Static Creep Micro-Macro Fracture Mechanics of Brittle Solids

Xiaozhao Li · Zhushan Shao · Chengzhi Qi

Static Creep Micro-Macro Fracture Mechanics of Brittle Solids

Springer

Xiaozhao Li
School of Civil and Transportation Engineering
Beijing University of Civil Engineering and Architecture
Beijing, China

Zhushan Shao
School of Civil Engineering
Xi'an University of Architecture and Technology
Xi'an, Shaanxi, China

Chengzhi Qi
School of Civil and Transportation Engineering
Beijing University of Civil Engineering and Architecture
Beijing, China

ISBN 978-981-99-8202-8 ISBN 978-981-99-8203-5 (eBook)
https://doi.org/10.1007/978-981-99-8203-5

This Springer imprint is published by the registered company Springer Nature Singapore Pte Ltd.
The registered company address is: 152 Beach Road, #21-01/04 Gateway East, Singapore 189721, Singapore

Paper in this product is recyclable.

Preface

Brittle solids (glass, ceramic, concrete, ice, and rock et al.) widely appear in the artificial engineering (deep underground engineering, architecture engineering, aerospace engineering, natural gas engineering, petroleum engineering, and mechanical manufacturing engineering et al.), and natural environment (earth crust, glacier, and mountains, et al.). Furthermore, numerous microcracks existing in brittle solids have a great influence on the macroscopic mechanical behaviors of brittle solids. The time-dependent creep mechanical behaviors of brittle solids strongly affect the long-term safety and stability of the engineering and environment and could trigger some engineering accidents (e.g., building component fracture, mechanical component damage, tunnel collapse et al.) and natural disasters (earthquake, landslide, and volcanic eruption et al.).

This book gathers a large amount of the recent research results on this topic to provide a better understanding of the static creep micro-macro fracture mechanics in brittle solids. To be precise, this is about to explore the effects of the external factors of stress paths, water content, seepage pressure, dynamic disturbance, and thermal treated temperature, and the internal factors of crack friction, angle, size, recovery, nucleation, and coalescence on the static creep fracture mechanical properties in brittle solids. This book provides important theoretical support in evaluation for long-term lifetime in brittle solid engineering and environment.

Chapter 1 presents an overview of the research background and research status of the static creep micro-macro fracture mechanics of brittle solids. Chapters 2 and 4 present a micro-macro method to evaluate the time-dependent deformation and shear strength during static creep fracture of brittle solids. Chapters 3, 5–12 introduce the effect factors of stress path, crack angle, crack nucleation, crack recovery, water chemistry reaction, seepage pressure, dynamic disturbance, and thermal treated temperature into the present micro-macro method of Chaps. 2 and 4.

Beijing, China Xiaozhao Li
Xi'an, China Zhushan Shao
Beijing, China Chengzhi Qi

Acknowledgments This work was supported by the National Natural Science Foundation of China (Grant Nos. 51708016, 51774018 and 12172036), the R&D program of Beijing Municipal Education Commission (Grant No. KM202110016014), and the Pyramid Talent Training Project of Beijing University of Civil Engineering and Architecture (Grant No. JDYC20200307).

Contents

Chapter 1
Introduction

Static creep fracture in brittle solids (e.g., glass, ceramic, concrete, ice and rock et al.) has a great significance for the evaluation of material lifetime in engineering application (e.g., deep underground engineering, architecture engineering, aerospace engineering, natural gas engineering, petroleum engineering, mechanical manufacturing engineering et al.). Numerous initial microcracks are being in brittle solids. The internal microcrack variations (e.g., friction, initiation, growth, recovery, nucleation, and coalescence) have great effect on the static creep fracture in brittle solids. The external factors of loading, chemical corrosion, seepage pressure, dynamic disturbance and temperature also strongly influence the static creep fracture in brittle solids. Thus, it is very necessary to establish a micro–macro method for evaluating the static creep fracture considering the effects of the external factors in brittle solids.

The research object of this book mainly focuses on the natural rock materials. The research results have a help for understanding the static creep fracture caused by microcrack variation in brittle solids.

Static creep failure of brittle rock has great significance for the prediction of important geohazards and stability of deep-buried excavations. During the long-term operation in upper crust and deep underground space, rock failure often occurs at stresses below its short-term failure strength. Such time dependent brittle deformation is generally induced by sub-critical crack growth which is caused by stress corrosion. The process of such rock deformation and failure caused by applied constant stress is defined as brittle creep. Long-term strength, steady-state creep rate and creep failure time play important roles in predicting the long-term creep failure of rock.

Brittle creep behaviors of rock attract the wide interest of scholars and engineers around the world. Based on the abundant experimental results, the static fatigue limit of granite was predicted by exponential function (Schmidtke and Lajtai 1985). The effects of stress state on failure time were also investigated (Sangha and Dhir 1972; Aubertin et al. 2000; Lau and Chandler 2004). According to relationship between steady-state creep rate and stress state, critical stress which transforms steady-state creep to accelerated creep was studied, and long-term strength was defined by stress at

X. Li et al., *Static Creep Micro-Macro Fracture Mechanics of Brittle Solids*,
https://doi.org/10.1007/978-981-99-8203-5_1

volume strain reversal. Rock damage increases constantly during steady-state creep. And rock failure will occur finally when applied stresses are larger than the long-term strength (Lajtai et al. 1991; Ma and Daemen 2006; Tsai et al. 2008; Brace et al. 1966; Chandler 2013). Using phenomenological approach, numerous viscoplastic creep models were presented. Parameters in proposed equations were obtained from a series of basic experiments, which predicted creep failure behaviors of rocks (Munson 1997; Bellenger and Bussy 2001; Challamel et al. 2005; Wu et al. 2015). All above publications presented well description of macroscopic mechanical behaviors for rock brittle creep.

Based on the crack growth during rock failure process, the micromechanical behaviors can be divided into three distinct stages: crack initiation, crack propagation and coalescence (Tang and Kou 1998; Park and Bobet 2010). The stress–strain curve was divided into five phases: crack closure, crack initiation, stable crack growth, unstable crack growth, and post-peak failure. It was experimental studied by using strain measurement and acoustic emission monitoring (Martin and Chandler 1994; Eberhardt et al. 1998; Chang and Lee 2004).

According to the cumulative number of acoustic emission events, the creep behavior of brittle rock was experimental investigated (Ohnaka 1983; Grgic and Amitrano 2009; He et al. 2012). Using scanning electron microscope test, the crack interaction increase with increment of the numbers of individual cracks is observed, and cracks coalescence play an important role in tertiary creep (Kranz 1979; Heap et al. 2009). Employing the mechanism of brittle creep caused by time-dependent extension of microcracks at low stresses, a transient creep law in nonhomogeneous brittle rocks was derived, and the influences of stress states and micro-parameters on creep behaviors of brittle rocks in compressive loadings were studied (Scholz 1968; Cruden 1970). Based on the micromechanical model of two interacting cracks and subcritical crack growth law, the accelerated creep crack growth and deformation brittle rocks are analyzed, and the effects of stress state, temperatures, and water contents on the creep failure time are discussed (Miura et al. 2003; Brantut et al. 2012). Nara et al. (2010, 2013) estimated parameters of subcritical crack growth by using of the Double-Torsion testing method. Based on experimental results, they calculated the long-term strength of rock under high temperature and humidity. Based on the mechanism of crack growth, the long-term creep failure of rock was investigated by using numerical simulation method e.g., the discrete element method (DEM), the finite difference program (i.e., FLAC), and the finite element method (FEM) (Yoshida and Horii 1992; Shao et al. 2006; Golshani et al. 2007; Caia et al. 2004; Diederichs et al. 2004; Li and Konietzky 2014).

As above mentioned, lots of experimental research of brittle creep failure in rock has been achieved. In the following studies of this book, we will try to establish a micro–macro method to investigate the brittle creep failure of rock in compression. The effect of external factors of loading path, temperature, seepage pressure, chemical corrosion on the microcrack variation and macroscopic deformation during brittle creep of rocks will be studied. This study will also provide an assistance in evaluation for the static creep fracture of other brittle solid materials containing initial microcracks.

References

Aubertin M, Li L, Simon R. A multiaxial stress criterion for short- and long-term strength of isotropic rock media. Int J Rock Mech Min Sci. 2000;37(8):1169–93.

Bellenger E, Bussy P. Phenomenological modeling and numerical simulation of different modes of creep damage evolution. Int J Solids Struct. 2001;38:577–604.

Brace WF, Paulding B, Scholz C. Dilatancy in the fracture of crystalline rocks. J Geophys Res. 1966;71:3939–53.

Brantut N, Baud P, Heap MJ, Meredith PG. Micromechanics of brittle creep in rocks. J Geophys Res. 2012;117:B08412.

Caia M, Kaisera PK, Tasakab Y, Maejima T, Morioka H, Minami M. Generalized crack initiation and crack damage stress thresholds of brittle rock masses near underground excavations. Int J Rock Mech Min Sci. 2004;41:833–47.

Challamel N, Lanos C, Casandjian C. Creep damage modelling for quasi-brittle materials. Eur J Mech A-Solids. 2005;24:593–613.

Chandler NA. Quantifying long-term strength and rock damage properties from plots of shear strain versus volume strain. Int J Rock Mech Min Sci. 2013;59:105–10.

Chang SH, Lee CI. Estimation of cracking and damage mechanisms in rock under triaxial compression by moment tensor analysis of acoustic emission. Int J Rock Mech Min Sci. 2004;41:1069–86.

Cruden DM. A theory of brittle creep in rock under uniaxial compression. J Geophys Res. 1970;75(17):3431–2.

Diederichs MS, Kaiser PK, Eberhardt E. Damage initiation and propagation in hard rock during tunneling and the influence of near-face stress rotation. Int J Rock Mech Min Sci. 2004;41:785–812.

Eberhardt E, Stead D, Stimpson B, Read RS. Identifying crack initiation and propagation thresholds in brittle rock. Can Geotech J. 1998;35:222–33.

Golshani A, Oda M, Okui Y, Takemura T, Munkhtogoo E. Numerical simulation of the excavation damaged zone around an opening in brittle rock. Int J Rock Mech Min Sci. 2007;44(6):35–845.

Grgic D, Amitrano, D. Creep of a porous rock and associated acoustic emission under different hydrous conditions. J Geophys Res. 2009;114(B10).

He M, Jia X, Coli M, Livi E, Sousa L. Experimental study of rockbursts in underground quarrying of Carrara marble. Int J Rock Mech Min Sci. 2012;52(6):1–8.

Heap MJ, Baud P, Meredith PG, Bell AF, Main IG. Time-dependent brittle creep in Darley Dale sandstone. J Geophys Res. 2009;114:B07203.

Kranz RL. Crack growth and development during creep of Barre granite. Int J Rock Mech Min Sci Geomech Abs. 1979;16(1):23–35.

Lajtai EZ, Scott Duncan EJ, Carter BJ. The effect of strain rate on rock strength. Rock Mech Rock Eng. 1991;24(2):99–109.

Lau JSO, Chandler NA. Innovative laboratory testing. Int J Rock Mech Min Sci. 2004;41(8):1427–45.

Li X, Konietzky H. Numerical simulation schemes for time-dependent crack growth in hard brittle rock. Acta Geotech. 2014;10(4):1–19.

Ma L, Daemen JJK. An experimental study on creep of welded tuff. Int J Rock Mech Min Sci. 2006;43(2):282–91.

Martin CD, Chandler NA. The progressive fracture of Lac du Bonnet granite. Int J Rock Mech Min Sci Geomech Abs. 1994;31(6):643–59.

Miura K, Okui Y, Horii H. Micromechanics-based prediction of creep failure of hard rock for long-term safety of high-level radioactive waste disposal system. Mech Mater. 2003;35(3):587–601.

Munson DE. Constitutive model of creep in rock salt applied to underground room closure. Int J Rock Mech Min Sci. 1997;34(2):233–47.

Nara Y, Takada M, Mori D, Owada H, Yoneda T, Kaneko K. Subcritical crack growth and long-term strength in rock and cementitious material. Int J Fract. 2010;164(1):57–71.

Nara Y, Yamanaka H, Oe Y, Kaneko K. Influence of temperature and water on subcritical crack growth parameters and long-term strength for igneous rocks. Geophys J Int. 2013;193(1):47–60.

Ohnaka M. Acoustic emission during creep of brittle rock. Int J Rock Mech Min Sci. 1983;20(3):121–34.

Park CH, Bobet A. Crack initiation, propagation and coalescence from frictional flaws in uniaxial compression. Eng Fract Mech. 2010;77:2727–48.

Sangha CM, Dhir RK. Influence of time on the strength, deformation and fracture properties of a lower Devonian sandstone. Int J Rock Mech Min Sci. 1972;9:343–54.

Schmidtke RH, Lajtai EZ. The long-term strength of Lac du Bonnet granite. Int J Rock Mech Min Sci Geomech Abs. 1985;6(22):461–5.

Scholz CH. Mechanism of creep in brittle rock. J Geophys Res. 1968;73(10):3295–302.

Shao JF, Chau KT, Feng XT. Modeling of anisotropic damage and creep deformation in brittle rocks. Int J Rock Mech Min Sci. 2006;43(4):582–92.

Tang CA, Kou SQ. Crack propagation and coalescence in brittle materials under compression. Eng Fract Mech. 1998;61:311–24.

Tsai LS, Hsieh YM, Weng MC. Time-dependent deformation behaviors of weak sandstones. Int J Rock Mech Min Sci. 2008;45(2):144–54.

Wu F, Liu JF, Wang J. An improved Maxwell creep model for rock based on variable-order fractional derivatives. Environ Earth Sci. 2015;73(11):6965–71.

Yoshida H, Horii H. A micromechanics-based model for creep behavior of rock. Appl Mech Rev. 1992;45(8):294–303.

Chapter 2
Static Creep Fracture Mechanism

Brittle rock deformation brings high hazards on the stability of rock mass during deep underground excavation and/or geological evolvement. Natural rocks intrinsically exist numerous microcracks. Microcrack growth has an important influence on the long-term creep behaviors of brittle rocks. A major challenge of this area is linking the microcrack growth behavior with the macroscopic mechanical behavior. In this chapter, a micro–macro method is proposed to calculate the relations of stress to axial, lateral, and volumetric strains, involving the long-term creep failure of brittle rocks.

2.1 Theory

2.1.1 Microcrack Growth Model Under Compression

A compressive stress-induced microcrack cracking model of brittle rocks (Ashby and Sammis 1990) is shown in Fig. 2.1a, which is used to evaluate the short-and long-term deformations by coupling sub-critical cracking rule and crack-strain correlation in brittle rocks. In the actual brittle rocks, the numerous random distributed initial microcracks exist without external loadings; some initial microcracks start to slide and wing microcracks start to extend under external loadings; the failure of brittle rocks often is caused by strain localization due to some microcracks accumulation and coalescence. However, in this model, the spatial effects of randomly distributed microcracks and the failure caused by strain localization cannot be considered. This spatial effect is approximately researched by an average and equivalent method. Rock failure from strain localization is equivalently investigated using the global failure by the model. This model is presumed to be a homogenous and isotropic elastic body. The growth direction of wing crack is assumed to be parallel to the direction of axial stress σ_1. Confining pressure σ_3 is posited to be equal σ_2. It is noted that the compressive stress is negative in the theoretical derivation. However, for convenience

X. Li et al., *Static Creep Micro-Macro Fracture Mechanics of Brittle Solids*,
https://doi.org/10.1007/978-981-99-8203-5_2

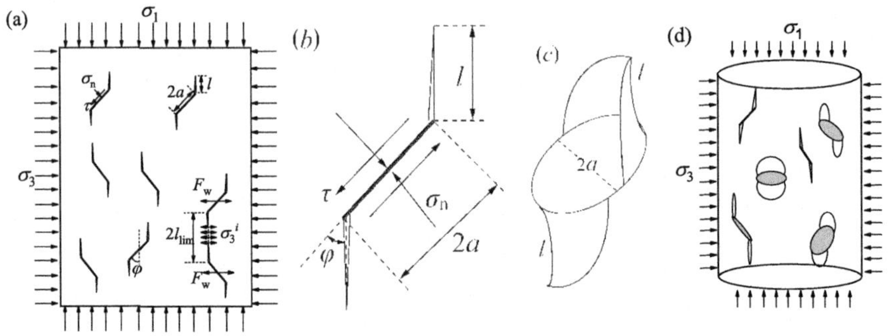

Fig. 2.1 Configurations of **a** overall microcrack model, **b** single total crack containing initial penny-shaped crack and wing crack, the three dimensions of **c** total crack and **d** overall microcrack model (Ashby and Sammis 1990)

of observation, the compressive stress is denoted as positive in the graphic results below.

The configuration of single total crack containing initial penny-shaped crack and wing crack is shown Fig. 2.1b, c. φ is the angle aligned to axial stress σ_1, and a is the radius of initial penny-shaped crack. The numerous initial crack size and angle in brittle materials are both assumed as an average value in this model, respectively. τ and σ_n is shear stress and normal stress acting on the surface of initial penny-shaped crack. l is wing crack length caused by the sliding of initial crack. F_w is the wedging force and $\sigma^i{}_3$ is internal stress, which causes the tensile effect on initial crack surfaces and wing crack tips, and plays an important part in prediction of mechanical properties at rock failure.

According to the microcracking model in Fig. 2.1, the stress intensity factor for mode I crack of the wing crack tips in brittle rocks is given as (Ashby and Sammis 1990):

$$K_I = \frac{F_w}{[\pi(l+\beta a)]^{\frac{3}{2}}} + \frac{2}{\pi}(\sigma_3 + \sigma_3^i)\sqrt{\pi l} \tag{2.1}$$

where

$$\sigma_3^i = \frac{F_w}{S - \pi(l+\alpha a)^2} \tag{2.2}$$

$$S = \pi^{1/3}[3/(4N_V)]^{2/3} \tag{2.3}$$

$$F_w = -(A_1\sigma_1 - A_2\sigma_3)a^2 \tag{2.4}$$

$$A_1 = \pi\sqrt{\frac{\beta}{3}}[\sqrt{1+\mu^2} - \mu] \tag{2.5}$$

$$A_2 = A_1 \frac{\sqrt{1+\mu^2}+\mu}{\sqrt{1+\mu^2}-\mu} \tag{2.6}$$

where N_V is the number of initial microcracks unit volume. μ is the friction coefficient on surface of initial penny-shaped crack. $\alpha = \cos\varphi$. β is a correction factor regulating the value of stress intensity factor K_I at $l = 0$. S is an average area of each crack. Assuming that each crack is assigned an average spherical volume ($4\pi r^3/3 = 1/N_V$) in unit volume, the average area is expressed as $S = \pi r^2 = \pi^{1/3}[3/(4N_V)]^{2/3}$. $\pi(l + \alpha a)^2$ is a projected area of total crack (initial crack and two wing cracks) to the plane, which is perpendicular to direction of σ_3. The limitation length for coalescence of two adjacent wing cracks can be solved by $S = \pi(l + \alpha a)^2$ as:

$$l_{\lim} = \left(D_o^{-1/3} - 1\right)\alpha a \tag{2.7}$$

where D_o is initial damage, which is defined as $D_o = 4\,N_V\,\pi(\alpha a)^3/3$.

Neglecting subcritical crack growth, cracks only start to grow at K_I reaching fracture toughness K_{IC}. Relation of stress state (σ_1, σ_3) and wing cracks growth l is expressed by using Eq. (2.1) as:

$$\sigma_1(l) = \frac{\sigma_3[c_3 + A_2(c_1 + c_2)] - K_{IC}/\sqrt{\pi a}}{A_1(c_1 + c_2)} \tag{2.8}$$

where

$$c_1 = \frac{1}{\pi^2}\left(\frac{l}{a} + \beta\right)^{-\frac{3}{2}} \tag{2.9}$$

$$c_2 = 2\frac{1}{(\pi\alpha)^2}\left(\frac{l}{a}\right)^{\frac{1}{2}} / \left[D_o^{-\frac{2}{3}} - \left(1 + \frac{l}{\alpha a}\right)^2\right] \tag{2.10}$$

$$c_3 = \frac{2}{\pi}\left(\frac{l}{a}\right)^{\frac{1}{2}} \tag{2.11}$$

Relation between axial stress and crack length is corresponding to stress–strain relation, and a peak value of function $\sigma_1(l)$ is equivalent to the short-term strength of brittle rocks (Ashby and Sammis 1990; Brantut et al. 2012). Furthermore, correlation of crack initiation stress σ_{1c} and confining pressure is expressed as (Ashby and Hallam 1986):

$$\sigma_{1c} = \frac{\sqrt{1+\mu^2}+\mu}{\sqrt{1+\mu^2}-\mu}\sigma_3 - \frac{\sqrt{3}}{\sqrt{1+\mu^2}-\mu}\frac{K_{IC}}{\sqrt{\pi a}} \tag{2.12}$$

Parameters a and μ are yielded by comparison of Eq. (2.12) and an experimental relation of confining pressure and axial crack initiation stress. Comparing Eq. (2.8) at $l = 0$ with Eq. (2.12), β is obtained as a constant value, i.e., 0.32.

2.1.2 Crack-Strain Relation for Brittle Solids

Due to random distribution of numerous microcracks in brittle rocks, following the Weibull probability distribution (Weibull 1951), the damage is defined as (Chen et al. 1997a):

$$D = 1 - \exp\left[-(\varepsilon_1/\varepsilon_o)^m\right] \tag{2.13}$$

where m, ε_0 are the material constant, ε_1 is axial strain. It is noted that the compressive strain is denoted as positive in theory derivation. Following Ashby and Sammis' (1990) definition, the damage induced by microcrck growth is expressed as:

$$D = \frac{4}{3}\pi N_V(l + \alpha a)^3 \tag{2.14}$$

Then, the initial damage of rock (i.e., when $l = 0$) is obtained by using Eq. (2.14) as:

$$D_o = 4\pi N_V(\alpha a)^3/3 \tag{2.15}$$

For the given rocks, the initial damage can be determined by experimental techniques of Acoustic Emission or Scanning Electron Microscope, and the average values of size a, inclination angle φ and friction coefficient μ for the initial crack plane can be solved by comparing theoretical and experimental data. In other words, a kind of solid material is corresponding to a set of the average parameters, which provides great help for understanding qualitatively the micro-mechanism of brittle solids under compressive loadings.

Substituting initial damage D_o into Eq. (2.13), the axial elastic strain ε_{1e} without considering wing crack growth could be defined as:

$$\varepsilon_{1e} = \varepsilon_o[-\ln(1 - D_o)]^{1/m} \tag{2.16}$$

In this study, assuming the value of damage from Eq. (2.13) is equivalent to damage from Eq. (2.14), the relation between microcrack growth and axial strain of brittle solids can be obtained as:

$$\varepsilon_1 = \varepsilon_o\left(-\ln\left[1 - \frac{(l + a\alpha)^3 D_o}{a^3\alpha^3}\right]\right)^{\frac{1}{m}} \tag{2.17}$$

$$l = a\alpha\left[D_o^{-1/3}\left(1 - \frac{1}{\exp\left[(\varepsilon/\varepsilon_o)^m\right]}\right)^{1/3} - 1\right] \tag{2.18}$$

Generally, the elastic relation between axial and lateral strains of rocks subjected to triaxial stress is expressed as:

$$\frac{\varepsilon_1}{\varepsilon_3} = \frac{\sigma_1 - 2\gamma\sigma_3}{-\gamma\sigma_1 + \sigma_3 - \gamma\sigma_3} \tag{2.19}$$

However, using the Eq. (2.19), the proposed analytical results of unified method for evaluating rock deformation hardly can be solved. For simplification of the proposed analytical method, the study mainly focuses on the rock mechanical properties in higher axial stress and lower confining pressure of brittle rocks. The parts including confining pressure σ_3 in Eq. (2.19) can be neglected. Thus, the Eq. (2.19) is approximately simplified as:

$$\varepsilon_3 = -\gamma\varepsilon_1 \tag{2.20}$$

where γ is Poisson rate. Furthermore, we assume that the inelastic relation of axial and lateral strains is also in accord with the description of Eq. (2.20). The rationality of these simplification and assumption is verified by the below study of comparison between theoretical and experimental results.

Based on axial and lateral strains, the elastic volumetric strain ε_{Ve} can be defined as (Martin and Chandler 1994):

$$\varepsilon_{\text{Ve}} = \varepsilon_{1\text{e}} + 2\varepsilon_{3\text{e}} \tag{2.21}$$

where $\varepsilon_{3\text{e}}$ is elastic lateral strain. Furthermore, the elastic axial strain at the initial condition is given in Eq. (2.16). Coupling Eqs. (2.16), (2.20) and (2.21), the elastic volumetric strain can be expressed as:

$$\varepsilon_{\text{Ve}} = \varepsilon_{\text{o}}[-\ln(1 - D_{\text{o}})]^{1/m}(1 - 2\gamma) \tag{2.22}$$

Volumetric strain is caused by microcrack growth in brittle rocks (Brace et al. 1966; Martin and Chandler 1994; Budiansky and O'Connel 1976). Thus, based on the microcrack growth mechanism (Budiansky and O'Connel 1976), the inelastic volumetric strain after wing cracking can be approximately defined as:

$$\varepsilon_{\text{V-ine}} = \frac{\Delta \text{V}}{\text{V}} \approx \frac{2N_{\text{V}}(l/2)^3}{1} = \frac{3D_o l^3}{16\pi(\alpha a)^3} \tag{2.23}$$

where V is the unit volume of rocks, and ΔV is the volumetric variation value. $N_{\text{V}}(l/2)^3$ is assumed to be approximately equal to the dilation volume of brittle rocks caused by the single wing crack growth.

Substituting the wing crack length l of Eq. (2.18) into Eq. (2.23), the relation between inelastic volumetric strain and axial strain can be derived as:

$$\varepsilon_{\text{V-ine}} = \frac{3D_{\text{o}}\left[D_{\text{o}}^{-1/3}\left(1 - \exp[-(\varepsilon_1/\varepsilon_{\text{o}})^m]\right)^{1/3} - 1\right]^3}{16\pi} \tag{2.24}$$

So, the total volumetric strain can be expressed as:

$$\varepsilon_V = \varepsilon_{Ve} - \varepsilon_{V\text{-}ine} = \varepsilon_o[-\ln(1 - D_o)]^{1/m}(1 - 2\gamma) - \frac{3D_o\left[D_o^{-1/3}\left(1 - \exp\left[-(\varepsilon_1/\varepsilon_o)^m\right]\right)^{1/3} - 1\right]^3}{16\pi} \tag{2.25}$$

Inserting Eq. (2.7) into Eq. (2.17), the axial failure strain can be predicted as:

$$\varepsilon_{1f} = \varepsilon_o\left(-\ln\left[1 - \frac{(l_{\lim} + a\alpha)^3 D_o}{a^3\alpha^3}\right]\right)^{\frac{1}{m}} \tag{2.26}$$

Coupling Eqs. (2.26) and (2.20), the lateral failure strain can be predicted as:

$$\varepsilon_{3f} = -\gamma\varepsilon_o\left(-\ln\left[1 - \frac{(l_{\lim} + a\alpha)^3 D_o}{a^3\alpha^3}\right]\right)^{\frac{1}{m}} \tag{2.27}$$

Coupling Eqs. (2.26) and (2.25), the volumetric failure strain can be predicted as:

$$\varepsilon_{Vf} = \varepsilon_o[-\ln(1 - D_o)]^{1/m}(1 - 2\gamma) - \frac{3D_o\left[D_o^{-1/3}\left(1 - \exp\left[-(\varepsilon_{1f}/\varepsilon_o)^m\right]\right)^{1/3} - 1\right]^3}{16\pi} \tag{2.28}$$

2.1.3 *Deformation Induced by Progressive Loadings*

Coupling Eqs. (2.17) and (2.8), the relation between axial stress and axial strain during progressive failure can be expressed as:

$$\sigma_1(\varepsilon_1) = \frac{\sigma_3[J_3 + A_2(J_1 + J_2)] - K_{IC}/\sqrt{\pi a}}{A_1(J_1 + J_2)} \tag{2.29}$$

where

$$J_1 = \pi^{-2}(\alpha J_4 + \beta)^{-3/2} \tag{2.30}$$

$$J_2 = 2\pi^{-2}\alpha^{-3/2}J_4^{1/2}/\left[D_0^{-2/3} - (J_4 + 1)^2\right] \tag{2.31}$$

$$J_3 = 2(\alpha J_4)^{1/2}/\pi \tag{2.32}$$

$$J_4 = D_o^{-1/3}\left[1 - \exp\left[-(\varepsilon_1/\varepsilon_o)^m\right]\right]^{1/3} - 1 \tag{2.33}$$

Coupling Eqs. (2.20) and (2.29), the relation between axial stress and lateral strain during progressive failure can be expressed as:

$$\sigma_1(\varepsilon_3) = \frac{\sigma_3[Z_3 + A_2(Z_1 + Z_2)] - K_{IC}/\sqrt{\pi a}}{A_1(Z_1 + Z_2)} \tag{2.34}$$

where

$$Z_1 = \pi^{-2}(\alpha Z_4 + \beta)^{-3/2} \tag{2.35}$$

$$Z_2 = 2\pi^{-2}\alpha^{-3/2}Z_4^{1/2}/\left[D_o^{-2/3} - (Z_4 + 1)^2\right] \tag{2.36}$$

$$Z_3 = 2(\alpha Z_4)^{1/2}/\pi \tag{2.37}$$

$$Z_4 = D_o^{-1/3}\left[1 - \exp\left[-\left(\frac{-\varepsilon_3}{\gamma\varepsilon_o}\right)^m\right]\right]^{1/3} - 1 \tag{2.38}$$

Coupling Eqs. (2.25) and (2.29), the relationship between axial stress and volumetric strain during progressive failure can be expressed as:

$$\sigma_1(\varepsilon_V) = \frac{\sigma_3[H_3 + A_2(H_1 + H_2)] - K_{IC}/\sqrt{\pi a}}{A_1(H_1 + H_2)} \tag{2.39}$$

where

$$H_1 = \pi^{-2}(\alpha H_4 + \beta)^{-3/2} \tag{2.40}$$

$$H_2 = 2\pi^{-2}\alpha^{-3/2}H_4^{1/2}/\left[D_o^{-2/3} - (H_4 + 1)^2\right] \tag{2.41}$$

$$H_3 = 2(\alpha H_4)^{1/2}/\pi \tag{2.42}$$

$$H_4 = \left[\frac{16\pi(\varepsilon_o(-\ln[1 - D_o])^{1/m}(1 - 2\upsilon) - \varepsilon_V)}{3D_o}\right]^{1/3} \tag{2.43}$$

Substituting Eq. (2.14) into Eq. (2.8), the relationship between damage and stress state (σ_1, σ_3) during progressive failure can be obtained.

$$\sigma_1(D) = \frac{2\sigma_3\pi\sqrt{B_1} + A_2(B_2 - B_3) - K_{IC}\pi^{3/2}a^{-1/2}}{A_1(B_2 - B_3)} \tag{2.44}$$

where

$$B_1 = \alpha\left[(D/D_o)^{1/3} - 1\right], \tag{2.45}$$

$$B_2 = (\beta + B_1)^{-3/2}, \tag{2.46}$$

$$B_3 = 2D_o^{2/3} B_1^{1/2} / \left[\alpha^2 \left(D^{2/3} - 1\right)\right]. \tag{2.47}$$

2.1.4 Subcritical Crack Growth

Crack growth could occur at $K_I < K_{IC}$ during subcritical crack growth. For a mode-I crack, the time-dependent crack growth is expressed as (Charles 1958; Atkinson 1984):

$$\frac{dl}{dt} = v\left(\frac{K_I}{K_{IC}}\right)^n \tag{2.48}$$

where n is stress corrosion index, and v is the characteristic crack velocity. The both parameters can be measured by subcritical crack growth test (Atkinson 1979; Wan et al. 2010).

Substituting Eq. (2.1) into Eq. (2.48), time-dependent wing crack growth of brittle rocks subjected to constant stress can be obtained as:

$$\frac{dl}{dt} = v(\pi a)^{\frac{n}{2}}\left[\frac{(\mathrm{A}_2\sigma_3 - \mathrm{A}_1\sigma_1)(c_1 + c_2) + \sigma_3 c_3}{K_{IC}}\right]^n \tag{2.49}$$

An initial equilibrium crack length l_o can be solved by substituting initial stress state (σ_1, σ_3) into Eq. (2.8). It is noted that the initial axial stress is between crack initiation stress and peak stress at given confining pressure. This value of initial equilibrium crack length is taken as the initial iteration value for numerical integration of Eq. (2.49). Numerical solution for evolution of wing crack length (i.e., l(t)) under different stress paths can be obtained.

2.1.5 Brittle Creeping Deformation

Substituting the evolution of wing crack length (i.e., l(t)) solved by Eq. (2.49) into Eq. (2.17), the time-dependent axial strain can be expressed as:

$$\varepsilon_1(t) = \varepsilon_o\left(-\ln\left[1 - \frac{(l(t) + a\alpha)^3 D_o}{a^3\alpha^3}\right]\right)^{\frac{1}{m}} \tag{2.50}$$

Coupling Eqs. (2.20) and (2.50), the time-dependent lateral strain can be expressed as:

$$\varepsilon_3(t) = -\gamma\varepsilon_o\left(-\ln\left[1 - \frac{(l(t) + a\alpha)^3 D_o}{a^3\alpha^3}\right]\right)^{\frac{1}{m}} \quad (2.51)$$

Coupling Eqs. (2.25) and (2.50), the time-dependent volumetric strain can be expressed as:

$$\varepsilon_V(t) = \varepsilon_o[-\ln(1 - D_o)]^{1/m}(1 - 2\gamma) - \frac{3D_o\left[D_o^{-1/3}\left(1 - \exp\left[-(\varepsilon_1(t)/\varepsilon_o)^m\right]\right)^{1/3} - 1\right]^3}{16\pi} \quad (2.52)$$

Contributions of mode-II and mode-III stress intensity factors to wing crack growth are much less than that of mode-I stress intensity factor. So these contributions are neglected in the presented method (Ashby and Sammis 1990; Brantut et al. 2012).

Crack growth at the closed crack is caused by the sliding of the initial crack after shear stress overcomes the friction force between crack interfaces. Furthermore, crack growth at the pore is caused by the stress concentrations around the pore (Ashby and Sammis 1990). If the closed crack is regarded as an open crack (which is approximately equivalent to the pore neglecting the effects of the geometrical shape of micro-defects), then the friction coefficient is neglected (i.e., $\mu = 0$), which provides an approximate understanding of the mechanical behavior of porous rocks on the basis of the proposed model.

2.2 Model Parameters

In the following analysis, all parameters are taken from the experimental results of Sanxia granite (Jiang et al. 2008; Zhu et al. 2007; Sun and Ling 1997) and Jinping marble (Wan et al. 2010; Wang et al. 2008; She and Cui 2010) in China. In order to determine the specific parameters used in the theoretical model, the macroscopic theoretical results (i.e. the axial stress–axial strain curve and evolution of creep curve) after each trial were used to calibrate the parameters by comparing the macroscopic experiment results. This process was repeated until the theoretical results achieved a good agreement with the experimental results. According to the triaxial test results, the empirical relationship between confining pressure and crack initiation stress for Sanxia granite was suggested as (Zhu et al., 2007) $\sigma_{1c} = 3.92\sigma_3 + 80.68$ MPa. Based on the triaxial test results, an empirical relationship of crack initiation stress states in Jinping marble was suggested as $\sigma_{1c} = 2.67\sigma_3 + 46$ MPa (Wang et al. 2008). The specific parameters are shown in Table 2.1.

Table 2.1 Specific parameters

Parameters	Value Sanxia granite	Value Jinping marble
Critical stress intensity factor K_{IC}	1.89 MPa·m$^{1/2}$	1.61 MPa·m$^{1/2}$
Index of stress corrosion n	21	57
Characteristic crack velocity v	0.17 m/s	0.16 m/s
Initial damage D_o	0.0409	0.048
Microcrack angle φ	59°	45°
Friction coefficient μ	0.74	0.51
Initial crack size a	2.05 mm	3.1 mm
Material constant m	1.8	1
Material constant ε_o	0.012	0.0147
Constant β	0.32	0.32
Poisson ratio γ	0.18	0.2

2.3 Progressive Fracture

The theoretical results of stress-induced axial, lateral and volumetric strains are shown in Fig. 2.2 ($\sigma_3 = 0$ MPa) and Fig. 2.3 ($\sigma_3 = 20$ MPa). The corresponding experimental results are also shown in Figs. 2.2 and 2.3. The theoretical results are in accord with the experimental results (Wang et al. 2008). The rationality of the proposed constitutive model for stress–strain relationship is verified. Comparing Figs. 2.2 and 2.3, it is also observed that the increment of confining pressure enhances the compressive strength of brittle rocks.

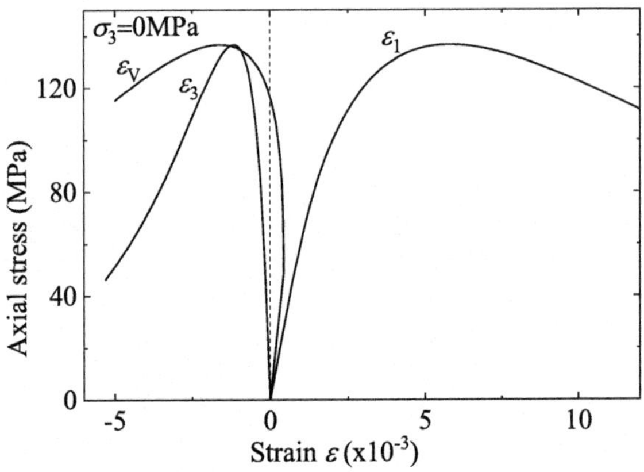

Fig. 2.2 Theoretical stress–strain results under confining pressure $\sigma_3 = 0$ MPa

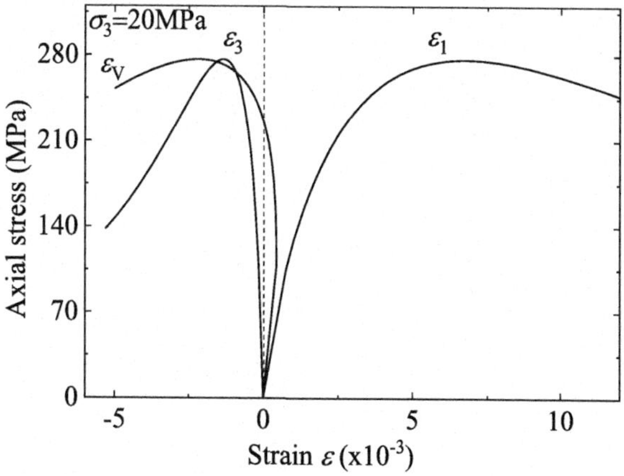

Fig. 2.3 Theoretical stress–strain results under confining pressure $\sigma_3 =$ 20 MPa

2.3.1 *Effects of Model Parameters on Stress-Crack Growth Relationship*

Figure 2.4 shows that the effect of fracture toughness on relationship between axial stress and crack length. It can be seen that the peak stress increase with increment of fracture toughness K_{IC}. Figure 2.5 shows that the effect of friction coefficient between microcracks on relationship between axial stress and crack length. It can be seen that the axial stress increases with increment of friction coefficient at the given crack length.

Due to the definition of initial damage $D_o = 4\pi N_V(\alpha a)^3/3$, the effect of initial microcrack size (i.e., a) on the relationship between axial stress and crack length under given initial damage D_o is different from the given initial microcrack number N_V.

For given initial damage, the effect of initial microcrack size on the relationship axial stress and crack length is shown in Fig. 2.6. It can be seen that the axial

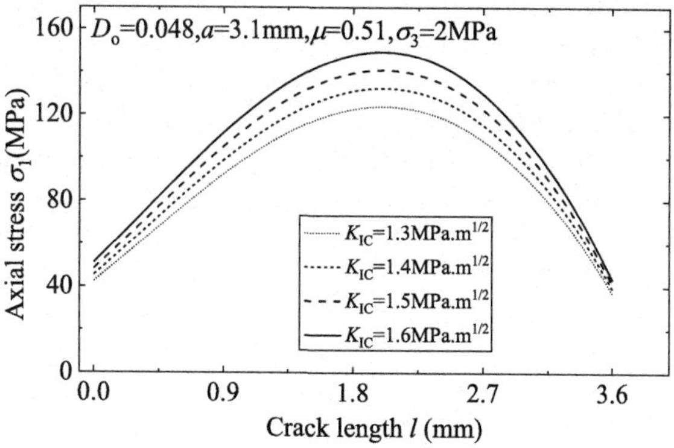

Fig. 2.4 Effect of fracture toughness on the relationship between axial stress and crack length

Fig. 2.5 Effect of friction coefficient on the relationship between axial stress and crack length

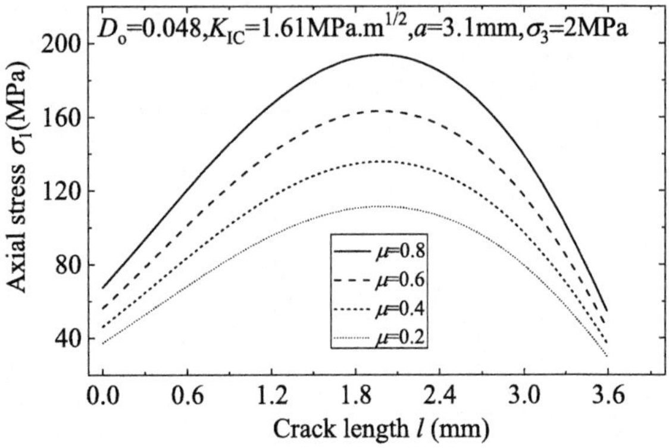

Fig. 2.6 Effect of initial crack size on the relationship between axial stress and crack length under given initial damage D_o

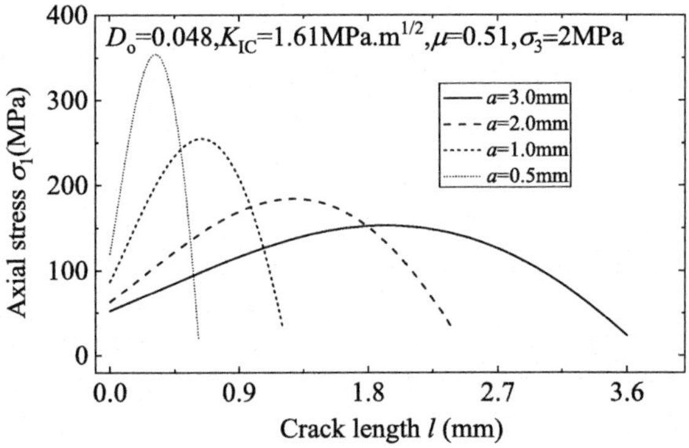

stress decreases with increment of initial microcrack size at the given crack length. Furthermore, it can be seen that the crack failure length increases with increment of initial crack size.

For given initial crack number N_V, the effect of initial microcrack size on relationship between axial stress and crack length is shown in Fig. 2.7. It also can be seen that the axial stress decreases with increment of initial microcrack size (i.e., initial damage) at the given crack length. Moreover, the crack failure length decreases increment of initial crack length. However, the sensitivity of initial microcrack size on peak axial stress for given initial crack number is larger than the given initial damage.

Theoretical results of axial stress and wing crack length are shown in Fig. 2.8. It can be seen that the axial stress increases with the increment of wing crack length before it reaches peak value. The axial stress decreases with the increment of wing crack length after peak value. The larger confining pressure causes the higher the axial stress at same wing crack length.

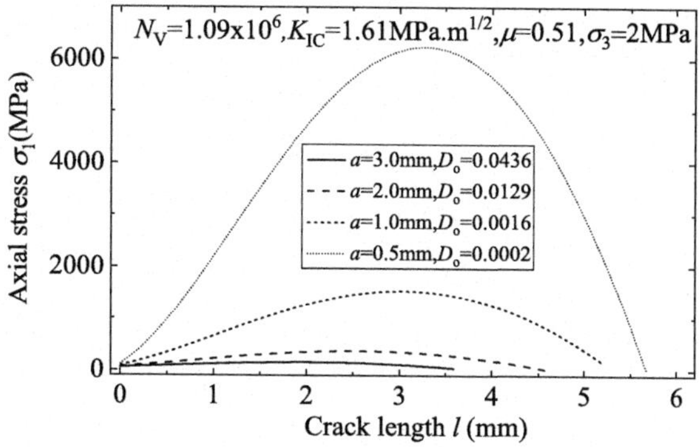

Fig. 2.7 Effect of initial crack size on the relationship between axial stress and crack length under given initial microcrack number N_V

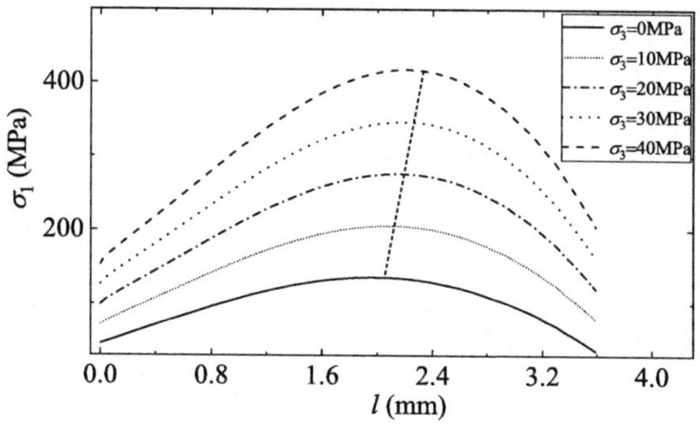

Fig. 2.8 Effect of confining pressure on relationship between axial stress and crack length

2.3.2 *Effects of Model Parameters on Stress–Strain Relationship*

The microcrack has great influence on the stress–strain relationship. Due to the random distribution of microcracks in brittle rock, the effects of microcrack on stress–strain relationship of brittle rock almost can't be investigated by common experiment directly. Thus, based on the micro–macro modelling method in this study, the effect of microcrack on stress–strain relationship will be theoretically predicted.

Figure 2.9 shows that the effect of fracture toughness on stress–strain relationship at $\sigma_3 = 2$ MPa. It can be seen that the peak stress (i.e., rock compressive strength) increase with increment of fracture toughness K_{IC}.

Figure 2.10 shows that the effect of friction coefficient between microcracks on stress–strain relationship at $\sigma_3 = 2$ MPa. It can be seen that the axial stress increases with increment of friction coefficient at the given strain. It means that the increment of friction coefficient enhances rock the compressive strength.

For given initial damage, the effect of initial microcrack size on stress–strain relationship at confining pressure $\sigma_3 = 2$ MPa is shown in Fig. 2.11. It can be seen that the axial stress decreases with increment of initial microcrack size at the given strain. For given initial crack number N_V, the effect of initial microcrack size on

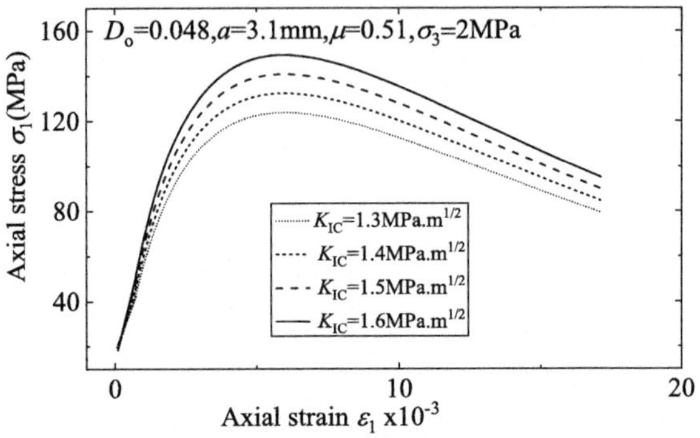

Fig. 2.9 Effect of fracture toughness on stress–strain relationship

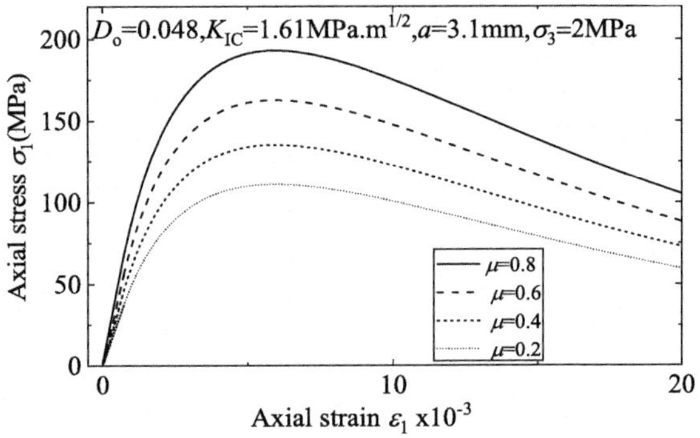

Fig. 2.10 Effect of friction coefficient on stress–strain relationship

stress–strain relationship is shown in Fig. 2.12. It also can be seen that the axial stress decreases with increment of initial microcrack size at the given strain. Furthermore, the rock compressive strength decreases with increment of initial damage. However, the sensitivity of initial microcrack size for given initial crack number is larger than the given initial damage.

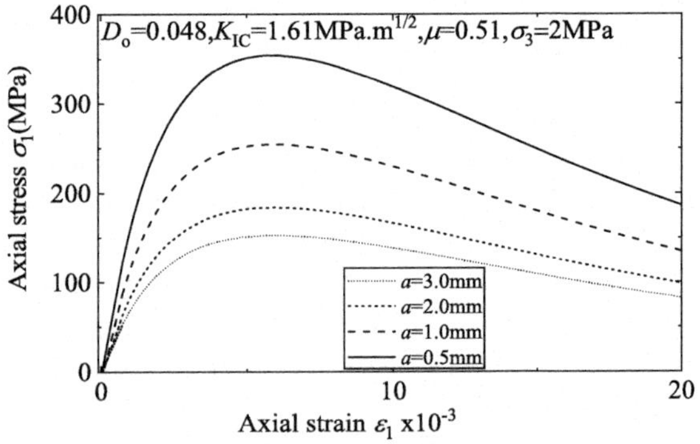

Fig. 2.11 Effect of initial crack size on the stress–strain relationship under given initial damage D_o

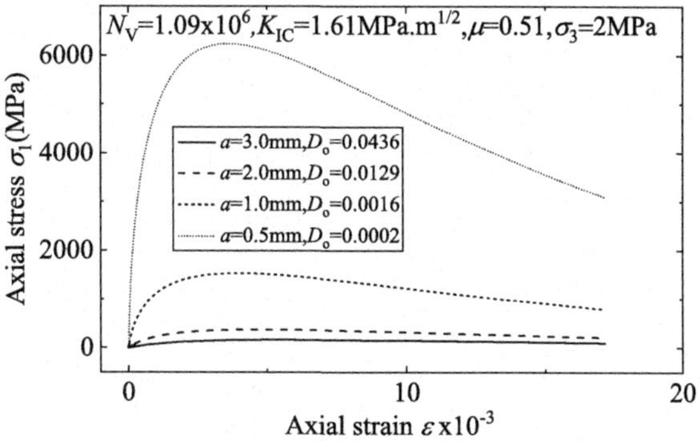

Fig. 2.12 Effect of initial damage on stress–strain relationship under given initial microcrack number N_V

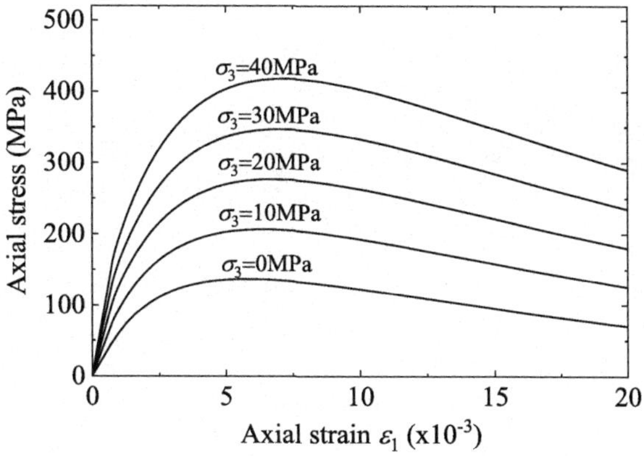

Fig. 2.13 Effect of confining pressure on relationship between axial stress and axial strain

The effects of confining pressure on stress-induced axial, lateral and volumetric strains are shown in Figs. 2.13, 2.14, 2.15 and 2.16. From Fig. 2.16, the correlation of stress-induced axial, lateral and volumetric strains can be seen directly under different confining pressures. It can be seen that peak stress increases with increment of confining pressure.

2.3.3 Correspondence Between Stress, Strain, Crack Length, and Damage

Theoretical relationships between axial stress and damage are shown in Fig. 2.17. It's worthy to note that the maximum theoretical value of damage is 1.0. This is an idealized theoretical state. In fact, rock failure would occur soon after the peak-value stress, and damage could not reach the value 1.0. According to the results shown in Fig. 2.17, the damage values at peak stress are about 0.280–0.339 corresponding to different confining pressure. The applied axial stress increases with the increment

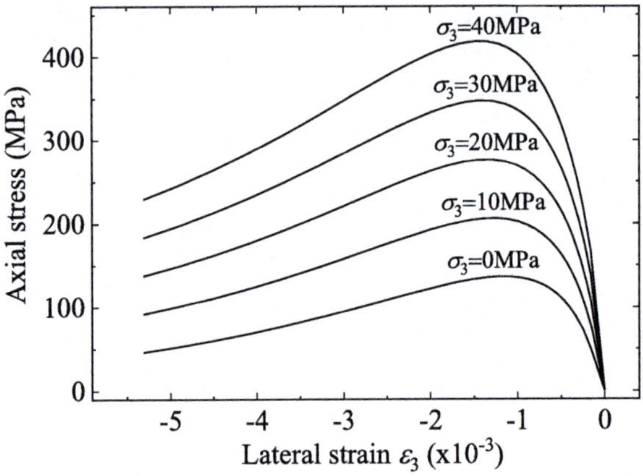

Fig. 2.14 Effect of confining pressure on relationship between axial stress and lateral strain

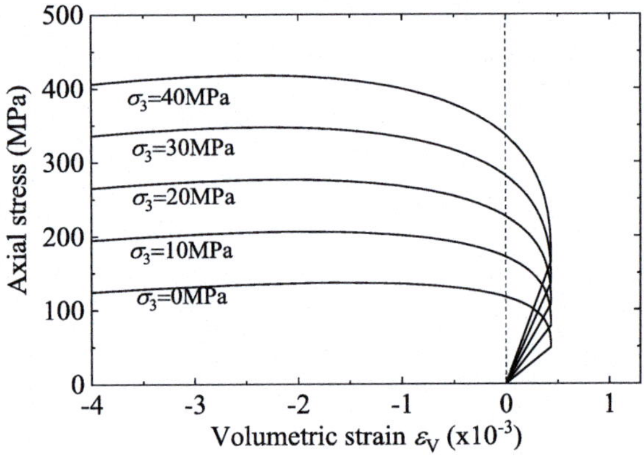

Fig. 2.15 Effect of confining pressure on relationship between axial stress and volumetric strain

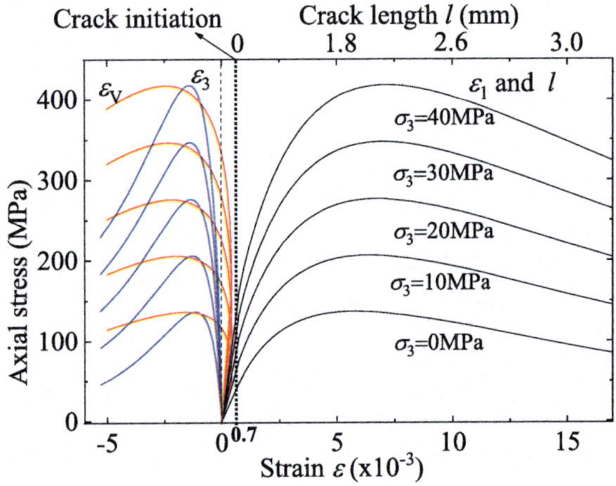

Fig. 2.16 Theoretical results between axial, lateral, volumetric strain and axial stress

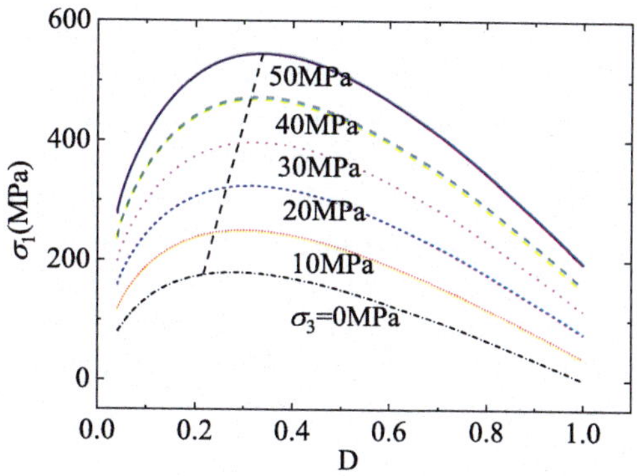

Fig. 2.17 Effects of confining pressures on relationship between axial stress and damage

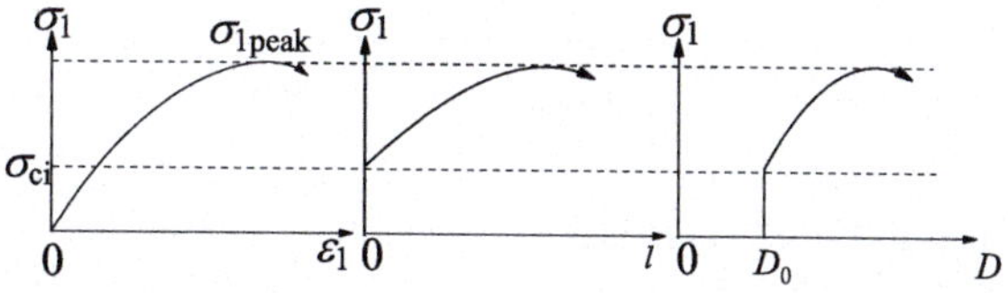

Fig. 2.18 Correspondence between axial stress, axial strain, crack length, and damage during progressive failure of rock

of damage before it reaches peak value. The axial stress would decrease with the increment of damage after peak value. The confining pressure also exhibits dramatic effect on damage evolution in rock. The larger the confining pressure, the higher the axial stress should be applied to reach to same damage value.

During the progressive failure of rock, the correspondence of axial stress, axial strain, crack length, and damage could be established by using of the theoretical method presented in this paper, as shown in Fig. 2.18. It can be seen that the suggested new method in this paper could not only describe the full stress–strain curve but also predict the damage variation. The axial strain is reversible when the axial stress is smaller than crack initiation stress. It could conclude that reversible strain occurs before crack growth, and damage keeps its initial value. It is crack growth that results in the irreversible strain in rock, and damage would increase with them.

2.3.4 *Model Parameter Effect on Rock Strength and Crack Initiation Stress*

Rock compressive strength and fracture toughness both have great significance for judge the mechanical behaviors. However, the relationship between rock strength and fracture toughness hardly is studied. Based on the stress–strain relationship studied above, the peak stress (i.e., rock compressive strength) and fracture toughness could be obtained. The effects of parameters in the equation of stress–strain relation on the relationship between rock strength and fracture toughness will be studied below.

As shown in Fig. 2.19, effect of friction coefficient on relationship between peak stress and fracture is obtained. It can be seen that the peak stress approximately linearly increases with increment of fracture toughness. Furthermore, the peak stress increases with increment of friction coefficient at given fracture toughness.

For given initial damage, effect of initial microcrack size on relationship between peak stress and fracture is obtained in Fig. 2.20. The peak stress increases with decrement of initial microcrack size at given fracture toughness.

For given initial microcrack number, effect of initial microcrack size (i.e., initial damage) on relationship between peak stress and fracture is obtained in Fig. 2.21. The peak stress also increases with decrement of initial damage at given fracture toughness.

The effects of confining pressure on crack initiation stress and peak stress are shown in Fig. 2.22. It could be seen that the crack initiation stress and peak stress increase linearly with increment of confining pressure. Moreover, these effects exhibit good agreement with experiment results (Wang et al. 2008).

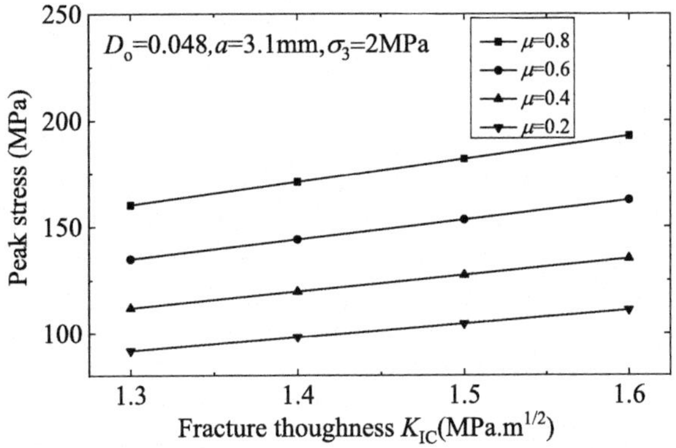

Fig. 2.19 Effect of friction coefficient on relationship between peak stress and fracture toughness

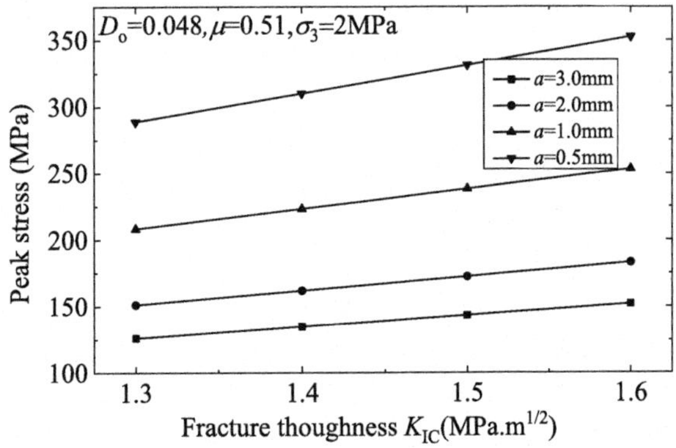

Fig. 2.20 Effect of initial microcrack szie on relationship between peak stress and fracture toughness under given initial damage D_o

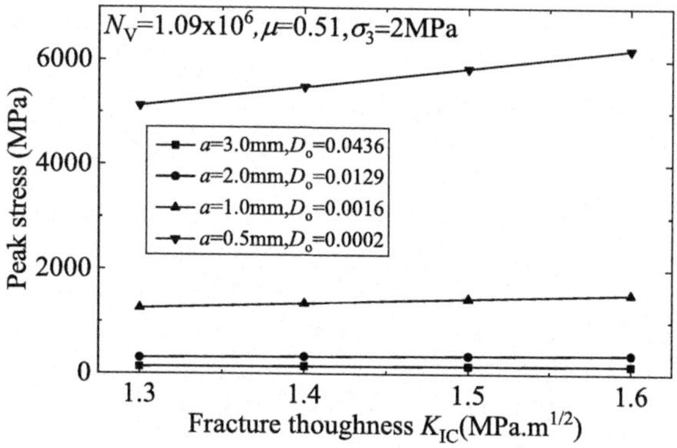

Fig. 2.21 Effect of initial damage on relationship between peak stress and fracture toughness under given initial microcrack number N_V

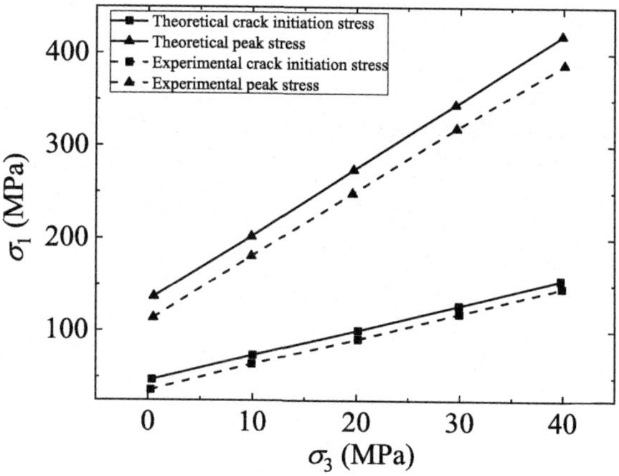

Fig. 2.22 Relations between crack initiation, peak stresses and confining pressure from test and theory

2.4 Creep Fracture

Above the mentioned progressive failure results, the crack initiation stress and peak stress (short-term strength) are studied. It is known that rock creep failure often occurs at stresses below its short-term failure strength. Thus, the progressive failure results provide a significant reference for selection of constant stress state which could lead to the creep failure.

Evolutions of axial, lateral and volumetric strains, and crack length under constant stress state (i.e., $\sigma_1 = 60$ MPa and $\sigma_3 = 2$ MPa) during creep are shown in Fig. 2.23. The axial and lateral strain have three stages: decelerated strain, steady-state strain, and accelerated strain, which is corresponding to the decelerated crack growth, steady-state crack length, and accelerated crack growth. Volumetric strain experience five stages: transient contraction, decelerated dilation, a transformation from contraction to dilation (see Fig. 2.24), steady-state dilation, and accelerated dilation. Figures 2.25 and 2.26 show another analytical result for evolutions of axial, lateral and volumetric strains under constant stress state (i.e., $\sigma_1 = 90$ MPa and

$\sigma_3 = 11$ MPa) during creep. The phenomena of analytical results are similar to the published experimental results (Chandler 2013).

The limitation length for coalescence of two adjacent wing cracks is solved by Eq. (2.7) (i.e., $l_{\text{lim}} = 3.8$ mm). The axial, lateral and volumetric failure strains are also predicted by Eqs. (2.26–2.28) as followed: $\varepsilon_{1\text{f}} = 0.059$, $\varepsilon_{2\text{f}} = 0.012$, and $\varepsilon_{\text{vf}} = 0.0145$. However, based on the analytical results (as shown in Figs. 2.23 and 2.25) solved by Eqs. (2.49–2.52), the actual failure crack length is about 2.5 mm, the actual axial failure strain $\varepsilon_{1\text{f}}$ is around 0.0094, the lateral failure strain $\varepsilon_{2\text{f}}$ is about 0.0019, and the volumetric failure strains ε_{vf} is around 0.0038. The actual failure crack length is not equal to the limitation length l_{lim} at crack coalescences solved by Eq. (2.7). The actual calculated failure strains solved by Eqs. (2.50–2.52) are smaller than that at crack coalescences solved by Eqs. (2.26–2.28). The causes of rock failure occur prior to crack coalescence could be summarized as followed: (1) rock is presumed to be a homogenous elastic body in wing crack model, (2) rock

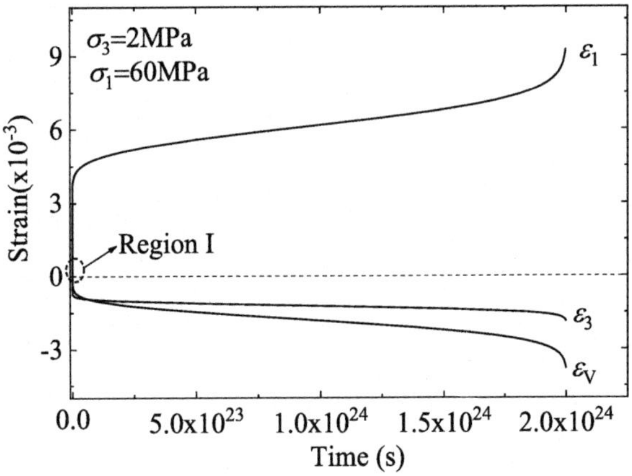

Fig. 2.23 Evolutions of strains and crack length under $\sigma_1 = 60$ MPa and $\sigma_3 = 2$ MPa

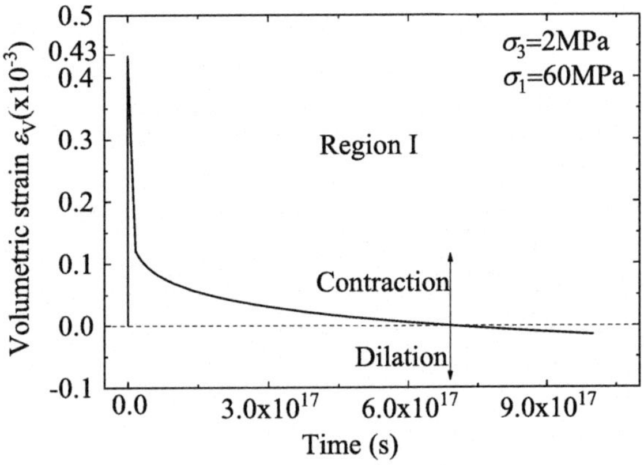

Fig. 2.24 Enlargement of volumetric strain for Region I of Fig. 2.23

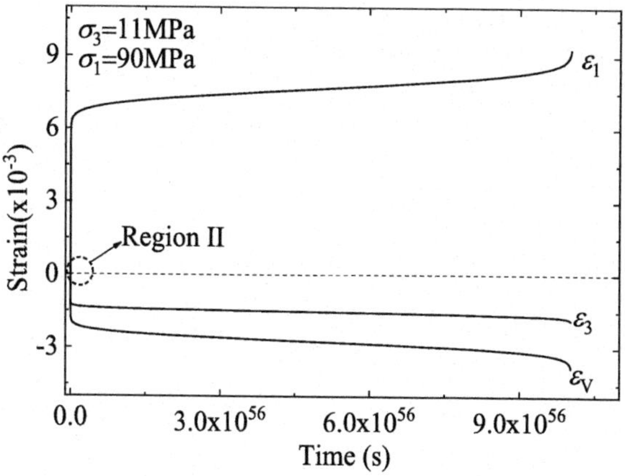

Fig. 2.25 Evolutions of strain and crack length under $\sigma_1 = 90$ MPa and $\sigma_3 = 11$ MPa

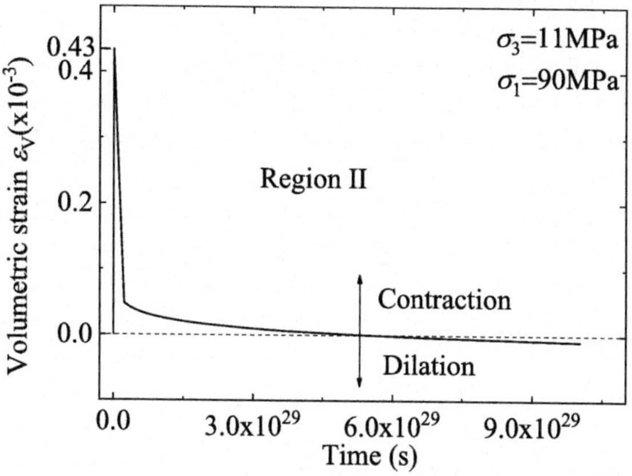

Fig. 2.26 Enlargement of volumetric strain for Region II of Fig. 2.25

failure from strain localization is equivalently investigated using the global failure by Ashby and Sammis' (1990) model.

A comparison of theoretical and experimental evolution for creep axial strain in Sanxia granite is also graphically presented in Fig. 2.27. The evolution of creep strain well agrees with the experiment result (Sun and Ling 1997). The comparison between theoretical and experimental results could present a good verification for the rationality of the theoretical method suggested in this paper.

2.4.1 Crack Growth, Strain, Stress Intensity Factor and Damage During Creep

In Fig. 2.28, the evolution of wing crack length is similar with the brittle creep process in rock. Wing crack growth during brittle creep can be divided into three

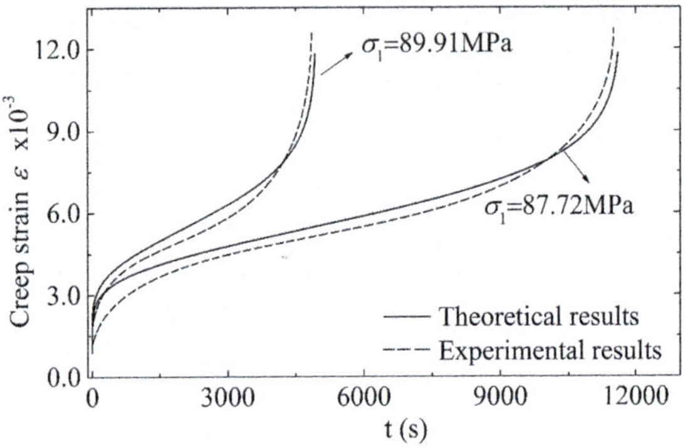

Fig. 2.27 Theoretical and experimental results of creep strain of Sanxia granite

phases: decelerating growth, steady-state growth, and accelerating growth. Crack velocity decreases from its initial value to a low value at first stage. Then it will keep an almost low constant value at second stage. Finally, it will accelerate from low value to a high value at the third stage.

As shown in Fig. 2.29, the creep strain of Sanxia granite exhibits three phases: decelerating strain, steady-state strain, and accelerating strain. Corresponding to the evolution of creep strain, the train rate decreases from a high value to low value at first stage. It will keep an almost low constant value at second stage. And it will increase from low value to high value at third stage.

In Fig. 2.30, the stress intensity factor at crack tip also exhibits three phases during brittle creep. It starts from a high value at crack initiation and decreases quickly to a low value at which the normalized stress intensity factor is smaller than 1.0. Then the stress intensity factor will keep an almost constant value until it increases transiently to a very high value in third stage. During brittle creep process, the normalized stress intensity factor is less than 1.0 in most of the time. This result could provide a theoretical support for the widely accepted acknowledge that brittle creep of rock is induced by sub-critical crack growth. Furthermore, it can be seen from Fig. 2.31 that the damage evolution is similar with the brittle creep process in Sanxia granite. Damage evolution during brittle creep can be also divided into three phases: decelerating damage, quasi-steady-state damage, and accelerating damage.

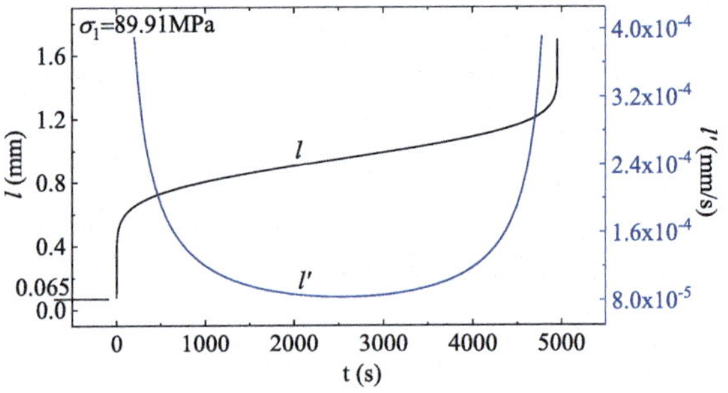

Fig. 2.28 Evolution of wing crack length and crack velocity during creep

Fig. 2.29 Evolution of axial creep strain and strain rate

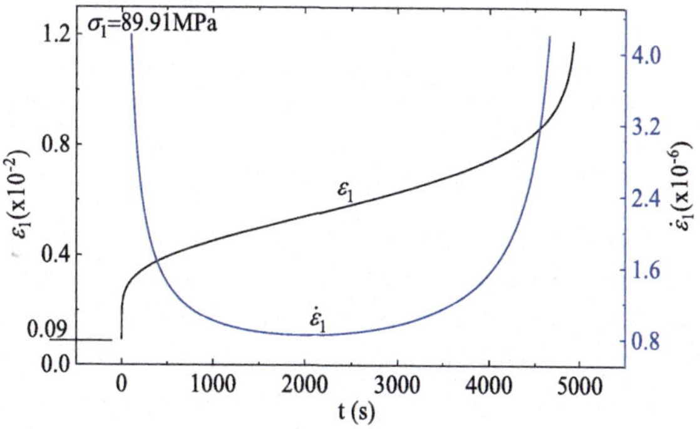

Fig. 2.30 Evolution of normalized stress intensity factor and creep strain rate

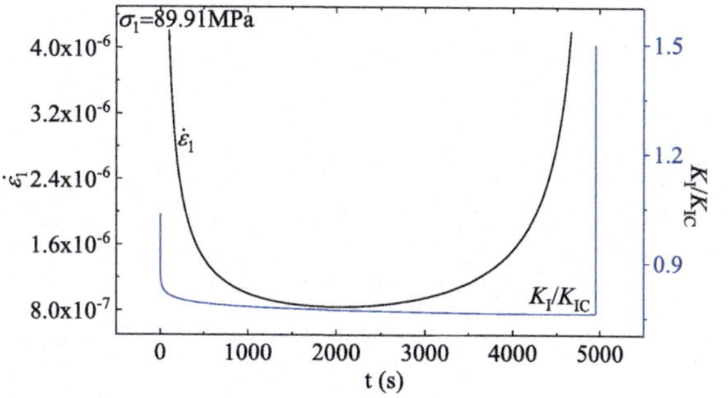

Fig. 2.31 Damage evolution during brittle creep

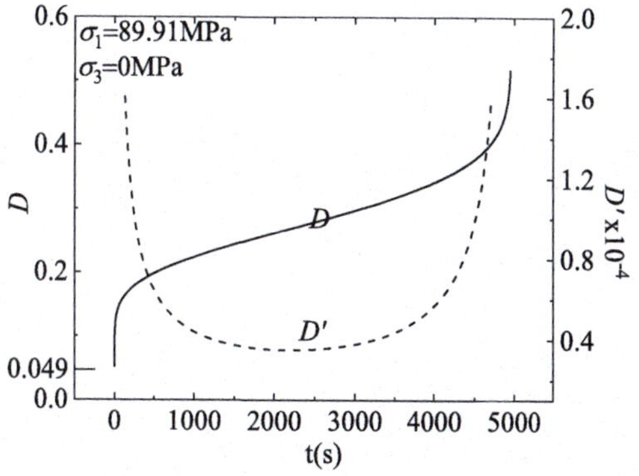

2.4.2 *Creep Failure Time*

The effect of axial stress on the evolution of creep strain is presented in Fig. 2.32. It can be seen that axial stress dramatically affect the creep failure time. A small variation of axial stress could result in huge difference for creep failure time. For the convenience of the presentation of all theoretical results, different time coordinates are used for different axial stresses. For example, 1.1×10^{24} s corresponds to 160 MPa, 3.4

$\times\ 10^{16}$ s corresponds to axial stress 180 MPa, and so on. It could be found that theoretical values of failure strain are almost identical under different axial stresses regardless of the huge difference of creep failure time.

(a) **Effect of Friction coefficient**

It can be seen from Fig. 2.33 that the effects of axial stress on creep failure time under different friction coefficients. The creep failure time decreases with increment axial stress. Furthermore, the creep failure time increases with increment of friction coefficient under the given axial stress. The difference of creep failure time on friction coefficient increases with decrement of axial stress. The variation rate of creep failure time on axial stress increases with increment of friction coefficient.

As shown in Fig. 2.34, the effects of confining pressure on creep failure time under different friction coefficients also obtained. The creep failure time increases with increment confining pressure. Furthermore, it can be seen that the creep failure time is very short under confining pressure $\sigma_3 = 0$ MPa. The creep failure time is

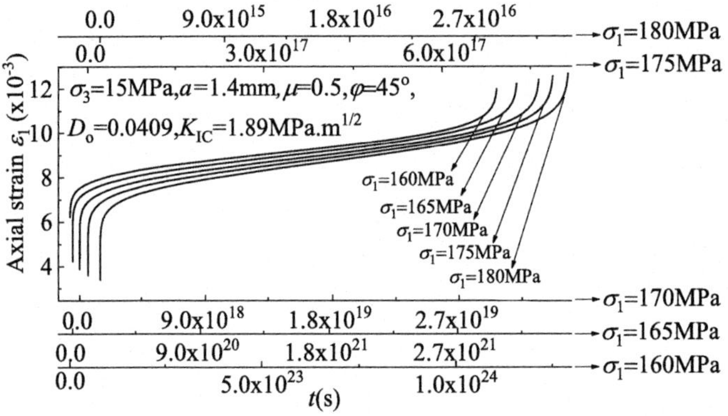

Fig. 2.32 Evolutions of creep strain under different axial stresses

Fig. 2.33 Effects of axial stress on creep failure time under different friction coefficients

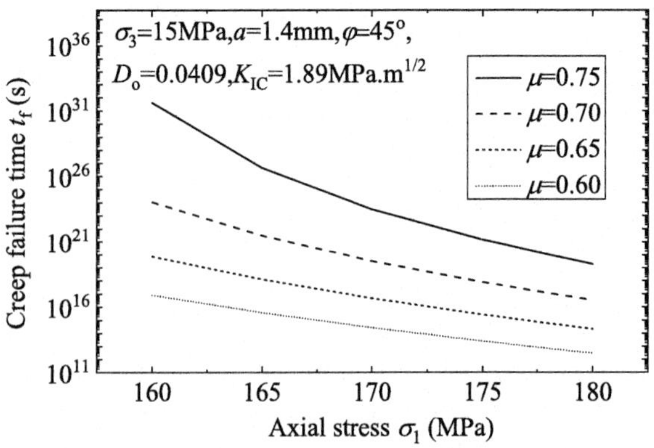

10^{32} or so (for $\mu = 0.75$) under confining pressure $\sigma_3 = 17$ MPa. In other word, the creep failure quickly appears when the confining pressure is unloaded.

The difference of creep failure time on friction coefficient increases with increment of confining pressure. The variation rate of creep failure time on confining pressure increases with increment of friction coefficient.

(b) Effect of Initial crack size

It can be seen from Fig. 2.35 that the effects of axial stress on creep failure time under different initial crack sizes. The creep failure time increases with decrement of initial crack size under the given axial stress. The difference of creep failure time on initial crack size increases with decrement of axial stress. The variation rate of creep failure time on axial stress increases with decrement of initial crack size.

As shown in Fig. 2.36, the effects of confining pressure on creep failure time under different initial crack sizes also obtained. The difference of creep failure time on initial crack size increases with increment of confining pressure. The variation rate of creep failure time on confining pressure increases with decrement of initial crack size.

(c) Effect of stress intensity factor

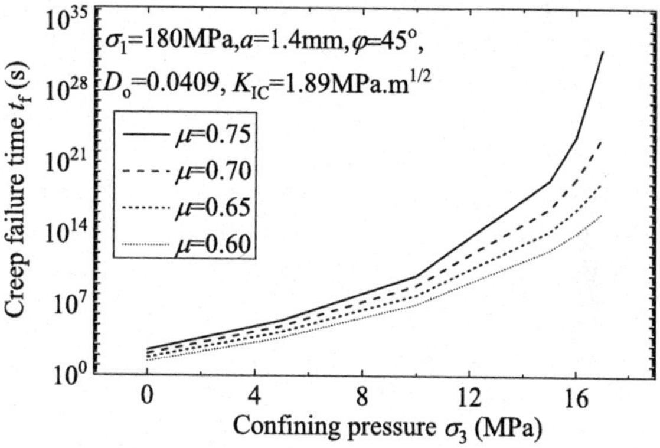

Fig. 2.34 Effects of confining pressure on creep failure time under different friction coefficients

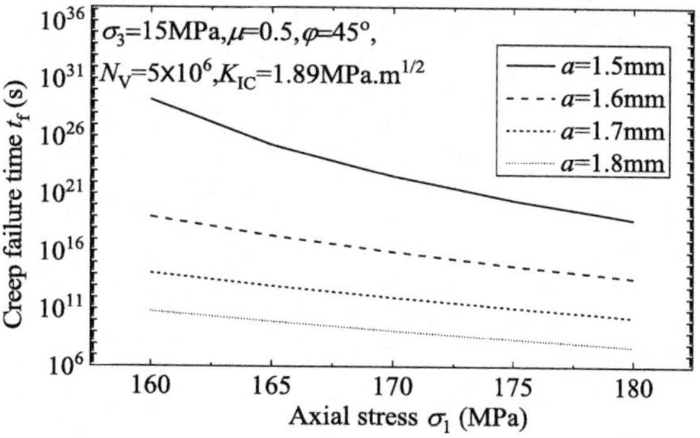

Fig. 2.35 Effects of axial stress on creep failure time under different initial crack sizes

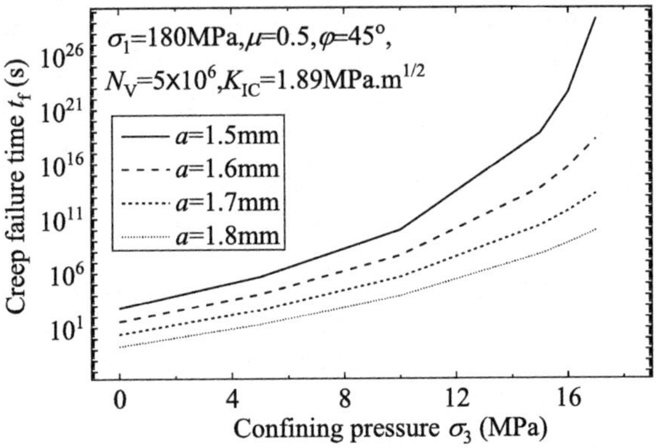

Fig. 2.36 Effects of confining pressure on creep failure time under different initial crack sizes

It can be seen from Fig. 2.37 that the effects of axial stress on creep failure time under different fracture toughness K_{IC}. The creep failure time increases with increment of fracture toughness under the given axial stress. The difference of creep failure time on fracture toughness approximately is same under different axial stresses. The variation rate of creep failure time on axial stress approximately is same under different fracture toughness.

As shown in Fig. 2.38, the effects of confining pressure on creep failure time under different fracture toughness also obtained. The difference of creep failure time on fracture toughness approximately is same under different confining pressures. The variation rate of creep failure time on confining pressure approximately is same under different fracture toughness.

In summary, it can be seen that the sensitivity of fracture toughness is smaller than the sensitivity of the initial crack size and friction coefficient on the variation rate of creep failure time on axial stress or confining pressure.

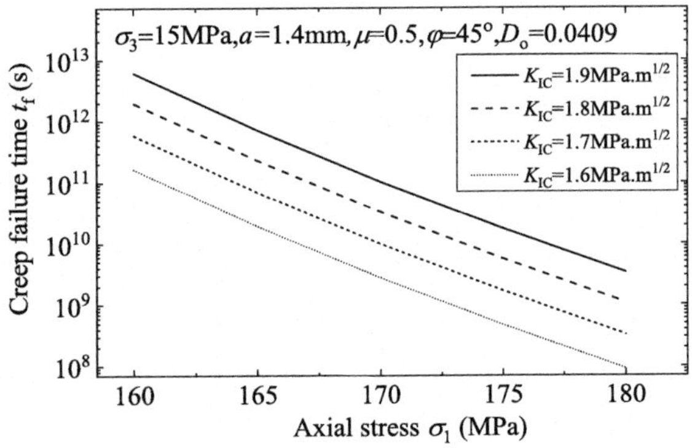

Fig. 2.37 Effects of axial stress on creep failure time under different fracture toughness

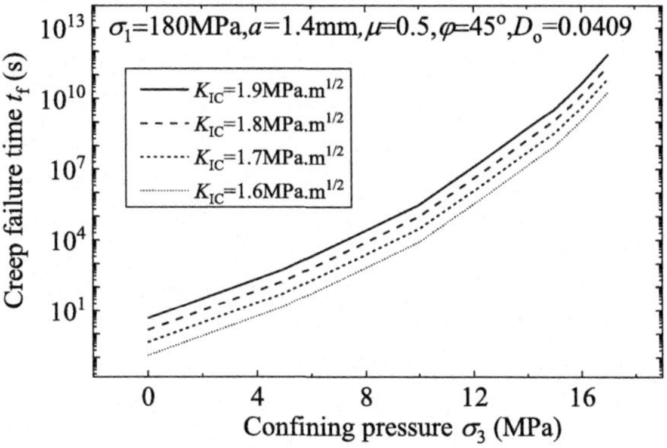

Fig. 2.38 Effects of confining pressure on creep failure time under different fracture toughness

2.4.3 Long-Term Strength

Theoretical results of steady-state creep rate and creep failure time are shown in Fig. 2.39. It is noted that the steady-state creep rate approximately equals to the minimum value of creep strain rate. Empirical equations are also presented corresponding to different confining pressures. The steady-state creep rate quickly increases with the increment of confining pressure, versus the quickly decreases of creep failure time. According to the empirical equations, the increment of confining pressure would results in the increment of creep failure time and the decrease of steady-state creep rate.

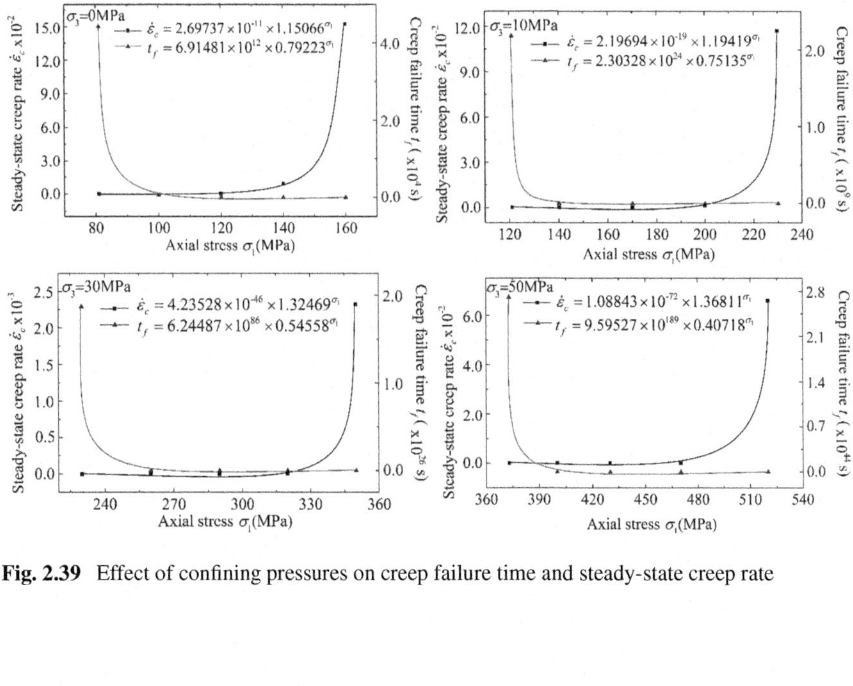

Fig. 2.39 Effect of confining pressures on creep failure time and steady-state creep rate

For deep-buried constructions, service life evaluation is very important for design and maintenance. The service life could be evaluated by creep failure time of rock. Using the empirical equations presented in Fig. 2.39, the basic stress states corresponding to different given service lives are shown in Fig. 2.40. It could be found that, in order to meet the given service life (creep failure time), the higher the axial pressure, the bigger the confining pressure is needed. For example, in order to meet 100-years service life, the confining pressure must be higher than 20 MPa corresponding to 224 MPa axial pressure. In other words, axial stress must be smaller than 224 MPa corresponding to 20 MPa confining pressure in order to meet 100-years service life. The basic relationship between axial stress and confining pressure is almost linear for a given service life. Here the axial stress and confining stress could be regarded as surrounding-rock pressure and bearing force of supporting structure, respectively.

Above the creep failure results, the stress states satisfying different creep failure time are obtained in Fig. 2.40. Thereinto, a critical stress that could transform steady-state creep to accelerating creep is defined as the long-term strength of rock. This critical stress has great significance for judging the long-term stability of rock. Thus, the long-term strength will be analyzed in detail below.

Using the theoretical model suggested in this paper, the long-term strength of Sanxia granite could be calculated by using Eqs. 2.49–2.52. It is noted that the accelerating creep stage can't appear at the creep curve by Eqs. 2.49–2.52 when the applied stress is larger than the long-term strength. All theoretical results of long-term strength are shown in Table 2.2. It could be found that the rates of initial cracking stress and long-term strength corresponding to peak stress increase with increment of confining pressure. In the case of smaller confining pressure (0, 10, and 20 MPa), stresses corresponding long-term strength are almost equal to crack initiation stress. For higher confining pressure (30, 40, and 50 MPa), long-term strength is different from crack initiation stress. It should be note that theoretical results of the rate of crack initiation stress (45–51%) agree with experiment results (40–50%) (Liu et al. 2009; Zhu et al. 2007). Otherwise, theoretical results of the rate of long-term strength

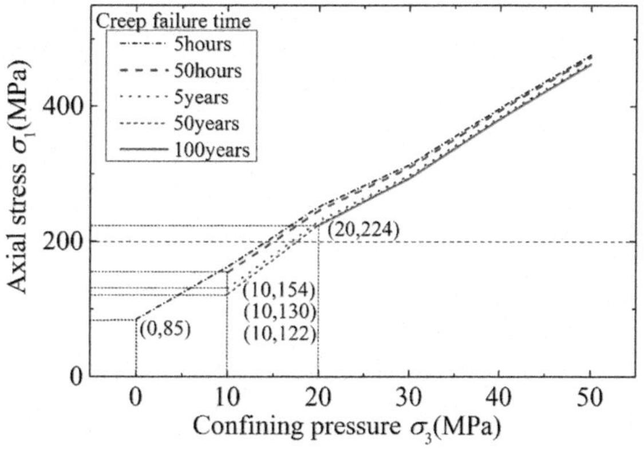

Fig. 2.40 Stress states (σ_1, σ_3) satisfying different creep failure time

Table 2.2 Critical values under different confining pressures

	Confining pressure (MPa)					
	0	10	20	30	40	50
Peak stress (MPa)	179.13	251.85	325.11	398.71	472.50	546.43
Crack initiation stress (MPa)	80.93	120.30	159.66	199.02	238.39	277.75
Long-term strength (MPa)	80.93	120.30	159.66	230.00	300.00	373.00
Crack initiation stress/peak stress (%)	45.18	47.77	49.11	49.92	50.45	50.83
Long-term strength/peak stress (%)	45.18	47.77	49.11	57.69	63.49	68.26
Wing crack length at peak stress (mm)	0.95	1.00	1.03	1.05	1.07	1.08
Damage at peak stress	0.280	0.302	0.315	0.324	0.334	0.339

(45–68%) are slightly below the experimental prediction value (70–80%) (Liu et al. 2009; Zhu et al. 2007). This difference could be due to the reason that long-term strength of rock is hard exactly determined by experimental method.

The confining pressure exhibits dramatic effect on the critical values during progressive and creep failure of rock. In order to observation and comparison well, the peak stress, crack initiation stress, long-term strength, wing crack length, damage, and so on, are shown in Table 2.2.

2.5 Conclusions

A micro–macro method is proposed to calculate the long-term creep deformations caused by microcrack growth under compression. This method is formulated by using wing crack model, subcritical growth law and a relation of crack growth and strain. During the progressive loading failure of rocks, axial stress is improved with the increment of confining pressure. The axial stress increases, and then decreases with the increasing of axial, lateral, volumetric strains and crack growth. The range between crack initiation stress and peak stress of rocks under different confining pressures is determined, which provides help for the stress states triggering rock creep failure. For long-term deformation under constant loadings, the proposed method reveals the evolution process of axial, lateral and volumetric strains caused by subcritical crack growth during creep. The actual calculated rock failure strains are smaller than that at crack coalescences, for the real rock failure generates from strain localization, not the global failure model developed by Ashby and Sammis. Limitation of crack growth $l_{\lim}$ at crack coalescence corresponding to axial, lateral and volumetric strains at rock failure are calculated, which provides a priori assumption and theoretical assistance for calculating the long-term creep deformations of rock failure caused by crack growth.

References

Ashby MF, Hallam SD. The failure of brittle solids containing small cracks under compressive stress states. Acta Metall. 1986;34(3):497–510.

Ashby MF, Sammis CG. The damage mechanics of brittle solids in compression. Pure Appl Geophy. 1990;133(3):489–521.

Atkinson BK. A fracture mechanics study of subcritical tensile cracking of quartz in wet environments. Pure Appl Geophys. 1979;117(5):1011–24.

Atkinson BK. Subcritical crack growth in geological materials. J Geophys Res. 1984;89(B6):4077–114.

Brace WF, Paulding B, Scholz C. Dilatancy in the fracture of crystalline rocks. J Geophys Res. 1966;71:3939–53.

Brantut N, Baud P, Heap MJ, Meredith PG. Micromechanics of brittle creep in rocks. J Geophys Res. 2012;117:B08412.

Budiansky B, O'Connel RJ. Elastic moduli of a cracked solid. Int J Solids Struct. 1976;12:81–97.

Chandler NA. Quantifying long-term strength and rock damage properties from plots of shear strain versus volume strain. Int J Rock Mech Min Sci. 2013;59:105–10.

Charles RJ. Static fatigue of glass I. J Appl Phys. 1958;29(11):1549–53.

Chen ZH, Tang CA, Huang RQ. A double rock sample model for rockbursts. Int J Rock Mech Min Sci. 1997;34(6):991–1000.

Jiang QQ, Li JT, Hu YF, Gui YL, Lai WM. Effects of water on subcritical crack growth. Chin J Rock Soil Mech. 2008;29(9):2527–30.

Liu QS, Hu YH, Liu B. Progressive damage constitutive models of granite based on experimental results. Chin J Rock Soil Mech. 2009;30(2):289–96.

Martin CD, Chandler NA. The progressive fracture of Lac du Bonnet granite. Int J Rock Mech Min Sci Geomech Abs. 1994;31(6):643–59.

She CX, Cui X. Influence of high pore water pressure on creep properties of rock. Chin J Rock Mech Eng. 2010;29(8):1603–9.

Sun J, Ling JM. On meso damage behaviour and long-term stability of high slope rock of the three gorges shiplocks. Chin J Rock Mech Eng. 1997;16(1):1–7.

Wan LH, Cao P, Huang YH, Wang YX. Study of subcritical crack growth of rocks and threshold values in different environments. Chin J Rock Soil Mech. 2010;31(9):2737–42.

Wang B, Zhu JB, Wu AQ, Hu JM, Xiong ZM. Experimental study on mechanical properties of Jinping marble under loading and unloading stress paths. Chin J Rock Mech Eng. 2008;27(10):2138–45.

Weibull W. A statistical distribution function of wide applicability. J Appl Mech. 1951;18:293–7.

Zhu ZQ, Sheng Q, Leng XL, Zhang ZR. Study on crack initiation mechanism of three gorges granite. Chin J Rock Mech Eng. 2007;26(12):2570–5.

Chapter 3
Loading and Unloading Path Effect on Creep Fracture

The stress path has great influence on creep properties of brittle rocks, which provides information on understanding the deformation behavior and time-delayed rockburst mechanism for surrounding rock during excavation of deep underground space. On basis of the proposed micro–macro method in Chap. 2, this chapter will introduce a stress path function to analyze the stress path of axial stress and confining pressure effect on the creep failure of brittle rocks.

3.1 Theory

The function of step loading is defined as $\sigma_1(\mathrm{t}) = \sigma_{1i} + \Delta\sigma[\mathrm{t/T}]$ (σ_{1i}, $\Delta\sigma$ are both positive), T is the cycle time, [t/T] denotes that t/T is integerized (e.g., if t/T = 0.4, [t/T] = 0, and if t/T = 1.9, [t/T] = 1). The function of step unloading is defined as $\sigma_3(\mathrm{t}) = \sigma_{3i} - \Delta\sigma[\mathrm{t/T}]$ (σ_{3i}, $\Delta\sigma$ are both positive), and the minimum of confining pressure σ_3 equals 0 (Fig. 3.1).

Inserting the stress-time function (i.e., $\sigma_1(t)$ and $\sigma_3(t)$) into Eq. (2.49), the expression of effects of stress path on evolution of subcritical crack growth is obtained as:

$$\frac{dl}{dt} = v(\pi a)^{n/2}\left\{\frac{[A_2\sigma_3(t) - A_1\sigma_1(t)](c_1 + c_2) + \sigma_3(t)c_3}{K_{\mathrm{IC}}}\right\}^n \tag{3.1}$$

X. Li et al., *Static Creep Micro-Macro Fracture Mechanics of Brittle Solids*,
https://doi.org/10.1007/978-981-99-8203-5_3

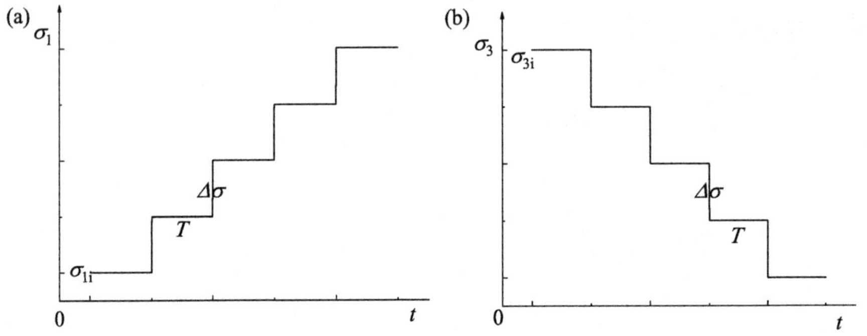

Fig. 3.1 Stepping **a** loading path of axial stress and **b** unloading path of confining pressure

3.2 Results and Discussions

3.2.1 Creep Axial, Lateral and Volumetric Strain Evolution Under Step Loading of Axial Stress

Evolutions of axial, lateral and volumetric strains are shown in Fig. 3.2 under multi-level loading (i.e., $\sigma_1 = 60$ MPa + 10 MPa [t/100,000 s] and $\sigma_3 = 2$ MPa). The corresponding stress–strain relation during multi-loading creep is shown in Fig. 3.3. The axial and lateral strains go through the decelerated and steady-state strain at each step for previous multilevel loadings, and the accelerated strain at the last step. It is noted that the positive volumetric strain represents the contraction, and negative volumetric strain represents the dilation compared with the initial volume of rocks. In Fig. 3.2, the volumetric deformation completes a transformation from contraction to dilation at the fifth-level loading, which is also observed in Fig. 3.3.

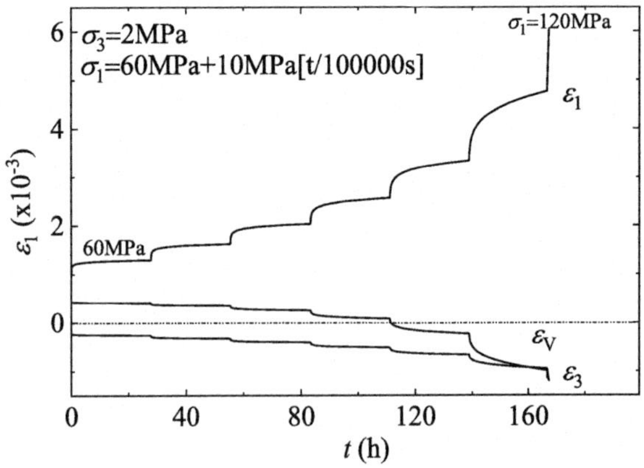

Fig. 3.2 Evolutions of strain and crack length under stepping axial stress and constant confining pressure $\sigma_3 = 2$ MPa

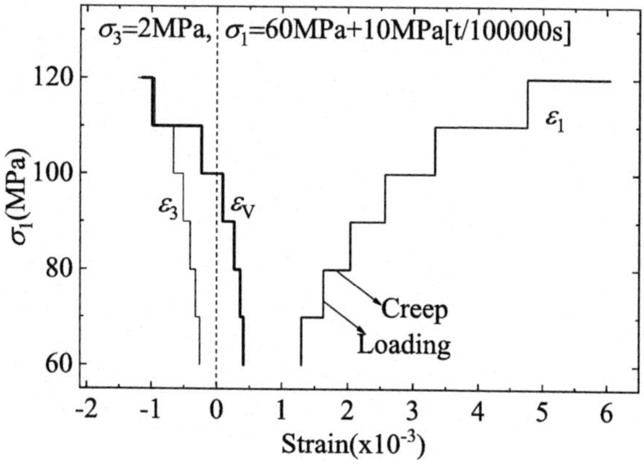

Fig. 3.3 Stress–strain relation during multi-loading creep of Fig. 3.2

As shown in Fig. 3.3, the variation value of strain at each step corresponding to the same $\Delta\sigma_1$ is increasing with increment of number of steps. Figure 3.4 shows another analytical result for evolutions of axial, lateral and volumetric strains under $\sigma_1 = 90$ MPa + 4 MPa [t/100,000 s] and $\sigma_3 = 11$ MPa. The corresponding stress-strain relation during multi-loading creep is shown in Fig. 3.5. The comparison of Figs. 3.2 and 3.4 has an understanding for effect of difference of $\Delta\sigma_1$ on multi-loading creep behaviors. The phenomena of theoretical results are also in good accord with the experimental results (Zhang et al. 2012, 2016) during multi-loading creep. Rationality of the proposed methods is verified further.

Comparing Figs. 3.2 and 2.23, or Figs. 3.4 and 2.25, the failure time for multilevel loading of axial stress is far less than the failure time for constant stress states. This phenomenon shows that the multilevel loading of axial stress strongly accelerates the crack growth to rock failure. Thus, the multi-loading creep provides a help for experimentally evaluating the long-term creep properties of brittle rocks at a relatively short time.

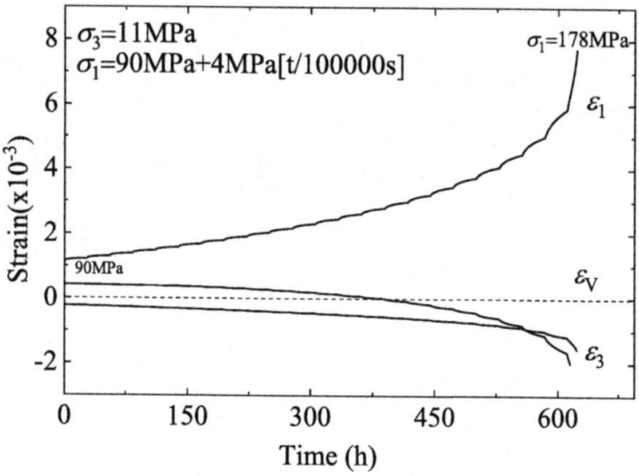

Fig. 3.4 Evolutions of strain and crack length under stepping axial stress and constant confining pressure $\sigma_3 = 11$ MPa

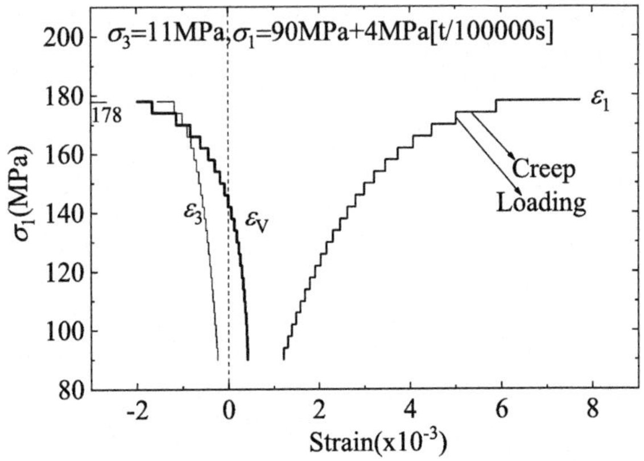

Fig. 3.5 Stress–strain relation during multi-loading creep of 3.10

3.2.2 *Creep Crack and Strain Rate Evolution Under Step Loading of Axial Stress*

The function of step loading of axial stress was selected as $\sigma_1 = 60$ MPa + 10 MPa [t/T] at $\sigma_3 = 2$ MPa, and $\sigma_1 = 90$ MPa + 4 MPa [t/T] at $\sigma_3 = 11$ MPa, respectively. T = 100,000 s ≈ 27.78 h. The initial value of step loading for axial stress was 60 MPa for $\sigma_3 = 2$ MPa and 90 MP for $\sigma_3 = 11$ MPa.

The evolutions of crack length were obtained by substituting the functions of step loading (i.e., $\sigma_1 = 60$ MPa + 10 MPa [t/100,000 s] for $\sigma_3 = 2$ MPa and $\sigma_1 = 90$ MPa + 4 MPa [t/100,000 s] for $\sigma_3 = 11$ MPa) into Eq. (3.1). Figure 3.6 shows that the crack length experienced transient and steady-state growth at low axial stress. Accelerated growth appeared at the last step when the axial stress reached 110 MPa (for $\sigma_3 = 2$ MPa) and 174 MPa (for $\sigma_3 = 11$ MPa).

Evolutions of axial strain under step loading of axial stress were shown in Fig. 3.7. The axial strain has two creep stages at low axial stress: transient and steady-state

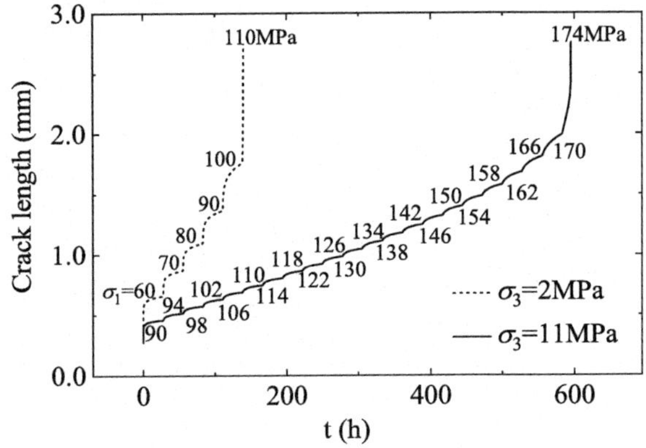

Fig. 3.6 Evolutions of crack length under step axial loading

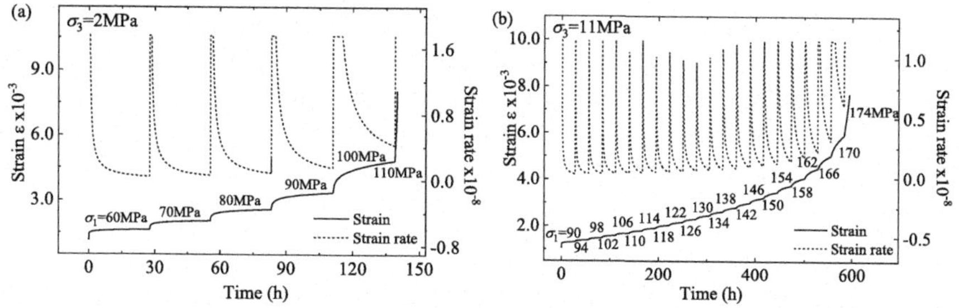

Fig. 3.7 Evolutions of axial strain and strain rate under step axial loading. **a** $\sigma_3 = 2$ MPa, **b** $\sigma_3 = 11$ MPa

creep. The accelerated creep appeared at the last step when the axial stress reached 110 MPa ($\sigma_3 = 2$ MPa) and 174 MPa ($\sigma_3 = 11$ MPa). The rock failed when the axial stress reached 110 MPa ($\sigma_3 = 2$ MPa) and 174 MPa ($\sigma_3 = 11$ MPa). The figure also shows that the increment of axial strain for each step increased with incremental steps. This phenomenon corresponds to the variation of steady-state creep strain rate. The axial creep strain rate for each step increased incrementally.

This stress-stepping test approach has been widely used and is reliable (Fabre and Pellet 2006; Ma and Daemen 2006; Zhang et al. 2012). The experimental and theoretical results of axial creep strain under different loading conditions are shown in Fig. 3.8. The figure shows that the theoretical results are consistent with the experimental results (She and Cui 2010).

The numerical results are not exactly the same as the experimental results. For the experimental results, the stress gradually increased after stabilization of the axial deformation. This deformation was considered to be stabilized when the strain rate became lower than a smaller value. Therefore, the step value $\Delta\sigma$ is different at each step in the stress-stepping creep test. However, because the stress path controlled by the strain rate can hardly be implemented in the analytical method, the step value $\Delta\sigma$ for the stress path is assumed to be constant. Thus, the numerical axial stress value could not accurately equal the experimental axial stress value in Fig. 3.8. The percentage of difference of axial stress for each step ranged from 0.7–13.4%. However, it worked well, and the experimental results would be well recovered.

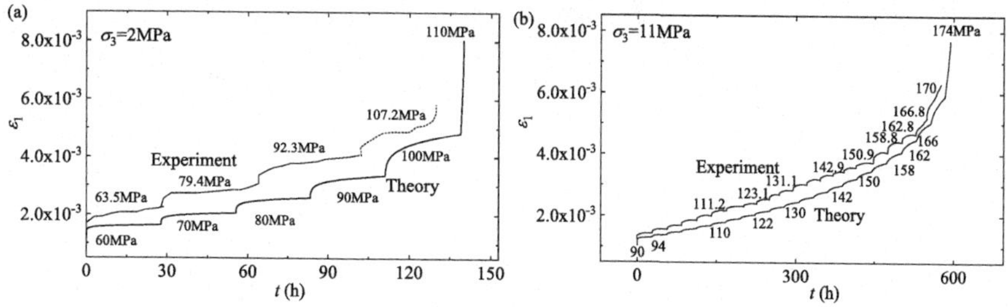

Fig. 3.8 Theoretical and experimental results of axial creep strain under loading of axial stress

Furthermore, the experimental axial stress value at each step was larger than the numerical stress, and the strain corresponding to the axial stress at each step in the experimental results was also larger than those in the numerical results. The larger axial stress corresponds to the larger strain. This phenomenon is known, and it shows that the difference between numerical and experimental results is acceptable.

3.2.3 Creep Strain, Crack and Strain Rate Evolution Under Step Unloading of Confining Pressure

The function of the step unloading of confining pressure was selected as $\sigma_3 = 20$–6 MPa [t/100 h]. The initial values of step unloading for confining pressure were both 20 MPa for $\sigma_1 = 110$ MPa and 130 MPa.

The evolutions of crack length under step unloading of confining pressure were identified as $\sigma_1 = 110$ MPa and 130 MPa by substituting the function of step unloading (i.e., $\sigma_3 = 20$–6 MPa [t/100 h]) into Eq. (3.1). Figure 3.9 shows that the creep failure time at $\sigma_1 = 110$ MPa was smaller than the creep failure time at $\sigma_1 = 130$ MPa. The crack length experienced a transient and steady-state growth at high confining pressure. The accelerated growth appeared at the last step when the confining pressure reached 2 MPa (for $\sigma_1 = 130$ MPa) and 0 MPa (for $\sigma_1 = 110$ MPa).

The evolution of the axial strain under step unloading of confining pressure was identified as $\sigma_1 = 110$ MPa and 130 MPa in Fig. 3.10. The axial strain showed two creep stages at high confining pressure: transient and steady-state creep. The accelerated creep appeared at the last step when the confining pressure reached 2 MPa (for $\sigma_1 = 130$ MPa) and 174 MPa (for $\sigma_3 = 11$ MPa). Furthermore, the increment of strain for each step increased incrementally. This result corresponds to the variation of the steady-state creep rate. The steady-state creep strain rate increased incrementally.

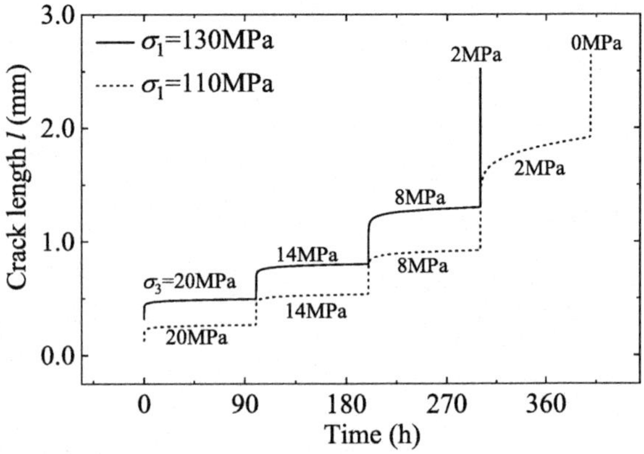

Fig. 3.9 Evolutions of crack length under step unloading of confining pressure

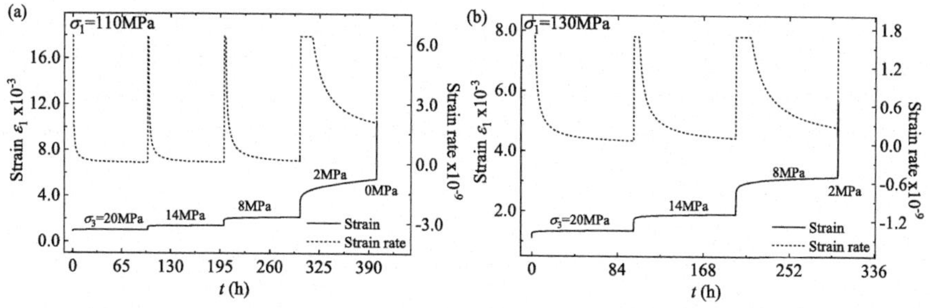

Fig. 3.10 Evolutions of axial strain and strain rate under step unloading of confining pressure **a** $\sigma_1 = 110$ MPa, **b** $\sigma_1 = 130$ MPa

Figure 2.22 shows that crack initiation stresses are 153 MPa for $\sigma_3 = 40$ MPa and 126 MPa for $\sigma_3 = 30$ MPa. Thus, the deformation behaved elastically when the applied axial stress was smaller than the crack initiation stress at a given confining pressure. As shown in Fig. 3.11, the deformation of rock (e.g., for $\sigma_3 = 40$ MPa and 30 MPa at $\sigma_1 = 110$ MPa, and $\sigma_3 = 40$ at $\sigma_1 = 130$ MPa) behaved elastically in the experimental results. However, the theoretical curves in Fig. 3.11 were only obtained from the stage of inelastic deformation.

During the stress-stepping test, the rock was observed to fail abruptly after the last decrement of the confining pressure and before the beginning of the accelerated creep phase (Fig. 3.11). This indicates that the applied axial stress was close to the rock compressive strength at this confining pressure. Furthermore, owing to the extremely short time of the rock failure stage under the unloading of confining pressure, the creep curves under the last step were difficult to obtain from the experiment in Fig. 3.11. The theoretical results show that the time was very short in the last step. Neglecting these effect factors and the diffidence of confining pressure between the theoretical and experimental value, the theoretical results are similar to the experimental results at the stage of inelastic deformation (Xia et al. 2009). This step provides theoretical assistance in studying the creep behavior under step unloading of confining pressure.

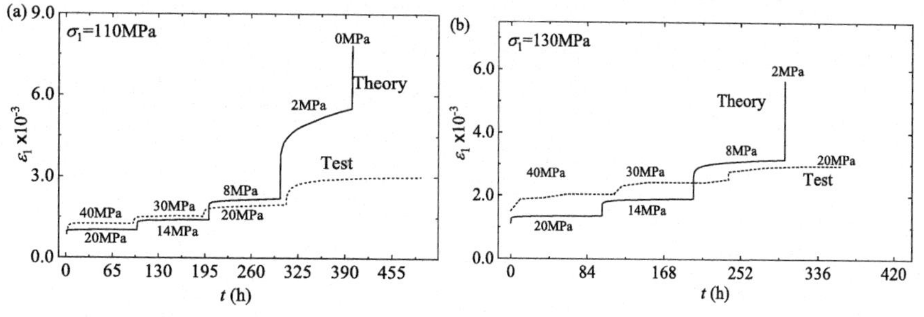

Fig. 3.11 Theoretical and experimental results of creep strain under step unloading of confining pressure

In the present study, the rationality of the theoretical creep results is only verified for brittle rocks. Actually, compressive creep behavior has also been widely studied for other brittle materials such as glass and concrete (Liu et al. 2002; Ranaivomanana et al. 2013; Mallet et al. 2015). Mechanisms such as the microcrack growth of brittle rocks were also observed. Thus, the theoretical model may also be applied to these brittle materials.

3.2.4 *Sensitivity of Model Parameters*

The effect of initial crack size on creep behavior under step loading of axial stress is shown in Fig. 3.12. Due to the initial damage $D_o = 4\pi N_V(\alpha a)^3/3$, the initial crack size is closely related to the initial damage. It other words, the initial crack size corresponds to the initial damage under given initial microcrack number N_V.

The creep failure time increased with decrements of initial crack size (i.e., initial damage). The creep failure time was 3676 s, and the creep failure appeared at the second step loading of axial stress when the initial crack size $a = 1.4$ mm. The creep failure time was 21,945 s, and the creep failure appeared at the seventh step loading of axial stress when the initial crack size $a = 1.3$ mm. This means that the number of steps during the creep failure also increases with decrements of initial crack size. The initial strain also increases with increments of initial crack size (i.e., initial damage). These results show that the larger the initial damage is, the larger the initial strain is.

The effects of friction coefficient, stress corrosion index, and fracture toughness on creep behavior under step loading of axial stress are shown in Figs. 3.13, 3.14 and 3.15, respectively. The creep failure time increased with increments of friction coefficients, stress corrosion index, or fracture toughness.

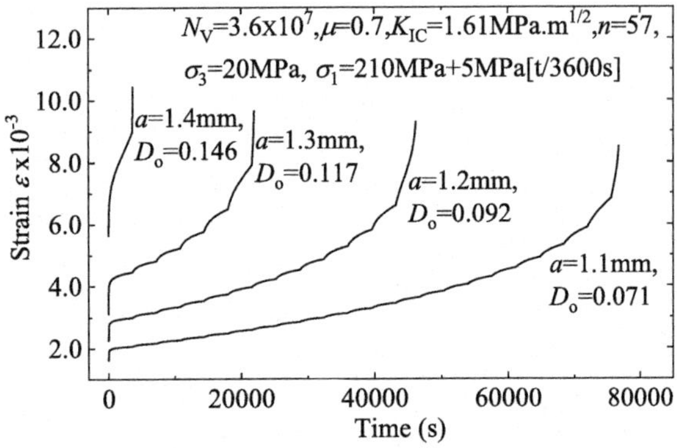

Fig. 3.12 Effect of initial crack size on creep behavior under step loading of axial stress

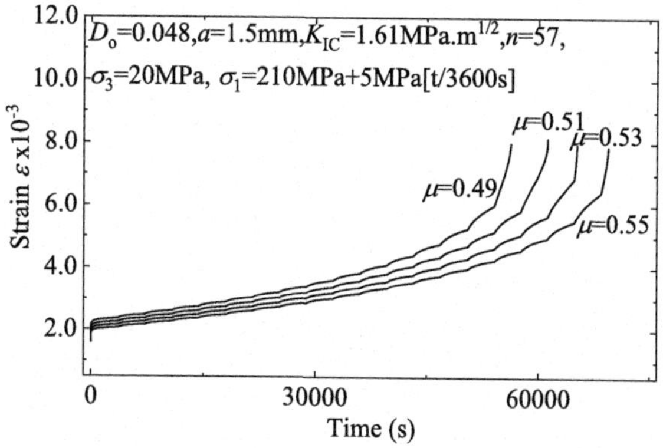

Fig. 3.13 Effect of friction coefficient on creep behavior under step loading of axial stress

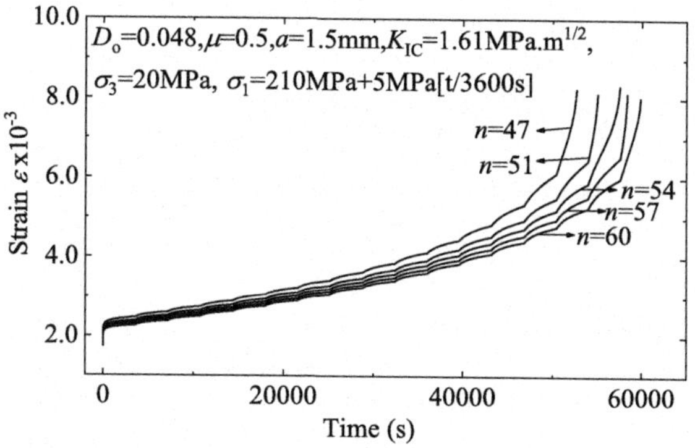

Fig. 3.14 Effect of stress corrosion index on creep behavior under step loading of axial stress

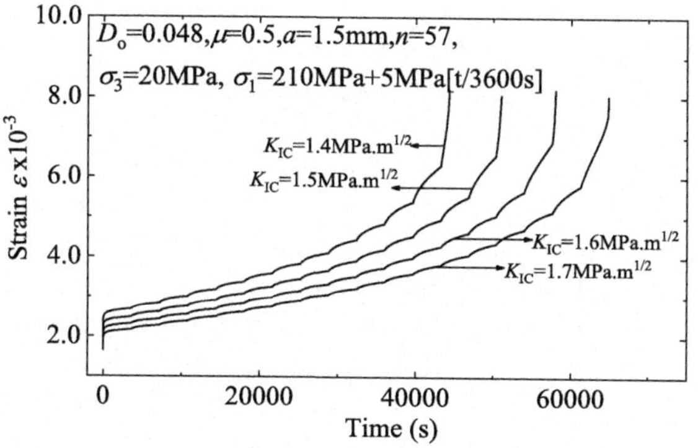

Fig. 3.15 Effect of fracture toughness on creep behavior under step loading of axial stress

3.2.5 *Effects of Unloading Rate of Confining Pressure*

The effects of the unloading rate of confining pressure on the strain and strain rate were obtained based on creep in Fig. 3.16. The constant axial stress $\sigma_1 = 110$ MPa,

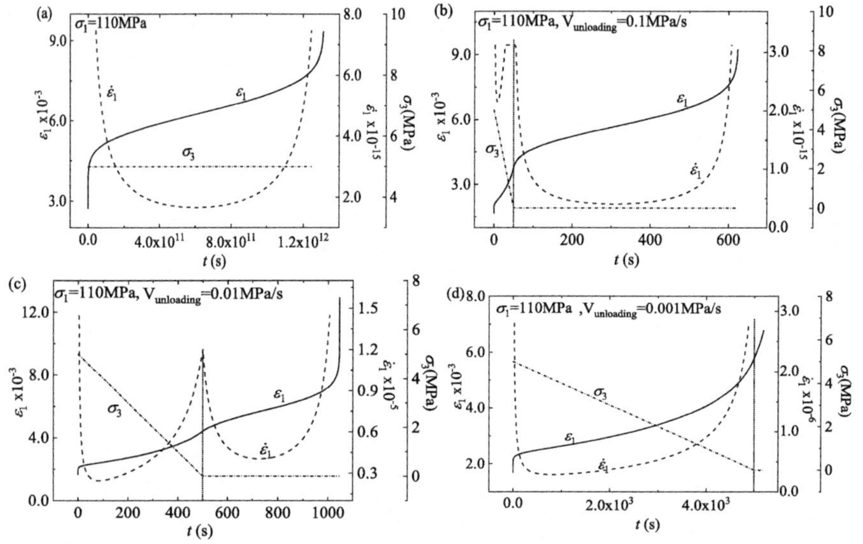

Fig. 3.16 Effects of unloading rate for confining pressure on axial strain and strain rate during creep

and the initial confining pressure $\sigma_3 = 5$ MPa. As shown in Fig. 3.16a, the creep failure time was 1.3×10^{12} s (i.e., 4122 years) when confining pressure $\sigma_3 = 5$ MPa.

As shown in Fig. 3.16b–d, the creep failure process was given when unloading rate of confining pressure $V_{\text{unloading}} = 0.1, 0.01$, and 0.01 MPa/s, respectively. The creep failure process had two stages: the evolutions of axial strain at $\sigma_3 = 5$ MPa-($V_{\text{unloading}}$)t and $\sigma_3 = 0$ MPa, respectively. Due to the break of the strain rate, an inflection point of axial strain appeared when the confining pressure decreased to 0 MPa. The time of the inflection point of axial strain was 50, 500, and 5000 s. The strain value of the inflection point was 4.0×10^{-3}, 4.6×10^{-3}, and 5.8×10^{-3}. The creep failure time was 650, 1050, and 5250 s or so throughout the process. The time was 570, 550, and 240 s or so at $\sigma_3 = 0$ MP. Furthermore, due to small variations of strain rate at t = 5000 s in Fig. 3.16d, the inflection point of axial strain was not apparent when the confining pressure decreases to 0 MPa.

A comparison of Figs. 3.16a–d shows that the creep failure time decreases with increments of unloading rate of confining pressure throughout the process. However, the time during $\sigma_3 = 0$ MPa increased with increments of unloading rate. A smaller unloading rate led to longer unloading time and then led to larger strain during the unloading stage and a shorter time at $\sigma_3 = 0$ MPa. Furthermore, the creep failure time at the unloading of confining pressure was significantly less than the creep failure time at constant confining pressure. This result shows that the unloading of confining pressure strongly influences the creep failure time.

3.2.6 *Coupling Effect of Step Axial Stress and Confining Pressure*

In the previous chapters, the stress path was only focused on studying either axial stress or confining pressure. However, in reality, the coupling of stress paths in step axial stress and confining pressure has more significance for the application of underground engineering. In this chapter, the coupling effect of step axial stress and confining pressure on creep failure is examined, thereby providing theoretical assistance to studies on rock mechanical behavior under loading and unloading during the underground excavation. The initial axial stress $\sigma_{1i} = 210$ MPa, initial confining pressure $\sigma_{3i} = 40$ MPa, and the step value $\Delta\sigma = 5$ MPa and below.

As shown in Fig. 3.17, the function of the stress path was $\sigma_1 = 210$ MPa + 5 MPa [t/3600 s] for $\sigma_3 = 40$ MPa under step loading of axial stress. The accelerated creep failure stage appeared at $\sigma_1 = 390$ MPa. The creep failure time was 130,312 s. The function of stress path was $\sigma_3 = 40$ MPa–5 MPa [t/3600 s] for $\sigma_1 = 210$ MPa under step unloading of confining pressure. The accelerated creep failure stage appeared at $\sigma_3 = 10$ MPa. The creep failure time was 21,601 s. The creep failure time for unloading of confining pressure was smaller than the loading of axial stress under the same step value and the initial stress state. These results show that the dependence of confining pressure on creep failure is larger than the dependence of axial stress on creep failure.

As shown in Fig. 3.18, the function of the stress path was $\sigma_1 = 210$ MPa + 5 MPa [t/3600 s] and $\sigma_3 = 40$ MPa–5 MPa [t/3600 s]. The accelerated creep failure stage appeared at $\sigma_1 = 235$ MPa and $\sigma_3 = 15$ MPa. The creep failure time was 18,001 s. The creep failure time in Fig. 3.18 was smaller than the creep failure time in Fig. 3.17, which shows that the coupling effect of step confining pressure and axial stress was larger than the single effect of axial stress or confining pressure on creep failure.

As shown in Fig. 3.19, the function of step loading for the axial stress path was $\sigma_1 = 210$ MPa + 5 MPa [t/3600 s]. The confining pressure was applied by the loading and unloading of the cycle step (i.e., $\sigma_3 = 40$ and 35 MPa). The accelerated creep failure stage appeared at $\sigma_1 = 355$ MPa and $\sigma_3 = 35$ MPa. The failure time

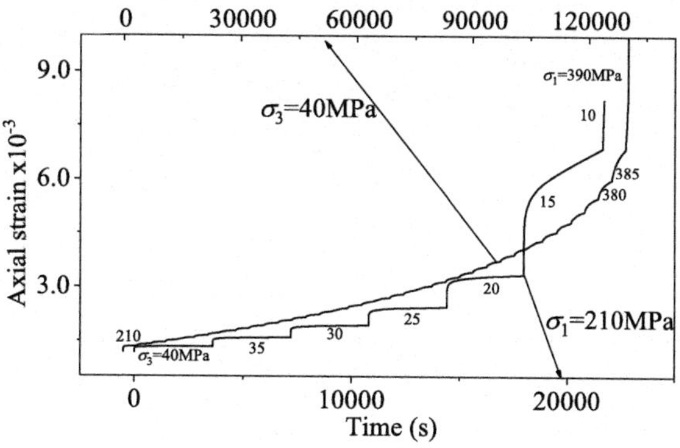

Fig. 3.17 Comparison of evolutions of axial strain under step loading of axial stress and step unloading of confining pressure

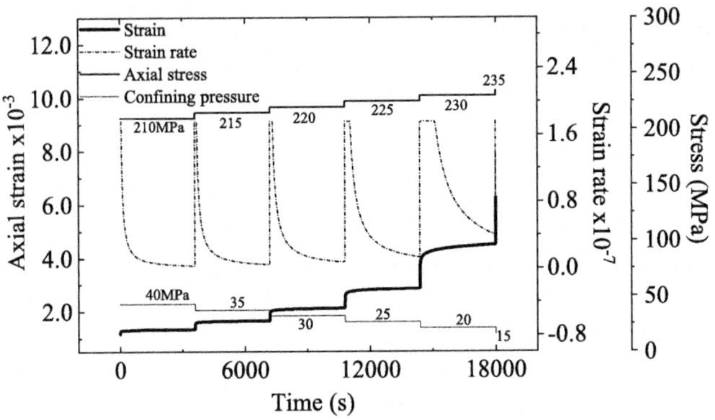

Fig. 3.18 Evolutions of axial strain and strain rate under step loading of axial stress and unloading of confining pressure

was 105,093 s, which is smaller than the failure time at step loading of axial stress in Fig. 3.17. These results show that the cycle loading and unloading of confining pressure weakens the mechanical property, increases the damage, and accelerates the failure rate of rock. The strain rate at $\sigma_3 = 40$ MPa was significantly smaller than $\sigma_3 = 35$ MPa. Thus, the strain rate at $\sigma_3 = 40$ MPa approximately equals a constant value as shown in Fig. 3.19. Furthermore, the minimum strain rate at $\sigma_3 = 35$ MPa increases incrementally.

The theoretical results are similar to triaxial unloading experiment results of studies in which the delayed rockburst of surrounding rock was studied during underground quarrying (He et al. 2010, 2012). The findings assist in understanding deformation behavior and time-delayed rockburst mechanism for surrounding rock during underground excavations.

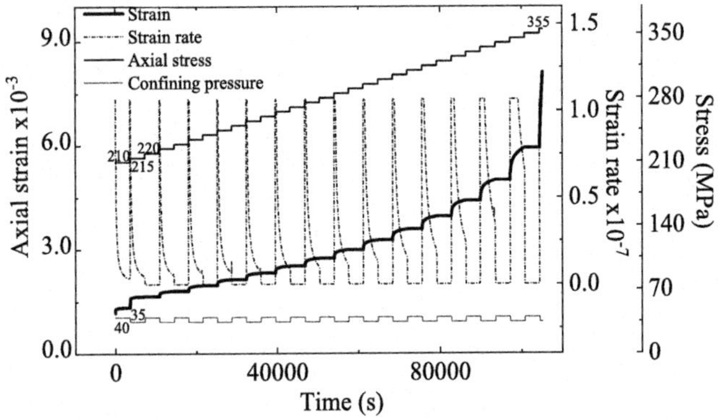

Fig. 3.19 Evolutions of axial strain and strain rate under step loading of axial stress and cycle step loading and unloading of confining pressure

3.2.7 Reloading Time Effect on Creep Failure After Unloading of Confining Pressure

In Fig. 3.20a–c, the creep evolutions of axial strain and strain rate under constant axial stress 110 MPa are shown when confining pressure unloads from 5 to 0 MPa during 500 s and reloads to 0.5 MPa after 100 s, 300 s and 500 s, respectively. After unloading the confining pressure to no confining pressure, the longer the time between the confining pressure reloading, the shorter the final creep time of the rock. This indicates that the stability time of brittle rock can be prolonged by timely reloading of confining pressure after unloading. For underground engineering, timely surrounding rock support can ensure good stability of surrounding rock for a long time after underground excavation.

3.2.8 Reloading Magnitude Effect on Creep Failure After Unloading of Confining Pressure

In Fig. 3.21, the creep evolutions of axial strain and strain rate under constant axial stress 110 MPa are shown when confining pressure unloads from 5 to 0 MPa during 500 s and then immediately reloads to 1 MPa after 500 s, respectively.

By comparing Figs. 3.20c and 3.21, it can be seen that the larger the confining pressure value of reloading, the longer the final creep failure time of rock after unloading. This shows that the increase of confining pressure has an important effect on the stability time of rock. It also indirectly shows that the resistance of surrounding rock supporting structure plays an important role in ensuring the stability of surrounding rock in underground engineering.

3.3 Conclusions

The crack growth, strain, and strain rate are investigated under step loading for axial stress at constant confining pressure or step unloading for confining pressure at constant axial stress. The two applied stress paths both accelerate the crack growth and shorten creep failure time.

Effects of the unloading rate of confining pressure on axial strain and strain rate are studied during creep. Creep failure time decreases with increment of unloading rate of confining pressure during the entire creep process. Coupling effects of step loading of axial stress and step unloading of confining pressure or step loading of axial stress and cycle loading and unloading of confining pressure on creep failure are also predicted. The coupling stress path weakens the mechanical property, increases the damage, and accelerates the failure of rock compared to the single stress path.

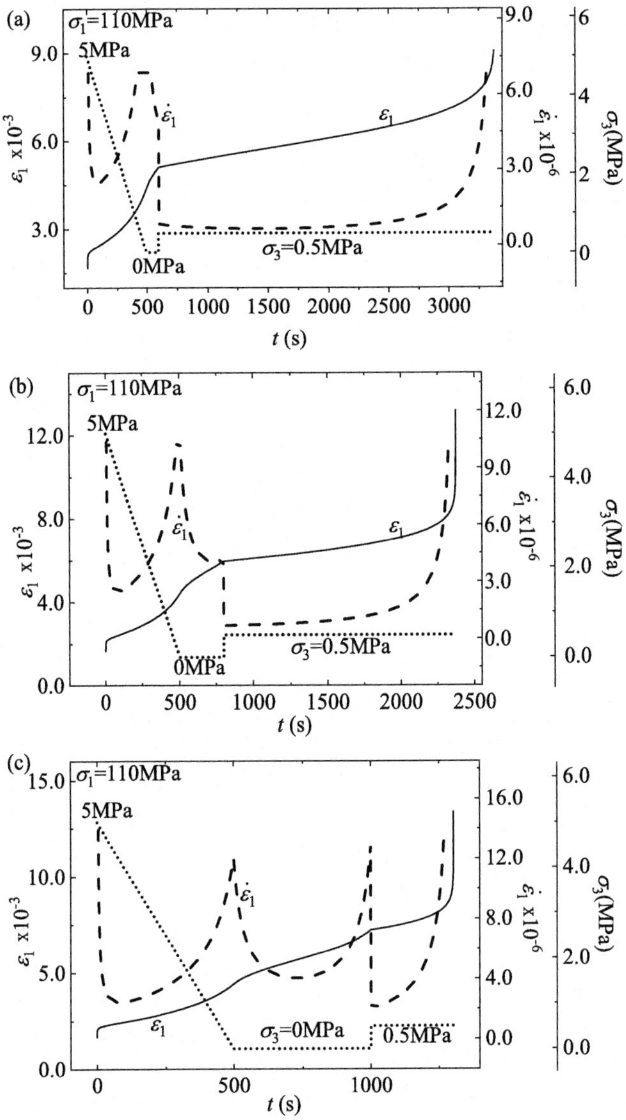

Fig. 3.20 The effect of reloading time **a** 600 s, **b** 800 s, and **c** 1000 s, after unloading of confining pressure on creep failure

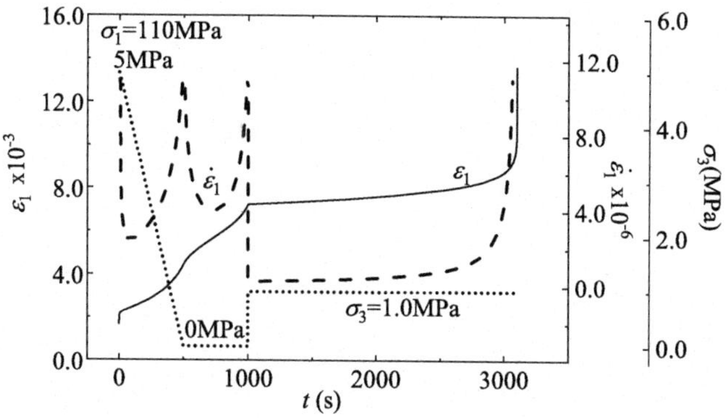

Fig. 3.21 Creep curve under the reloading magnitude 1 MPa at 1000 s after unloading of confining pressure

References

Fabre G, Pellet F. Creep and time-dependent damage in argillaceous rocks. Int J Rock Mech Min Sci. 2006;43(6):950–60.

He MC, Miao JL, Feng JL. Rock burst process of limestone and its acoustic emission characteristics under true-triaxial unloading conditions. Int J Rock Mech Min Sci. 2010;47(2):286–98.

He M, Jia X, Coli M, Livi E, Sousa L. Experimental study of rockbursts in underground quarrying of Carrara marble. Int J Rock Mech Min Sci. 2012;52(6):1–8.

Liu GT, Gao H, Chen FQ. Microstudy on creep of concrete at early age under biaxial compression. Cem Concr Res. 2002;32(12):1865–70.

Ma L, Daemen JJK. An experimental study on creep of welded tuff. Int J Rock Mech Min Sci. 2006;43(2):282–91.

Mallet C, Fortin J, Guéguen Y, Bouyer F. Brittle creep and subcritical crack propagation in glass submitted to triaxial conditions. J Geophys Res. 2015;120(2):879–93.

Ranaivomanana N, Multon S, Turatsinze A. Tensile, compressive and flexural basic creep of concrete at different stress levels. Cem Concr Res. 2013;52(10):1–10.

She CX, Cui X. Influence of high pore water pressure on creep properties of rock. Chin J Rock Mech Eng. 2010;29(8):1603–9.

Xia CC, Yan ZJ, Wang XD, Zhang CS, Zhao X. Research on elasto-viscopastic constitutive relation of marble under unloading condition. Chin J Rock Mech Eng. 2009;28(3):459–66.

Zhang ZL, Xu WY, Wang W, Wang RB. Triaxial creep tests of rock from the compressive zone of dam foundation in Xiangjiaba hydropower station. Int J Rock Mech Min Sci. 2012;50(35):133–9.

Zhang Y, Shao J, Xu W, Jia Y. Time-dependent behavior of cataclastic rocks in a multi-loading triaxial creep test. Rock Mech Rock Eng. 2016;49:3793–803.

Chapter 4
Shear Strength Evolution During Creep Fracture

Shear fracture triggered by subcritical crack extension in intact brittle rocks under long-term compressive loading plays a significant role in the evaluation of earthquake mechanisms. Numerous studies into shear mechanical properties of rocks in earth crusts focused on shear friction in faults (Dieterich 1979a, b; Ruina 1983) or shear fracture in intact rocks during creep (Ohnaka et al. 1997; Brantut and Viesca 2015), which provides an important help for evaluating earthquake nucleation. Thus, this chapter attempts to establish a method to link the proposed micro–macro model of Chaps. 2 and 3 to shear fracture properties. This method will introduce Mohr–Coulomb failure criterion and Mohr–Coulomb strain-softening model on basis of the Chaps. 2 and 3 to explore the relationship between the microcrack behaviors and shear fracture properties in brittle rocks during progressive and creep failure.

4.1 Theory

4.1.1 *Relationships Between Micromechanical Parameters and Shear Strength*

Due to the existence of a maximum $\sigma_{1\text{peak}}$ in Eq. (2.8), a peak point (l_{peak}, $\sigma_{1\text{peak}}$) can be achieved by the curve of axial stress and crack growth. The relationship between axial stress and crack length at peak point can be expressed as:

$$\sigma_{1\text{peak}}(l_{\text{peak}}) = \frac{\sigma_3[c_3 + A_2(c_1 + c_2)]}{A_1(c_1 + c_2)} - \frac{K_{\text{IC}}}{A_1\sqrt{\pi a}(c_1 + c_2)} \tag{4.1}$$

where l_{peak} is the wing crack length corresponding to peak stress $\sigma_{1\text{peak}}$. It is noted that the values of parameters c_1, c_2 and c_3 in Eq. (4.1) are achieved at $l = l_{\text{peak}}$.

X. Li et al., *Static Creep Micro-Macro Fracture Mechanics of Brittle Solids*,
https://doi.org/10.1007/978-981-99-8203-5_4

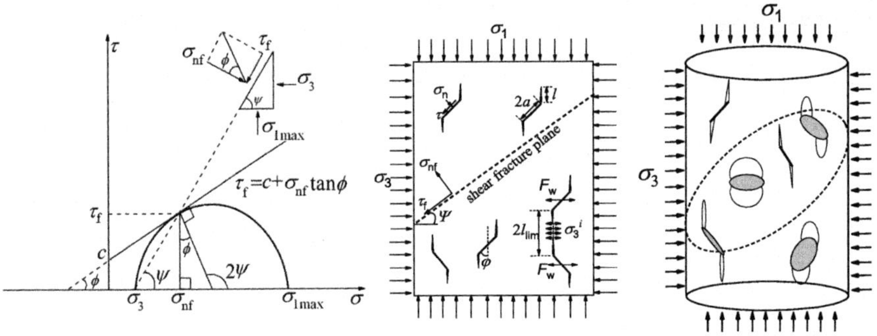

Fig. 4.1 Mohr–Coulomb failure criterion. (σ_{nf} and τ_{f} are the normal stress and shear strength on shear failure surface)

As shown in Fig. 4.1, the Mohr–Coulomb failure criterion is given as:

$$\sigma_{1\,\text{max}} = \frac{\sigma_3}{\tan^2(45^\circ - \phi/2)} - \frac{2c}{\tan(45^\circ - \phi/2)} \tag{4.2}$$

where c is cohesion, ϕ is internal friction angle, and $\sigma_{1\text{max}}$ is the maximum principal stress.

For the conventional triaxial compressive tests, the peak stress in the measured stress–strain curve equals the maximum principal stress (i.e., $\sigma_{1\text{peak}} = \sigma_{1\text{max}}$). Furthermore, for the single rock sample, the material parameters (i.e., a, α, β and μ) in Eq. (4.1) could be held constant. Thus, it can be seen that Eq. (4.1) is in agreement with Mohr–Coulomb failure criterion in Eq. (4.2). The axial stress σ_1 (i.e., $\sigma_{1\text{peak}}$ or $\sigma_{1\text{max}}$) and confining pressure σ_3 in Eqs. (4.1) and (4.2) both fall on a linear failure envelope. Combining Eqs. (4.1) and (4.2), the expressions of the relationships between micromechanical parameters and shear strength, cohesion, internal friction angle can be derived as:

$$\tau_{\text{f}}(l_{\text{peak}}) = \sigma_{\text{nf}} \tan\phi - c \tag{4.3}$$

where

$$\sigma_{\text{nf}}(l_{\text{peak}}) = \frac{\sigma_{1\text{peak}} + \sigma_3}{2} + \frac{\sigma_3 - \sigma_{1\text{peak}}}{2}\cos 2\psi \tag{4.4}$$

$$\phi(l_{\text{peak}}) = 90^\circ - 2arc\tan\sqrt{\frac{A_1(c_1 + c_2)}{c_3 + A_2(c_1 + c_2)}} \tag{4.5}$$

$$c(l_{\text{peak}}) = \frac{K_{\text{IC}}}{2\sqrt{A_1 \pi a[c_3 + A_2(c_1 + c_2)](c_1 + c_2)}} \tag{4.6}$$

and ψ is the macroscopic angle for shear failure plane in rocks. The correlation of shear failure plane angle and internal friction angle is $90^\circ + \phi = 2\psi$.

The peak axial strain can be easily obtained by the stress–strain curves in a compressive test. Substituting Eq. (2.18) into Eqs. (4.3–4.6), the shear strength, internal friction angle and cohesion related to macroscopic strain can be predicted as:

$$\tau_f(\varepsilon_{1\text{peak}}) = \sigma_{\text{nf}} \tan \phi - c \tag{4.7}$$

where

$$\sigma_{\text{nf}}(\varepsilon_{1\text{peak}}) = \frac{\sigma_{1\text{peak}} + \sigma_3}{2} + \frac{\sigma_3 - \sigma_{1\text{peak}}}{2} \cos 2\psi \tag{4.8}$$

$$\phi(\varepsilon_{1\text{peak}}) = 90^\circ - 2arc\tan\sqrt{\frac{A_1(J_1 + J_2)}{J_3 + A_2(J_1 + J_2)}} \tag{4.9}$$

$$c(\varepsilon_{1\text{peak}}) = \frac{K_{IC}}{2\sqrt{A_1 \pi a[J_3 + A_2(J_1 + J_2)](J_1 + J_2)}} \tag{4.10}$$

It is noted that the values of parameters J_1, J_2 and J_3 in Eq. (4.1) are achieved at $\varepsilon_1 = \varepsilon_{1\text{peak}}$.

4.1.2 Evolution of Shear Strength During Progressive and Creep Failure

Equations (4.3) and (4.7) only explain the relationship between shear strength in Mohr–Coulomb's failure criterion and microcrack parameters at the peak point of Eq. (4.1). However, in the Mohr–Coulomb strain-softening law (Zhao and Cai 2010) in Fig. 4.2, the locus of the peak point at the cyclic compressive test is approximately equivalent to the post-peak curve of the stress–strain constitutive curve from the conventional compressive loading test, which helps in the evaluation of continuous shear properties at the post-peak phase of the stress–strain relation. The invariant l_{peak} and $\sigma_{1\text{peak}}$ in Eqs. (4.3–4.6) are replaced respectively with the variables l and $\sigma_1(l)$ at the case of $l \geq l_{\text{peak}}$ (which corresponds to the case of $\varepsilon_1 \geq \varepsilon_{1\text{peak}}$). In addition, the cracking-dependent shear strength at the post-peak relation between stress and crack extension in Eq. (4.1) (i.e., $l \geq l_{\text{peak}}$) can be obtained as

$$\tau_f(l) = \sigma_{\text{nf}}(\tan \phi(l)) - c(l), \quad (l \geq l_{\text{peak}}) \tag{4.11}$$

where

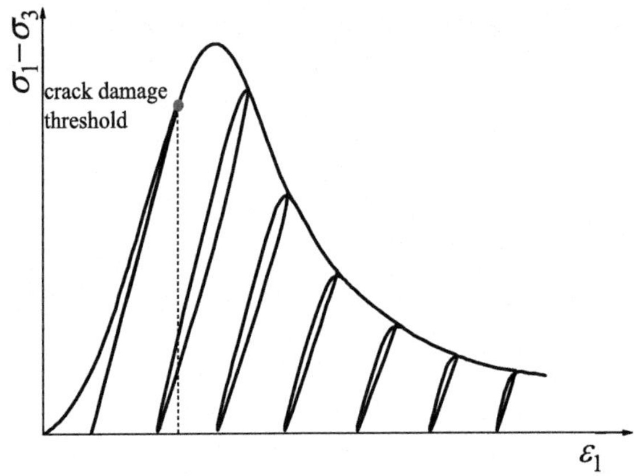

Fig. 4.2 Stress–strain curve under triaxial loading–unloading cycle test which is related to Mohr–Coulomb strain-softening model

$$\sigma_{\text{nf}}(l) = \frac{\sigma_1(l)+\sigma_3}{2} + \frac{\sigma_1(l)-\sigma_3}{2}\cos 2\psi(l) \tag{4.12}$$

$$\phi(l) = 90^{\circ} - 2arc\tan\sqrt{\frac{A_1(c_1+c_2)}{c_3+A_2(c_1+c_2)}} \tag{4.13}$$

$$c(l) = \frac{K_{\text{IC}}}{2\sqrt{A_1\pi a[c_3+A_2(c_1+c_2)](c_1+c_2)}} \tag{4.14}$$

where the inclination angle of the shear fracture plane $\psi(l) = \phi(l)/2 + 45^{\circ}$.

The strain-related shear strength, cohesion, and internal friction angle are also derived as follows:

$$\tau_{\text{f}}(\varepsilon_1) = \sigma_{\text{nf}}(\tan\phi(\varepsilon_1)) - c(\varepsilon_1), \quad (\varepsilon_1 \geq \varepsilon_{1\text{peak}}) \tag{4.15}$$

where

$$\sigma_{\text{nf}}(\varepsilon_1) = \frac{\sigma_1(\varepsilon_1)+\sigma_3}{2} + \frac{\sigma_1(\varepsilon_1)-\sigma_3}{2}\cos 2\psi(\varepsilon_1) \tag{4.16}$$

$$\phi(\varepsilon_1) = 90^{\circ} - 2arc\tan\sqrt{\frac{A_1(J_1+J_2)}{J_3+A_2(J_1+J_2)}} \tag{4.17}$$

$$c(\varepsilon_1) = \frac{K_{\text{IC}}}{2\sqrt{A_1\pi a(J_3+A_2(J_1+J_2))(J_1+J_2)}} \tag{4.18}$$

where the inclination angle of the shear fracture plane $\psi(\varepsilon_1) = \phi(\varepsilon_1)/2 + 45^{\circ}$.

In the proposed Eqs. (4.13) and (4.14), it is noted that internal friction angle and cohesion are related to initial crack size a, friction coefficient μ, crack number N_{V}, fracture toughness K_{IC}, wing crack length l, crack angle φ, and constant correction factor β. Moreover, parameters a, μ, N_V, K_{IC}, φ, and β all are material constants in this proposed model, which is determined by the types of materials. Thus, cohesion and internal friction angle vary with wing crack growth. Furthermore, in Eq. (4.12), the crack growth length l is caused by the applied variable axial stress. If crack growth is caused by stress corrosion, i.e., subcritical crack growth (Charles 1958; Atkinson 1984), the applied stress keeps constant. The variable $\sigma_1(l)$ is replaced by constant stress σ_1. In other words, due to the close correlations of cohesion, internal friction angle and microcrack growth, the applied variable stress causing crack growth is replaced by stress corrosion causing crack growth. The normal stress acting on shear failure plane from Eq. (4.12) is modified as:

$$\sigma_{nf}(l) = \frac{\sigma_1 + \sigma_3}{2} + \frac{\sigma_1 - \sigma_3}{2}\cos 2\psi(l) \tag{4.19}$$

For the case of $l \geq l_{peak}$, the time-dependent crack extension [i.e., l(t)] can be solved by introducing l_{peak} as the initial iteration value of crack extension (i.e., $l_o = l_{peak}$) for the numerical integration of Eq. (3.1). l_{peak} is obtained by the crack–stress relation of Eq. (4.1) under the given value of initial stress state. Substituting this time-dependent crack extension into Eqs. (4.11), (4.19), (4.13)and (4.14) can achieve the evolution of shear strength under constant stresses as follows:

$$\tau_f(t) = \sigma_{nf}(\tan\phi(t)) - c(t), \tag{4.20}$$

where

$$\sigma_{nf}(t) = \frac{\sigma_1 + \sigma_3}{2} + \frac{\sigma_1 - \sigma_3}{2}\cos 2\psi(t), \tag{4.21}$$

$$\phi(t) = 90^\circ - 2arc\tan\sqrt{\frac{A_1(c_1(t) + c_2(t))}{c_3(t) + A_2(c_1(t) + c_2(t))}}, \tag{4.22}$$

$$c(t) = \frac{K_{IC}}{2\sqrt{A_1\pi a[c_3(t) + A_2(c_1(t) + c_2(t))](c_1(t) + c_2(t))}}, \tag{4.23}$$

$$\psi(t){=}45^\circ{+}\frac{\phi(t)}{2}. \tag{4.24}$$

The stress path effect on evolution of shear strength can be further analyzed by replacing the constant axial stress and confining pressure with time-dependent axial stress $\sigma_1(t)$ and confining pressure $\sigma_3(t)$ in Eq. (4.20).

4.2 Results and Discussions

4.2.1 Effects of Microcrack Parameters on Shear Strength

Using the proposed relationships between micromechanical parameters and shear strength, the effects of initial damage, friction coefficient and confining pressure on shear strength will be studied in detail.

Due to the random distribution of initial microcracks in brittle rocks, it is impossible to find two same rock samples. Thus, the cyclic experiment for one rock sample could provide us the more precise and comparable results to study the effects of initial damage on shear strength. It is known that the damage increases with number of cycles in a cyclic loading experiment. We can examine each cycle as an independent test, which shows that the damage is increasing with each cycle. The experimental results between two independent cycles could be compared to obtain the effect of initial damage on mechanical properties.

The initial damage D_o is proportional to initial microcrack size a at the given number of initial microcracks per unit volume N_V. It is also seen that the initial damage is consistent with the crack density. Thus, the value of initial damage in real rock is not larger than 1. In the following analysis of this section, parameter N_V is selected as 9×10^7 per cubic meter, which guarantees that the value of initial damage D_o is less than 1 when initial microcrack size in the range of $a = 0.6$–1.0 mm. It is noted that the parameter N_V is determined as a constant value in order to provide a comparable result for discussing the effects of initial crack size a on shear strength, for other parameters referring the Jinping marble.

In this study, the cyclic loading experimental results from Martin and Chandler (1994) adopting a damage-controlled method are used to verify our theoretical results. It is worthy to note that a special care was taken to avoid fast failure when axial stress is close to rock strength, which aims to keep on the test in the post-failure region.

As shown in Fig. 4.3, it can be seen that the internal friction angle decreases with increment of initial microcrack size for given confining pressure and friction coefficient. In other words, the internal friction angle decreases with increment of initial damage. This result agrees well with published experiment results (Martin and Chandler 1994).

It also can be seen from Fig. 4.3a that the internal friction angle increases with increment of friction coefficient. The dependence of internal friction angle on initial microcrack size decreases with increment of friction coefficient. As shown in Fig. 4.3b, the effect of confining pressure on internal friction angle is very small. The slight difference of internal friction angle may be caused by slight increment of peak crack length l_{peak} or axial strain ε_{peak} with increment of confining pressure (see Sect. 4.2.3).

It can be seen from Fig. 4.4 that the cohesion decreases with increment of initial microcrack size for certain value of confining pressure and friction coefficient. This result agrees well with experiment results (Martin and Chandler 1994). It also can be seen from Fig. 4.4a that the cohesion increases with increment of friction coefficient.

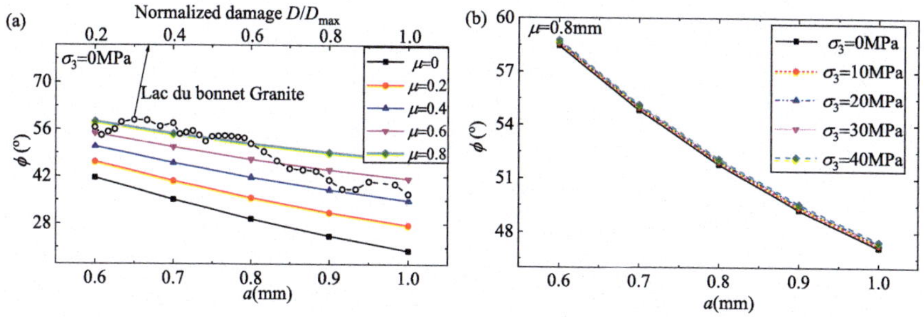

Fig. 4.3 Effects of **a** friction coefficient and **b** confining pressure on internal friction angle. (Experimental data of Lac du bonnet Granite is taken from Martin and Chandler (1994))

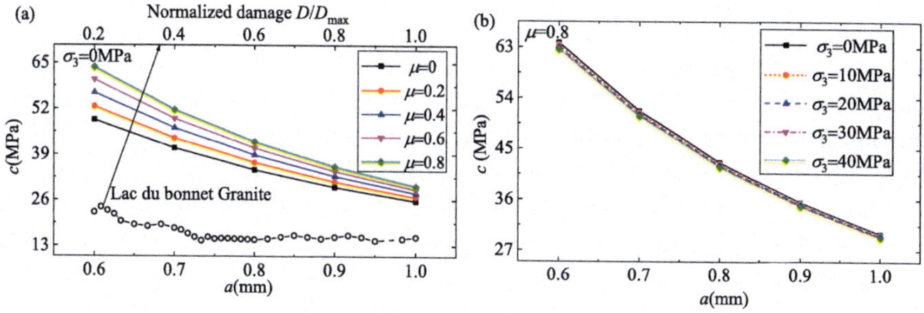

Fig. 4.4 Effects of **a** friction coefficient and **b** confining pressure on cohesion

The dependence of cohesion on initial microcrack size increases with increment of friction coefficient. As shown in Fig. 4.4b, the effect of confining pressure on cohesion is very small.

The shear strength decreases with increment of initial microcrack size (see Fig. 4.5). This result agrees well with experiment results (Martin and Chandler 1994). As shown in Fig. 4.5a, the shear strength increases with increment of friction coefficient for given confining pressure and initial microcrack size. The dependence of shear strength on initial microcrack size increases with increment of friction coefficient. The dependence of shear strength on initial microcrack size increases with increment of confining pressure in Fig. 4.5b.

4.2.2 Quasi-static Variations of Shear Properties with Crack Growth or Axial Strain During Post-peak Failure

The variations of shear strength, cohesion, and internal friction angle from microcrack growth of brittle rock appearing at the post-peak phase of quasi-static failure are shown in Fig. 4.6a. The figure provides an understanding of shear behavior

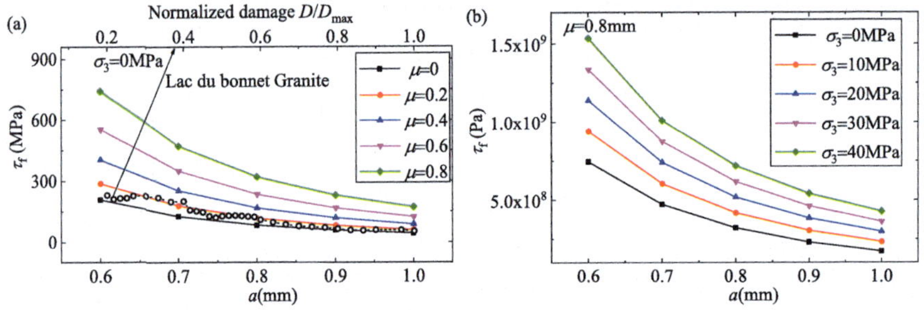

Fig. 4.5 Effects of **a** friction coefficient and **b** confining pressure on shear strength

caused by microcrack growth. However, crack growth-dependent shear strength, cohesion, and internal friction angle are not macroscopically observed and measured in conventional compressive tests. To clearly and directly verify the rationality of the presented analytical method, the macroscopic description of the relationship among shear strength, internal friction angle, cohesion, and axial strain is shown in Fig. 4.6b. At the post-peak phase of the axial stress–strain curve, shear strength and cohesion continuously weaken with increasing axial strain caused by crack growth; and internal friction angle initially strengthens and then weakens with increasing axial strain caused by crack growth, consistent with the experimental results under triaxial compressive loadings (Martin and Chandler 1994).

Furthermore, many studies explained that the onset of shear fracture in intact rocks has a similar phenomenon with the onset of rapid slip along a preexisting fault in the rate- and state-dependent laws (Dieterich 1979a, b; Ruina 1983). The onset of shear fracture has the same underlying physical process as the onset of earthquakes (Brace and Byerlee 1966; Ohnaka 1995; Beeler 2004). Thus, the variation of shear strength with deformation (i.e., axial strain) in Fig. 4.6b is similar to the variation of shear stress with deformation (slip displacement) during fast slip in Figs. 4.7 and 4.8. A direct comparison of the proposed and published data (Dieterich 1979a, b; Ruina 1983; Ohnaka 1995) on the relationship between normalized shear stress

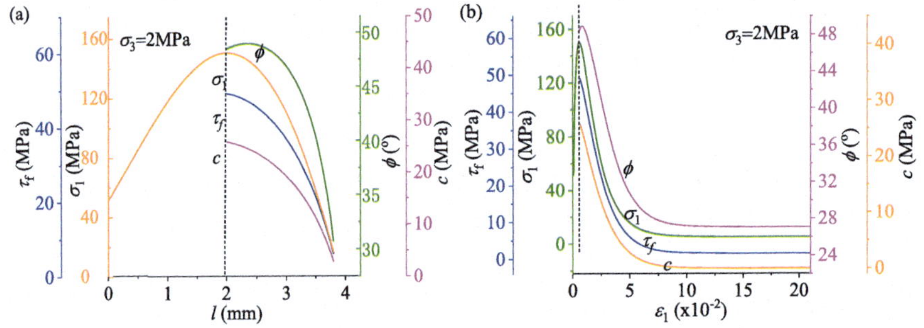

Fig. 4.6 Variations of axial stress, shear strength, internal friction angle, and cohesion along with **a** wing crack length and **b** axial strain

and normalized deformation is also shown in Fig. 4.9. The difference between the proposed and published data could be due to some average assumptions for the case of the microcrack geometries and distributions in this proposed model. However, the rate- and state-dependent laws in Fig. 4.7 can explain the effect of sudden variation of slip velocity on shear stress (i.e., the strengthening phase, where shear stress rapidly increases with slip displacement or time). However, in this section, our proposed model only explains shear fracture in intact rocks and cannot explain the effect of slip rate variation on shear strength.

These variations of shear properties are closely related to the stress-induced quasi-static crack growth and axial strain. Such sudden variations of applied stress cause the variations of deformation and deformation rate, which may approximately explain the effect of sudden variation of slip velocity on shear stress in the rate- and state-dependent laws (Dieterich 1979a, b; Ruina 1983).

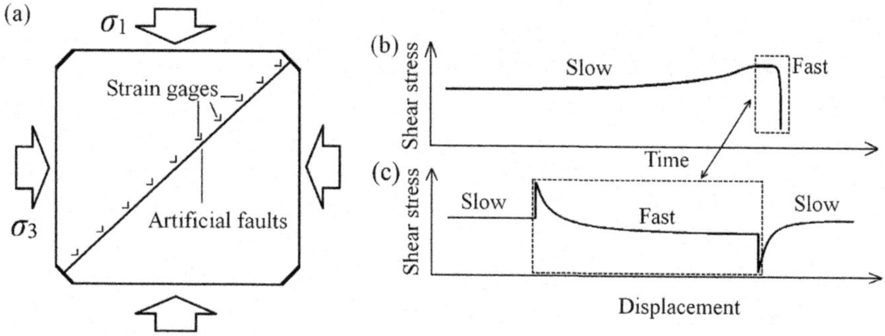

Fig. 4.7 **a** Schematic of rock friction experiment along artificial faults, **b** relationship between shear stress and time along the fault under compression, and **c** relationship between shear stress and displacement along the fault under compression (which is related to the rate- and state-dependent friction laws) (Dieterich 1979a, b)

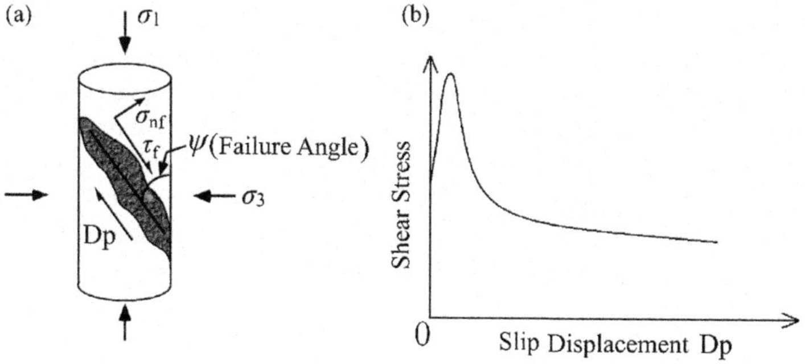

Fig. 4.8 **a** Configuration of shear fracture in an intact rock sample and **b** relationship between shear stress and slip displacement of intact brittle rocks under lithospheric conditions (Ohnaka 1995)

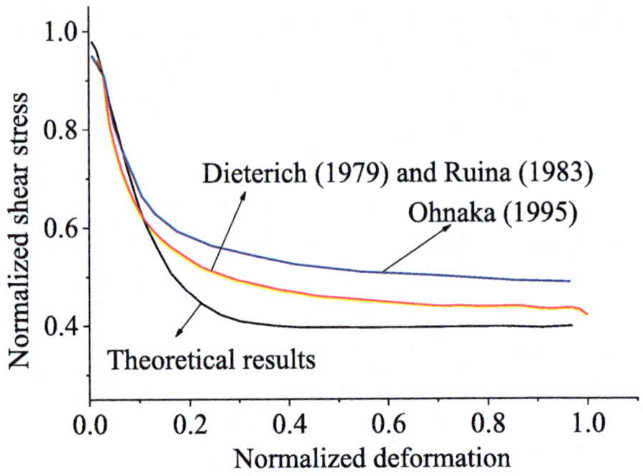

Fig. 4.9 Comparisons of the relationship between normalized shear stress and normalized deformation in this study, Dieterich (1979a, b), Ruina (1983), and Ohnaka (1995)

4.2.3 Time-Dependent Static Crack Growth, Strain, and Shear Properties Under Constant Stress State

The evolutions of crack growth, strain, and shear properties (i.e., internal friction angle, cohesion, and shear strength) of brittle rocks under constant stress state (i.e., creep) are shown in Fig. 4.10. The figure also illustrates the corresponding rates of crack extension, strain, and shear properties. Crack initially experiences a slow accelerated growth, and then a rapid accelerated growth in Fig. 4.10b, which causes slow and rapid accelerated strain, respectively in Fig. 4.10c. Shear strength undergoes a slow and rapid accelerated weakening stage in Fig. 4.10f, which is attributed to cohesion that has an accelerated weakening process in Fig. 4.10d and the internal friction angle that initially has a decelerated strengthening and subsequent accelerated weakening process in Fig. 4.10e.

Rock failure occurs at the rapid accelerated stage of crack growth and strain and rapid accelerated weakening stage of internal friction angle, cohesion, and shear strength. Previous studies rarely focused on the evolutions of cohesion and internal friction angle in brittle rocks. Figure 4.10e shows that the transformation between the strengthening and weakening stages of internal friction angle is clearly judged by the rate of internal friction angle and shear strength. The positive rate of internal friction angle represents the strengthening stage, and the negative rate of the internal friction angle represents the weakening stage. Furthermore, the time elapsed in the transformation between the strengthening and weakening stages of the internal friction angle is approximately 1.23×10^9 s in Fig. 4.10e.

Under quasi-static compression, the similarity of the onset of shear fracture in intact rocks in this proposed model with the onset of rapid slip along a preexisting fault in the rate- and state-dependent laws (Dieterich 1979a, b) is verified by comparing the variations of shear stress with deformation in intact rocks and faults in Fig. 4.9. Figure 4.7 also shows the variations of time-dependent shear stress during fast slip in the rate- and state-dependent laws for the quasi-static condition. The shear stress

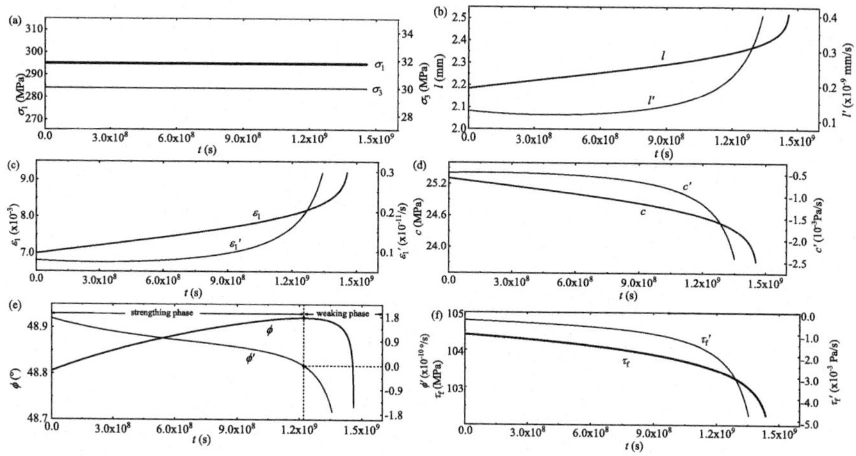

Fig. 4.10 Evolution and rate of **b** crack growth, **c** strain, **d** cohesion, **e** internal friction angle, and **f** shear strength under (**a**) constant axial stress and confining pressure

during fast slip decreases over time. However, if this quasi-static condition is replaced with the static condition, that is, the stress state maintains constant, then the time-dependent shear strength is obtained in Fig. 4.10f.

The proposed results of the evolution of shear strength caused by microcrack growth in Fig. 4.10f are similar to the phenomenon of time-dependent shear stress during the fast slip of Dieterich (1979a, b) in Fig. 4.7b. The normalized relation between shear stress and time is shown in Fig. 4.11 owing to the difference at the scale of coordinate values. The initial state of time-dependent mechanical behavior is selected by the peak point of the stress–strain relationship in Fig. 4.6b. The applied stress at the post-peak phase of the stress–strain curve is also equivalent to the rock strength (Zhao and Cai 2010). Thus, the shear strength of this study is equivalent to the shear stress obtained experimentally by Dieterich (1979a).

Thus, our study results for the evolution of shear strength in intact rocks have great implications for judging the process of earthquake nucleation in the earth crust on the basis of micromechanics. In the process of earthquake nucleation, the time-dependent stable and unstable deformations from rock fracture are controlled by the physical process of crack growth (Beeler 2004; Svetlizky and Fineberg 2014). Stable deformation generates a small inelastic strain caused by subcritical crack growth, which corresponds to the steady-state crack growth (Fig. 4.10b), creep strain (Fig. 4.10c), evolution of cohesion (Fig. 4.10d), evolution of internal friction angle (Fig. 4.10e), and evolution of shear strength (Fig. 4.10f). Unstable deformation (i.e., rapid slip) generates a large strain caused by crack coalescence and supercritical crack growth, which corresponds to the accelerated crack growth (Fig. 4.10b) and creep strain (Fig. 4.10c), and the drop of cohesion (Fig. 4.10d), internal friction angle (Fig. 4.10e), and shear strength (Fig. 4.10f).

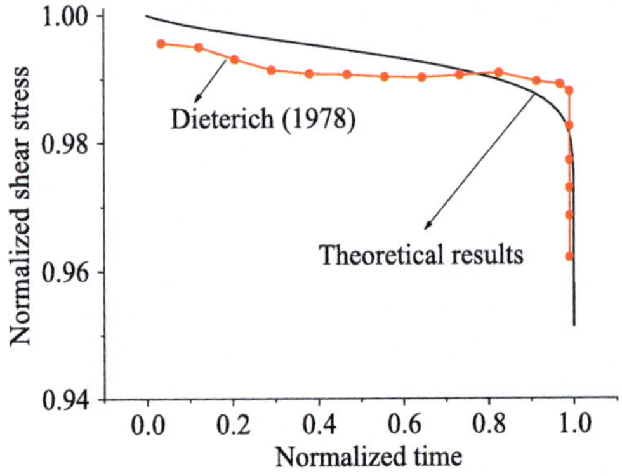

Fig. 4.11 Comparison of study results for time-dependent normalized shear stress in this study and in Dieterich (1979a). The normalized shear stress (or time) is defined by the division of the values between shear stress (or time) and maximum shear stress (or time)

The environment of rocks subjected to constant stress state is approximately equivalent to the case of lithospheric conditions. This finding provides an implication of regular earthquakes on the basis of the shear failure of intact rocks caused by microcrack growth.

Furthermore, the sensitivities of initial crack size, friction coefficient, confining pressure, and initial damage on the evolutions of cohesion, internal friction angle, and shear strength are studied and shown in Fig. 4.12. As shown in Fig. 4.12a–c, when initial crack number N_V is given as 1.3×10^6, the final time at the drop of cohesion, internal friction angle, and shear strength decreases with the increment of initial crack size. As shown in Fig. 4.12d–f, the final time at drop of cohesion, internal friction angle, and shear strength increases with the increment of friction coefficient between crack interfaces. As shown in Fig. 4.12g–i, the final time at the drop of cohesion, internal friction angle, and shear strength increases with the increment of confining pressure. The internal friction angle and cohesion slightly change under different confining pressures in a given time because of the little relevance of internal friction angle, cohesion, and confining pressure in Eqs. (4.13) and (4.14). Shear strength increases with the increment of confining pressure in a given time. In Fig. 4.12j–l, the final time at the drop of cohesion, internal friction angle, and shear strength increases with the decrement of initial damage. Internal friction angle, cohesion, and shear strength decrease with the increment of initial damage in a given time.

The small change of model parameters (i.e., initial crack size, friction coefficient, confining pressure, or initial damage) likewise causes the time variation over several orders of magnitudes. This phenomenon shows that the changes in rock internal properties or external effect factors largely influence the rock fracture process during creeping deformation.

The evolutions of strain, cohesion, internal friction angle, and shear strength are closely related to the subcritical crack growth of brittle rocks. Furthermore, the initial

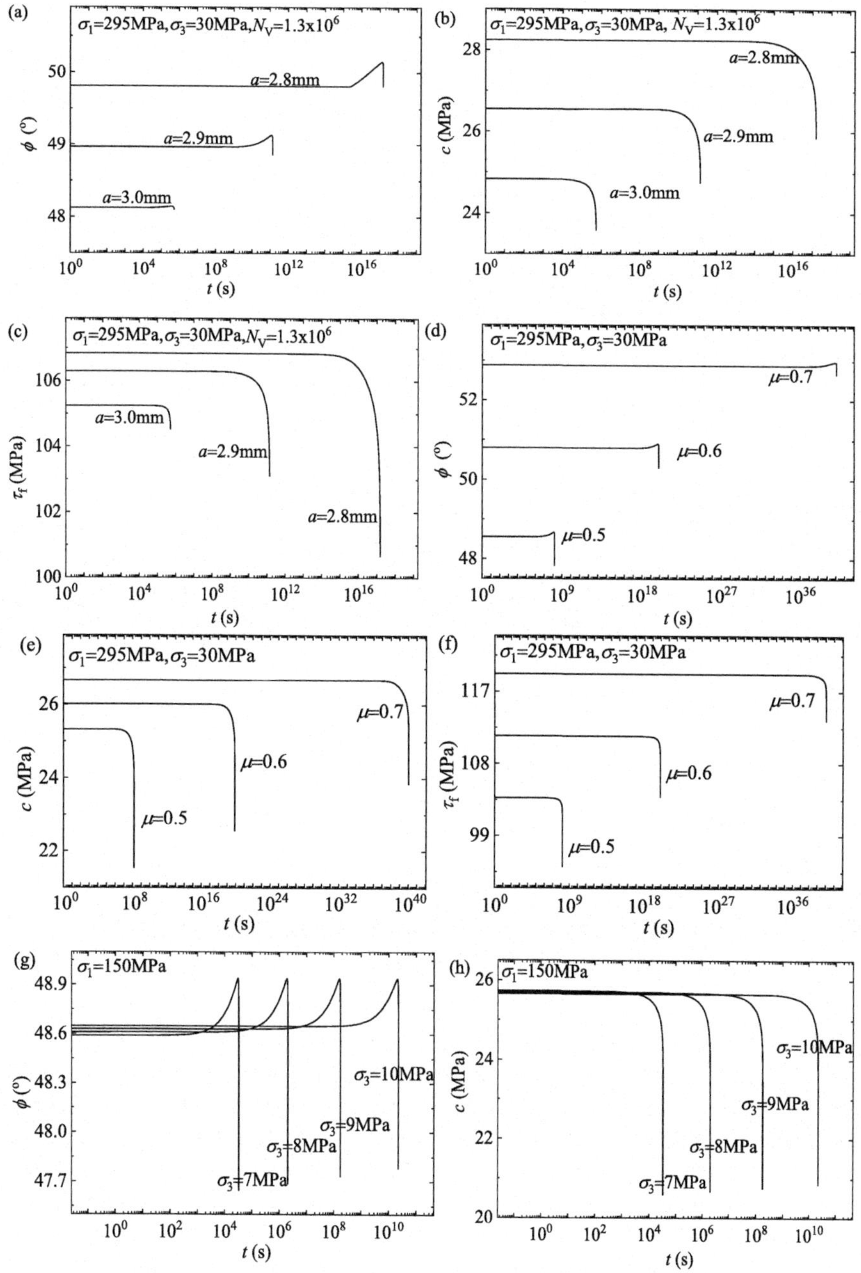

Fig. 4.12 Effects of **a–c** initial crack size a, **d–f** crack friction coefficient μ, **g–i** confining pressure σ_3, and **j–l** initial damage D_0 on the evolutions of cohesion, internal friction angle, and shear strength

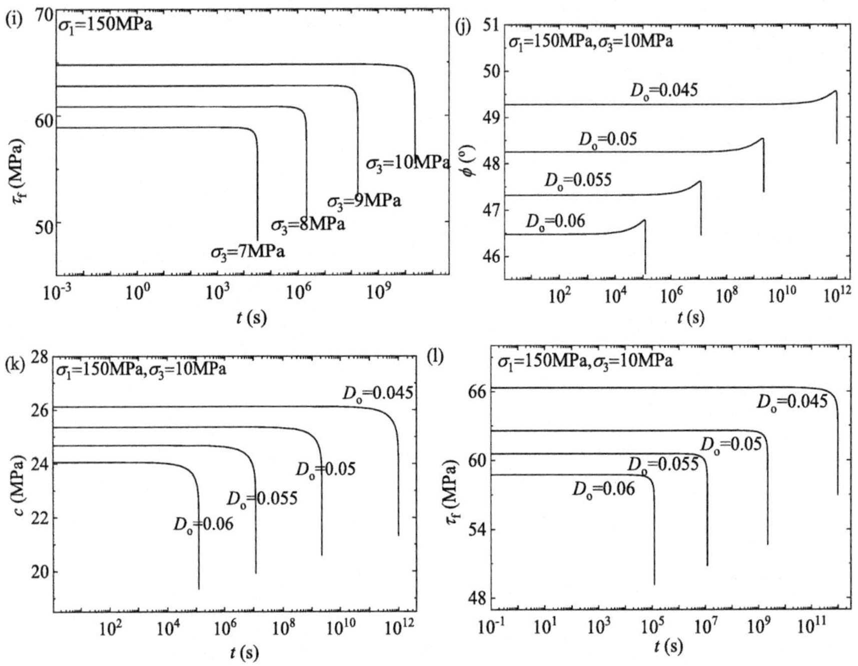

Fig. 4.12 (continued)

state of the rocks is selected by the peak point of the stress–strain relationship (which corresponds to the relationship between crack growth and stress) in Fig. 4.6.

4.2.4 Effects of Stress Changes on the Evolutions of Crack Growth, Strain, and Shear Properties

Stress applied to rocks in earth crust changes due to the geological tectonic development induced by different factors (e.g., ocean or solid Earth tides), which influence the slow earthquake mechanism. This mechanism, in which the episodic tremor recorded by weak seismic signals accompanies slow slip on faults, is presented (Shelly et al. 2011; Crampin et al. 2015; Leeman et al. 2016). Tremor and slow stick–slip are caused by minor stress variations from the ocean or solid Earth tides (Houston 2015). Thus, evolutions of deformation and shear strength of rocks triggered by stress changes have important implications in the evaluation of the slow earthquake mechanism (Hardebeck et al. 1998; Belardinelli et al. 2003). In the analyses, the initial axial stress is selected as 295 MPa, and the initial confining pressure is selected as 30 MPa, thereby providing a qualitative comparison of the effects of different stress paths on time-dependent compressive-shear failure.

Figure 4.13 shows the effects of step loading of axial stress on time-dependent crack growth, axial strain, cohesion, internal friction angle, and shear strength as well as illustrates the corresponding rates. Crack length and axial strain increase at steady state at the earlier two steps and accelerate at the last step. Cohesion and shear strength decrease at steady state at the earlier two steps and rapidly drop at the last step. Internal friction angle increases at steady state at the earlier two steps and decelerates and is then accompanied by an accelerated drop at the last step. The steady-state rates of crack growth, axial strain, cohesion, and shear strength increase with an increment of step numbers. The sudden rise of axial stress at each step in Fig. 4.13a causes the step variations in crack velocity, axial strain rate, rate of cohesion, rate of internal friction angle, and rate of shear strength.

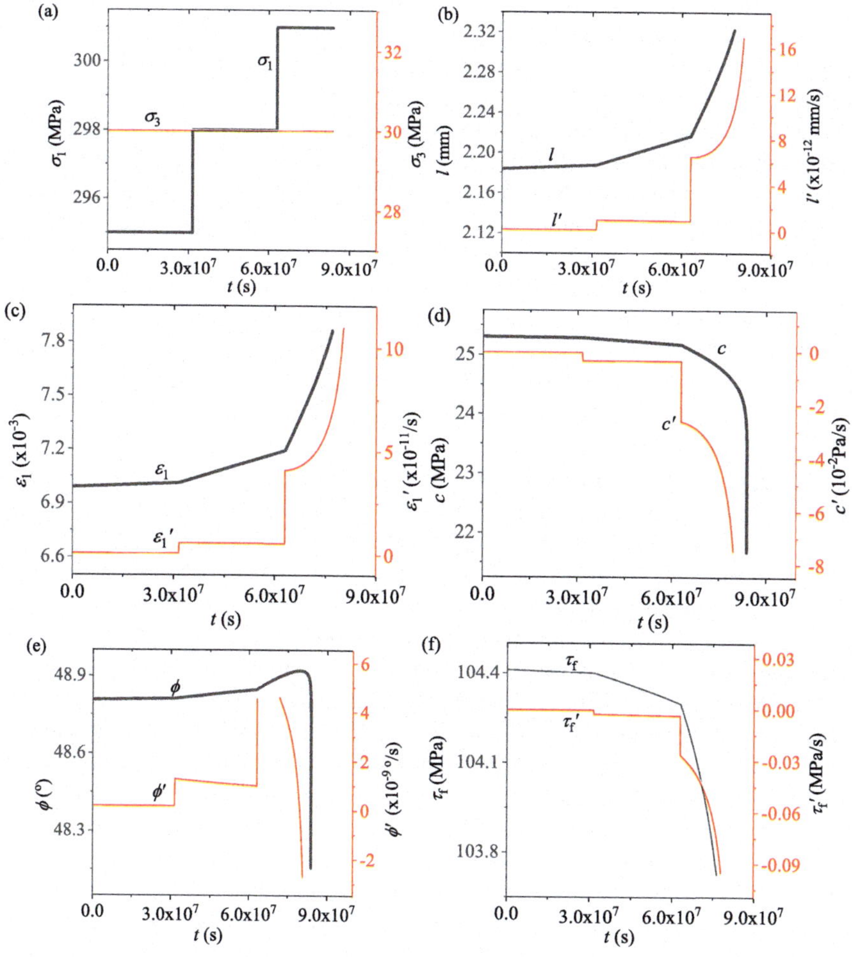

Fig. 4.13 Evolution and rate of **b** crack growth, **c** strain, **d** cohesion, **e** internal friction angle, and **f** shear strength under (**a**) step loading of axial stress and constant confining pressure

Figure 4.14 shows the effects of step unloading of confining pressure on the time-dependent crack growth, axial strain, cohesion, internal friction angle, and shear strength. The figure also shows the corresponding rates. Crack length and axial strain increase at a steady state at the earlier six steps and accelerate at the last step. Cohesion decreases at a steady state at the earlier six steps and rapidly drops at the last step. Internal friction angle increases at a steady state at the earlier six steps and decelerates initially and then rapidly drops at the last step. Shear strength decreases initially at a steady state and then slightly drops at the earlier each step, and finally shows a remarkable drop at the last step. The slight decrease in shear strength can cause a small earthquake, and the large decrease in shear strength can cause a large earthquake. The sudden drop of confining pressure at each step in Fig. 4.14a causes the step variations in crack velocity, axial strain rate, rate of cohesion, and rate of internal friction angle. However, in Fig. 4.14f, the sudden rise of confining pressure causes a sudden rise in shear strength, leading to a large variation in the rate of shear strength. Therefore, the sudden change of confining pressure largely influences shear strength.

The comparison between Figs. 4.13 and 4.14 and 4.10 shows that the final failure time at step loading of axial stress and confining pressure is shorter than that at constant stress state ($\sigma_1 = 295$ MPa, $\sigma_3 = 30$ MPa). Thus, the step loading of axial stress and unloading of confining pressure accelerate rock failure and earthquake occurrence.

The evolution and rate of crack, strain, cohesion, internal friction angle, and shear strength subjected to the increasing repeated loading and unloading of axial stress are shown in Fig. 4.15 under constant confining pressure. The function in the path of axial stress is $\sigma_1 = 295$ MPa $+ 0.1$ t $-$ 3 MPa [t/T], cyclic time T $= 60 \times 60 \times 24 \times 365$ s, and [t/T] represents the integralization for t/T (if t/T $= 0.6$, then [t/T] $= 0$; if t/T $= 8.9$, then [t/T] $= 8$).

Except for the phases of sudden drop of axial stress in Fig. 4.15a, crack undergoes a slow accelerated growth at earlier cycles and a rapid accelerated growth at the last cycle. In addition, the minimum crack velocity at each cycle increases with the increment of cyclic number in Fig. 4.15b. The phenomenon of strain evolution in Fig. 4.15c is similar to crack growth. Except for the phases of the sudden drop of axial stress, cohesion has a slow accelerated weakening process at earlier cycles, and a rapid drop of cohesion appears in the last cycle. The minimum rates of cohesion at each cycle increases with the increment of cyclic number in Fig. 4.15d. Except for the phases of sudden drop of axial stress, internal friction angle has a slow accelerated strengthening process at most earlier cycles and an accelerated weakening process at the subsequent less cycles. A rapid drop of internal friction angle appears at the last cycle. Before the internal friction angle enters the weakening stage (i.e., rate of internal friction angle is negative), it experiences transformation from the accelerated strengthening stage to the decelerated strengthening stage in Fig. 4.15e. In Fig. 4.15f, except for the phases of sudden drop of axial stress, shear strength has a slow accelerated weakening process at most earlier cycles and a rapid accelerated weakening process accompanied by a large drop of shear strength at the last cycle.

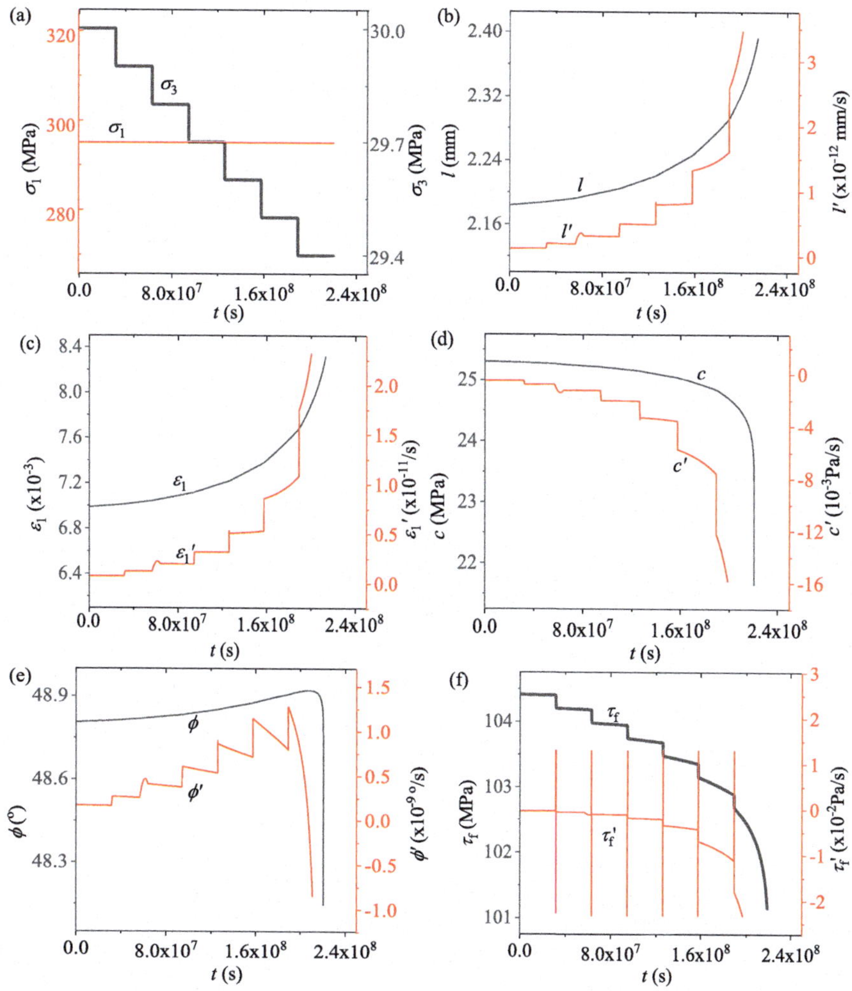

Fig. 4.14 Evolution and rate of **b** crack growth, **c** strain, **d** cohesion, **e** internal friction angle, and **f** shear strength under (**a**) step unloading of confining pressure and constant axial stress

Figure 4.16 shows the evolution and rate of crack, strain, shear properties under decreasing repeated unloading and loading of confining pressure, and constant axial stress. The function in the path of confining pressure is σ_3 = 30 MPa − 0.004 t + 0.1 MPa [t/(60 × 60 × 24 × 365 s)]. The evolutions of crack, strain, cohesion, internal friction angle, and shear strength have the same phenomenon as the evolutions of crack, strain, and shear properties under the increasing repeated loading and unloading of axial stress in Fig. 4.15. This finding shows that the mechanisms induced by increasing repeated loading and unloading of axial stress and decreasing repeated unloading and loading of confining pressure are similar. The sudden drop of

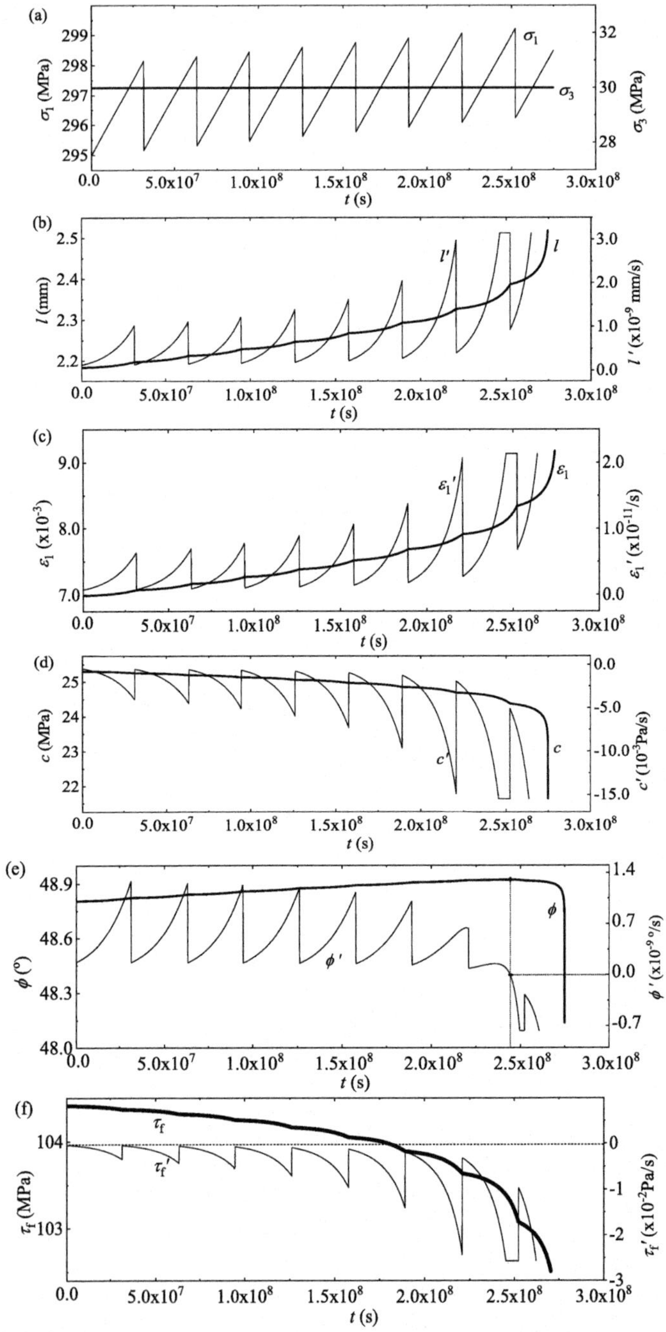

Fig. 4.15 Evolution and rate of **b** crack growth, **c** strain, **d** cohesion, **e** internal friction angle, and **f** shear strength under (**a**) increasing repeated loading and unloading of axial stress and constant confining pressure

axial stress in Fig. 4.15a and the sudden rise of confining pressure in Fig. 4.16a decelerate the evolutions of crack length, axial strain, cohesion, internal friction angle, and shear strength. The continuous slow increment of axial stress in Fig. 4.15a and the continuous slow decrement of confining pressure in Fig. 4.16a accelerate the evolutions of crack length, axial strain, cohesion, internal friction angle, and shear strength.

The comparison between Figs. 4.10, 4.15 and 4.16 shows that the time of rock failure under repeated loading and unloading of axial stress or confining pressure is shorter than the time of rock failure under constant stress state. The stress paths of increasing repeated loading and unloading of axial stress and decreasing repeated loading and unloading of confining pressure accelerate the crack growth. For earthquake mechanisms, the effect of increasing repeated loading and unloading of axial stress or decreasing repeated loading and unloading of confining pressure applied to rocks under lithospheric conditions accelerates the occurrence of an earthquake.

Figures 4.15 and 4.16 show the effect of increasing repeated loading and unloading of axial stress or decreasing unloading and loading of confining pressure, respectively. The coupling effect of increasing repeated loading and unloading of axial stress and decreasing unloading and loading of confining pressure is shown in Fig. 4.17. The path function of axial stress and confining pressure is equivalent to the function of Figs. 4.15 and 4.16, respectively. The evolutions of crack, strain, and shear properties are similar to the phenomena in Figs. 4.15 and 4.16. The comparison between Figs. 4.17, 4.15 and 4.16 shows that the time of rock failure about the coupling effect of axial stress and confining pressure is shorter than the time of rock failure about the single effect of axial stress or confining pressure. Therefore, the coupling effect of increasing repeated loading and unloading of axial stress and decreasing repeated loading and unloading of confining pressure applied to rocks under lithospheric conditions accelerates the crack growth to fracture of rocks and the occurrence of a regular earthquake.

Figure 4.18 shows the coupling effect of increasing repeated loading and unloading of axial stress and confining pressure on crack growth, strain, and shear properties. The function of the path of axial stress is $\sigma_1 = 295$ MPa + 0.1 t – 3 MPa [t/(60 × 60 × 24 × 365 s)], and the function of the path of confining pressure is σ_3 = 30 MPa + 0.004 t – 0.1 MPa [t/(60 × 60 × 24 × 365 s)]. Except for the phases of sudden variation of axial stress and confining pressure, shear strength is always in the decelerated strengthening phase (i.e., rate of shear strength decreases and is larger than 0 at each cycle) at earlier cycles and has an accelerated weakening phase at subsequent cycles. In addition, shear strength rapidly drops in the last cycle. The coupling effect of the sudden drops of confining pressure and axial stress causes a large drop in shear strength, which may cause a small earthquake. For each cycle, the increasing axial stress dominates the accelerated variations of crack growth, axial strain, cohesion, and internal friction angle, and the sudden drop of axial stress dominates the sudden drop of crack velocity, strain rate, cohesion rate, and rate of internal friction angle. The slow increasing confining pressure dominates the decelerated strengthening shear strength, and the sudden drop of confining pressure dominates the sudden drop of shear strength.

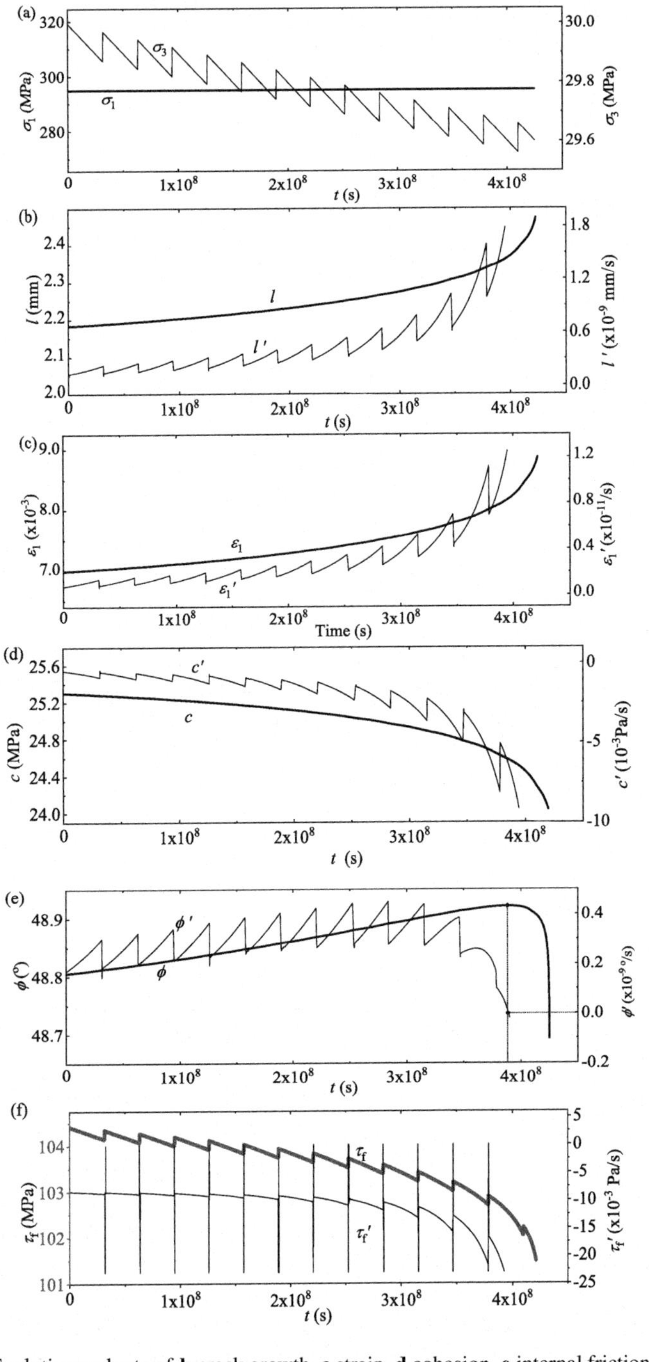

Fig. 4.16 Evolution and rate of **b** crack growth, **c** strain, **d** cohesion, **e** internal friction angle, and **f** shear strength under (**a**) decreasing repeated unloading and loading of confining pressure, and constant axial stress. The dashed circle is the enlargement area

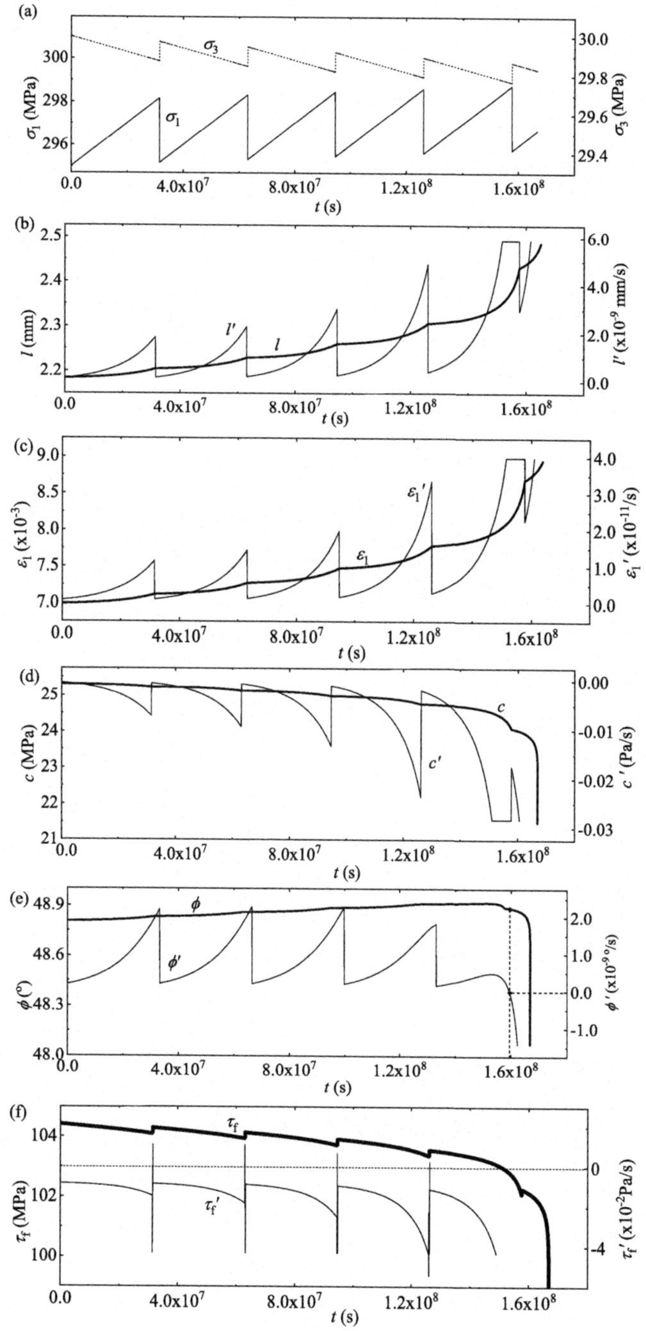

Fig. 4.17 Evolution and rate of **b** crack growth, **c** strain, **d** cohesion, **e** internal friction angle, and **f** shear strength under (**a**) increasing repeated loading and unloading of axial stress, and decreasing repeated unloading and loading of confining pressure

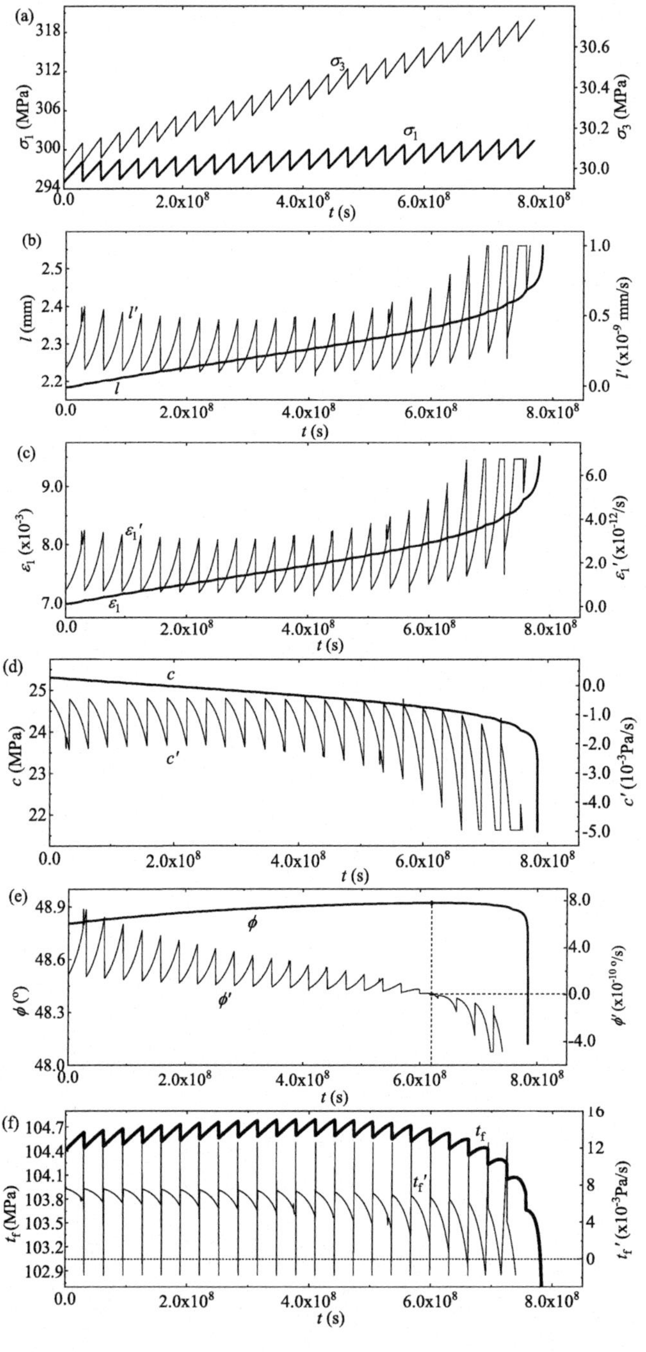

Fig. 4.18 Evolution and rate of **b** crack growth, **c** strain, **d** cohesion, **e** internal friction angle, and **f** shear strength under (**a**) increasing repeated loading of axial stress and confining pressure

Compared with Fig. 4.16, the increasing repeated loading and unloading of confining pressure of Fig. 4.18 decelerate the crack growth, increase the time of rock failure, and slow down the occurrence of earthquake.

4.3 Conclusions

An analytical approach is proposed to study the relationship microcrack geometry and shear strength, the evolution of shear strength during creep failure.

(1) Increment of initial microcrack size, and decrement of friction coefficient and confining pressure weaken the shear strength of brittle rocks.
(2) In the post-peak phase of quasi-static stress–strain curves for brittle rocks, the internal friction angle initially increases up to a maximum value and then decreases with further crack growth or strain increment. The cohesion and shear strength decrease with crack growth or strain in increment.
(3) For brittle rocks subjected to constant stress state, the cohesion and shear strength have slow accelerated weakening stage and a rapid drop, and internal friction angle has decelerated strengthening and accelerated weakening stages. The time-dependent shear strength is similar to the evolution of shear stress of brittle rocks during earthquake nucleation, which has an implication for the evaluation of the regular earthquake mechanism.
(4) Under step loading of axial stress and step unloading of confining pressure, the internal friction angle has a steady-state strengthening phase at earlier steps and a decelerated strengthening phase initially and then an accelerated weakening phase at the last step. Cohesion and shear strength have a steady-state weakening phase at earlier steps and an accelerated weakening phase at the last step. Under the step unloading of confining pressure, shear strength initially decreases at a steady state and is then accompanied by a slight drop at an earlier step, and finally a large accelerated drop at the last step.
(5) Subjected to the increase in repeated loading and unloading of axial stress or the decrease in repeated unloading and loading of confining pressure, internal friction angle has a slow accelerated strengthening process at most early cycles and a rapid accelerated weakening process at subsequent few cycles. In addition, cohesion has an accelerated weakening process at each cycle. Shear strength also has a slow accelerated weakening process at most early cycles and a rapid accelerated weakening process accompanied by a drop of shear strength at the last cycle.
(6) The sensitivity of sudden variation of confining pressure on shear strength is higher than that of the sudden variation of axial stress. The sudden rise or drop of confining pressure cause the sudden rise or drop of shear strength, thereby causing the large sudden rise and drop in the rate of shear strength. These phenomena of drops of shear strength caused by stress changes are similar to

the slow earthquake mechanism caused by minor variations of stress from the different factors (e.g., ocean or solid Earth tides).

Note: In the time-dependent shear properties caused by subcritical crack growth, the initial state starts from the peak point of the stress–strain relation measured by the standard post-failure test. The rock can undergo a long-time subcritical crack growth during the post-peak phase of the stress–strain relation.

References

Atkinson BK. Subcritical crack growth in geological materials. J Geophys Res. 1984;89(B6):4077–114.

Beeler NM. Review of the physical basis of laboratory-derived relations for brittle failure and their implications for earthquake occurrence and earthquake nucleation. Pure Appl Geophys. 2004;161:1853–76.

Belardinelli ME, Bizzarri A, Cocco M. Earthquake triggering by static and dynamic stress changes. J Geophys Res. 2003;108(B3):2135.

Brace WF, Byerlee JD. Stick-slip as a mechanism for earthquakes. Science. 1966;153(3739):990–2.

Brantut N, Viesca RC. Earthquake nucleation in intact or healed rocks. J Geophys Res. 2015;120(1):191–209.

Charles RJ. Static fatigue of glass I. J Appl Phys. 1958;29(11):1549–53.

Crampin S, Gao Y, Bukits J. A review of retrospective stress-forecasts of earthquakes and eruptions. Phys Earth Planet Inter. 2015;245:76–87.

Dieterich JH. Modeling of rock friction 1. Experimental results and constitutive equations. J Geophys Res. 1979a;84(B5):2161−68.

Dieterich JH. Modeling of rock friction 2. Simulation of preseismic slip. J Geophys Res. 1979b;84(B5):2169−45.

Hardebeck JL, Nazareth JJ, Hauksson E. The static stress change triggering model: constraints from two southern California aftershock sequences. J Geophys Res. 1998;103(B10):24427–37.

Houston H. Low friction and fault weakening revealed by rising sensitivity of tremor to tidal stress. Nat Geosci. 2015;8(5):409–15.

Leeman JR, Saffer DM, Scuderi MM, Marone C. Laboratory observations of slow earthquakes and the spectrum of tectonic fault slip modes. Nat Commun. 2016;7:11104.

Martin CD, Chandler NA. The progressive fracture of Lac du Bonnet granite. Int J Rock Mech Minng Sci Geomech Abs. 1994;31(6):643–59.

Ohnaka M. A shear failure strength law of rock in the brittle-plastic transition regime. Geophys Res Lett. 1995;22(1):25–8.

Ohnaka M, Akatsu M, Mochizuki H, Odedra A, Tagashira F, Yamamoto Y. A constitutive law for the shear failure of rock under lithospheric conditions. Tectonophysics. 1997;277:1–27.

Ruina A. Slip instability and state variable friction laws. J Geophys Res. 1983;88(B12):10359–70.

Shelly DR, Peng Z, Hill DP, Aiken C. Triggered creep as a possible mechanism for delayed dynamic triggering of tremor and earthquakes. Nat Geosci. 2011;4(6):384–8.

Svetlizky I, Fineberg J. Classical shear cracks drive the onset of dry frictional motion. Nature. 2014;509(7499):205.

Zhao XG, Cai M. A mobilized dilation angle model for rocks. Int J Rock Mech Min Sci. 2010;47(3):368–84.

Chapter 5
Crack Angle Effect on Creep Fracture

In brittle rock, inclination of microcracks greatly influences its mechanical behavior. Due to the complex distribution of numerous initial microcracks in brittle rocks, effects of microcrack inclination on strength and failure of brittle rock hardly is investigated directly. Thus, effect of crack was experimentally investigated by use of brittle specimen which contains minor pre-existing open crack (see Fig. 5.1). Mechanism of microcrack initiation, propagation and coalescence during fracture process under uniaxial or biaxial compression has been widely investigated. However, the analytical method to explore the mechanism of progressive and creep failure considering effect of crack angle is necessary. Based on the theoretical derivation method of Chaps. 2 and 4, employing an improved micromechanical wing crack model considering effect of crack angle, an analytical investigation of progressive and creep failure for brittle rock containing initial cracks under compression is presented in this chapter. Effect of crack angle, initial crack size, confining pressure and friction coefficient on crack initiation stress, strength, stress-crack relation, and stress–strain relation will be studied in details.

5.1 Theory

5.1.1 Stress-Crack-Strain Relation Considering Crack Angle Effect

In Ashby and Sammis' study, $F_w = (\tau + \mu\sigma_n)\pi a^2 \sin\varphi = -(A_1\sigma_1 - A_2\sigma_3)a^2$, where $A_1 = \pi(\beta/3)^{1/2}[(1+\mu^2)^{1/2} - \mu]$, $A_2 = A_1[(1+\mu^2)^{1/2} + \mu]/[(1+\mu^2)^{1/2} - \mu]$. This expression of F_w is derived by some assumptions and simplifications (e.g., $\tan 2\varphi = 1/\mu$). Effect of crack angle cannot be considered. In some published studies,

X. Li et al., *Static Creep Micro-Macro Fracture Mechanics of Brittle Solids*,
https://doi.org/10.1007/978-981-99-8203-5_5

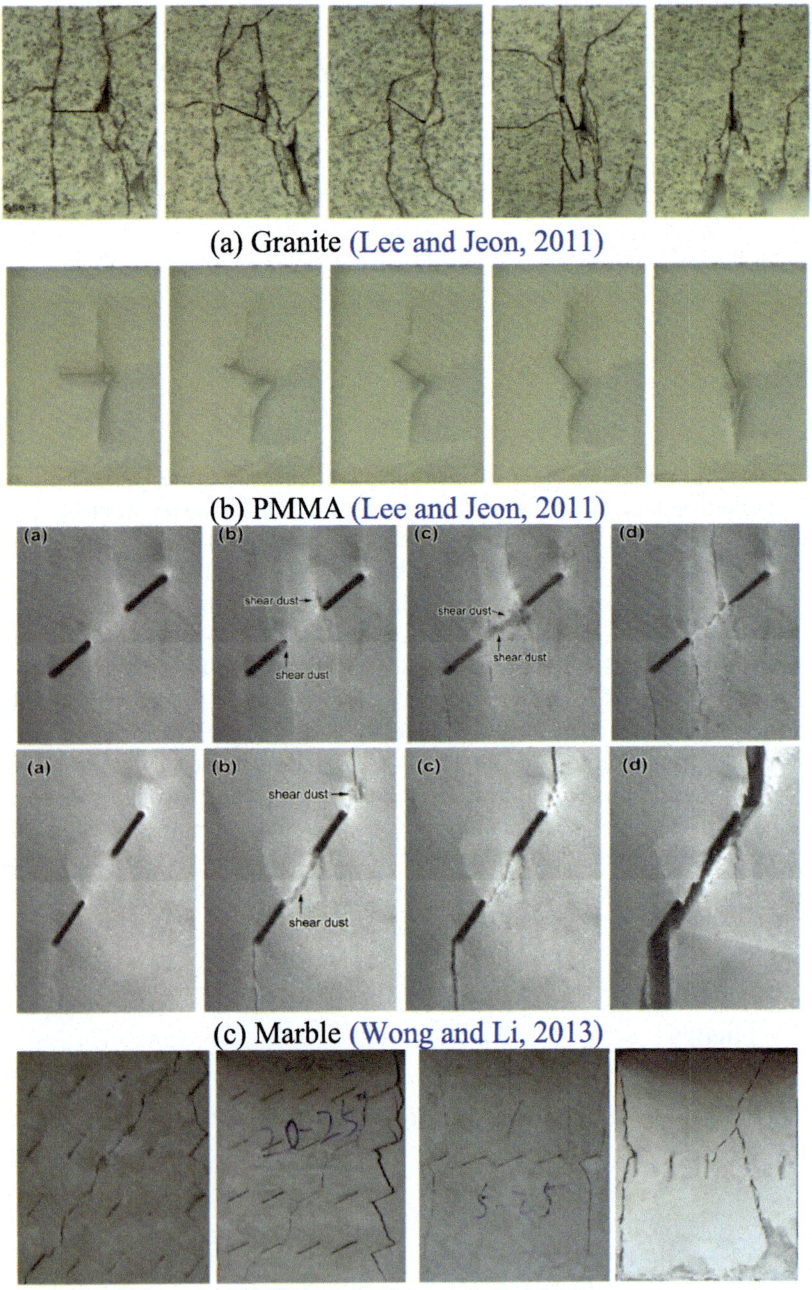

Fig. 5.1 Compressed cracking process under different crack inclinations for brittle materials of **a** Granite, **b** PMMA, **c** Marble, and **d** Concrete

crack angle is often assumed as $\varphi = 45°$ to maximize the effect of crack angle. For considering the effect of crack angle on rock mechanics, F_w is directly expressed as:

$$F_w = (\tau + \mu\sigma_n)\pi a^2 \sin\varphi \quad (5.1)$$

where

$$\tau = \left[(\sigma_3 - \sigma_1)/2\right] \sin 2\varphi \quad (5.2)$$

$$\sigma_n = (\sigma_1 + \sigma_3)/2 + \left[(\sigma_3 - \sigma_1)/2\right] \cos 2\varphi \quad (5.3)$$

where τ and σ_n are shear stress and normal stress at initial crack surface respectively.

Referring Eq. (2.1), the improved mode-I crack stress intensity factor considering crack angle effect is:

$$K_I = \sigma_1 \pi a^2 B_3 B_1 + \sigma_3\left(\pi a^2 B_3 B_2 + 2\sqrt{l/\pi}\right) \quad (5.4)$$

where

$$B_1 = [(\mu - \mu\cos 2\varphi - \sin 2\varphi)\sin\varphi]/2 \quad (5.5)$$

$$B_2 = [(\sin 2\varphi + \mu + \mu\cos 2\varphi)\sin\varphi]/2 \quad (5.6)$$

$$B_3 = \frac{1}{[\pi(l + \beta a)]^{\frac{3}{2}}} + \frac{2\sqrt{l}}{\sqrt{\pi}[S - \pi(l + \alpha a)^2]} \quad (5.7)$$

The relationship between stress and crack growth is rewritten as:

$$\sigma_1(l) = \frac{K_{IC} - \sigma_3\left(\pi a^2 B_3 B_2 + 2\sqrt{l/\pi}\right)}{\pi a^2 B_3 B_1} \quad (5.8)$$

The parameter β was assumed to be a constant value $\beta = 0.45$ (Ashby and Sammis 1990) or 0.32 (Bhat and Sammis 2011). However, the purpose of parameter β is to adjust the value of stress intensity factor K_I when the wing crack length l is equal to zero. Thus, it can be concluded that crack initiation stress is also influenced by correction factor β. Based on Eq. (5.8) at $l = 0$, this correction factor β is derived as:

$$\beta = \left\{K_{IC}^{-1}\sqrt{a/\pi}[(\sigma_1(l = 0))B_1 + \sigma_3 B_2]\right\}^{2/3} \quad (5.9)$$

Correction factor β and crack angle φ can be predicted by comparing Eq. (5.9) and the experimental relation of crack initiation stress and confining pressure. Furthermore, If the experimental relation between crack initiation stress and confining pressure is selected as $\sigma_{1c} = 2.67\sigma_3 + 46$ MPa $= \sigma_1$ $(l = 0)$ of Jinping Marble in China, inserting these crack initiation stresses under different confining pressures into Eq. (5.9), correction factor β and crack angle φ are both obtained in Fig. 5.2 (i.e., $\varphi = 32.03°$, $\beta = 0.276$). Crack angle can be assumed as an average value of numerous and random distributed initial microcrack angles in natural rocks, which provides an assistance in application of the microcrack growth-based model to predicting the mechanical properties of rocks.

Based on the density and size of microcracks, the damage can also be expressed as (Budiansky and O'Connel 1976):

$$D = N_V(l + a)^3 \tag{5.10}$$

Combining Eq. (5.10) and the damage relating to strain of Eq. (2.13), the relationship between crack growth and axial strain at low confining pressure is:

$$\varepsilon_1 = \varepsilon_o\left[-\ln\left[1 - \frac{D_o(l + a)^3}{a^3}\right]\right]^{1/m} \tag{5.11}$$

$$l = a\left\{D_{\circ}^{-1/3}\left[1 - \exp\left[-(\varepsilon/\varepsilon_o)^m\right]\right]^{1/3} - 1\right\} \tag{5.12}$$

where ε is axial strain, m is material constant, and $D_{\text{o}} = N_V a^3$.

Inserting Eq. (5.12) into Eq. (5.8), the theoretical expression of constitutive relation of axial stress and strain is derived as:

$$\sigma_1(\varepsilon_1) = \frac{K_{IC} - \sigma_3\left(\pi a^2 B_{3\varepsilon} B_2 + 2\sqrt{B_4/\pi}\right)}{\pi a^2 B_{3\varepsilon} B_1} \tag{5.13}$$

where

$$B_{3\varepsilon} = \frac{1}{[\pi(B_4 + \beta a)]^{\frac{3}{2}}} + \frac{2\sqrt{B_4}}{\sqrt{\pi}[S - \pi(B_4 + \alpha a)^2]} \tag{5.14}$$

$$B_4 = a\left\{D_{\circ}^{-1/3}\left[1 - \exp\left[-(\varepsilon/\varepsilon_o)^m\right]\right]^{1/3} - 1\right\} \tag{5.15}$$

5.1.2 *Time Dependent Crack Growth and Deformation Considering Crack Angle Effect*

During subcritical crack growth, the mode-I crack velocity is $dl/dt = v(K_{\mathrm{I}}/K_{\mathrm{IC}})^n$. Substituting Eq. (5.4) into this equation of crack velocity, the theoretical expression of wing crack growth over time under constant stress state is:

$$\frac{dl}{dt} = v\left\{\left[\sigma_1 \pi a^2 B_3 B_1 + \sigma_3\left(\pi a^2 B_3 B_2 + 2\sqrt{l/\pi}\right)\right] \Big/ K_{IC}\right\}^n \tag{5.16}$$

where v is characteristic crack velocity, n is stress erosion index. Coupling this solved time-dependent crack growth $l(t)$ and Eq. (5.11), the evolution of axial strain during creep is obtained.

$$\varepsilon_1(t) = \varepsilon_o\left[-\ln\left[1 - \frac{D_o(l(t)+a)^3}{a^3}\right]\right]^{1/m} \tag{5.17}$$

5.1.3 *Linkage of Shear Strength and Micro-Parameters*

Using Eq. (5.8), the relationship between peak stress $\sigma_{1\mathrm{peak}}$ and peak crack length l_{peak} is expressed as:

$$\sigma_1(l_{peak}) = \frac{K_{IC}}{\pi a^2 B_3 B_1} - \sigma_3 \frac{\left(\pi a^2 B_3 B_2 + 2\sqrt{l_{peak}/\pi}\right)}{\pi a^2 B_3 B_1} \tag{5.18}$$

Comparing Eqs. (5.18) and (4.2), cohesion and internal friction angle are solved. Shear strength is achieved by combining the Mohr–Coulomb failure envelope as:

$$\tau_{\mathrm{f}}(l_{peak}) = \sigma_{\mathrm{nf}} \tan\phi(l_{peak}) - c(l_{peak}) \tag{5.19}$$

where

$$\sigma_{nf}(l_{peak}) = \frac{\sigma_{1peak} + \sigma_3}{2} + \frac{\sigma_{1peak} - \sigma_3}{2}\cos 2\psi(l_{peak}) \tag{5.20}$$

$$\phi(l_{peak}) = 90^\circ - 2\arctan\left(\sqrt{-\frac{\pi a^2 B_3(l_{peak}) B_1}{\pi a^2 B_3 B_2 + 2\sqrt{l_{peak}/\pi}}}\right) \tag{5.21}$$

$$c(l_{peak}) = -K_{IC}\left(2\pi a^2 B_3(l_{peak}) B_1\right)^{-1} \tan\left(45^\circ - \phi(l_{peak})/2\right) \tag{5.22}$$

where c and ϕ represent for cohesion and internal friction angle respectively, ψ is the failure angle of shear plane, $2\psi = 90° + \phi$.

5.1.4 Evolution of Shear Strength During Progressive and Creep Failure

Referring the same theoretical derivation of Chap. 4, the relationship between shear strength and crack growth at $l \geq l_{\text{peak}}$ can be expressed as:

$$\tau_{\text{f}}(l) = \sigma_{\text{nf}} \tan\phi(l) - c(l) \tag{5.23}$$

where

$$\sigma_{\text{nf}}(l) = \frac{\sigma_1(l) + \sigma_3}{2} + \frac{\sigma_1(l) - \sigma_3}{2}\cos 2\psi(l) \tag{5.24}$$

$$\phi(l) = 90° - 2\arctan\left(\sqrt{-\frac{\pi a^2 B_3 B_1}{\pi a^2 B_3 B_2 + 2\sqrt{l/\pi}}}\right) \tag{5.25}$$

$$c(l) = -K_{IC}\left(2\pi a^2 B_3 B_1\right)^{-1}\tan\left(45° - \phi(l)/2\right) \tag{5.26}$$

where the shear failure angle $\psi(l) = 45° + \phi(l)/2$.

Inserting the relationship between crack growth and axial strain of Eq. (5.12) into Eqs. (5.23–5.26), the macroscopic descriptions for correlations of internal friction angle, cohesion, shear strength and axial strain are obtained. For $\varepsilon \geq \varepsilon_{\text{peak}}$, the strain-dependent shear properties can be expressed as:

$$\tau_{\text{f}}(\varepsilon_1) = \sigma_{\text{nf}} \tan\phi(\varepsilon_1) - c(\varepsilon_1) \tag{5.27}$$

where

$$\sigma_{\text{nf}}(\varepsilon_1) = \frac{\sigma_1 + \sigma_3}{2} + \frac{\sigma_1 - \sigma_3}{2}\cos 2\psi(\varepsilon_1) \tag{5.28}$$

$$\phi(\varepsilon_1) = 90° - 2\arctan\left(\sqrt{-\frac{\pi a^2 B_{3\varepsilon} B_1}{\pi a^2 B_{3\varepsilon} B_2 + 2\sqrt{B_4/\pi}}}\right) \tag{5.29}$$

$$c(\varepsilon_1) = -K_{IC}\left(2\pi a^2 B_{3\varepsilon} B_1\right)^{-1}\tan\left(45° - \phi(\varepsilon_1)/2\right) \tag{5.30}$$

Referring the same theoretical derivation of Chap. 4 and substituting the time-dependent wing crack growth obtained by Eq. (5.16) into Eqs. (5.23–5.26), the evolutions of shear strength, internal friction angle and cohesion under constant stress state

(σ_1, σ_3) at $l \geq l_{\text{peak}}$ can be obtained as:

$$\tau_{\text{f}}(t) = \sigma_{n\text{f}} \tan\phi(t) - c(t) \tag{5.31}$$

where

$$\sigma_{\text{nf}}(t) = \frac{\sigma_1(t) + \sigma_3(t)}{2} + \frac{\sigma_1(t) - \sigma_3(t)}{2}\cos 2\psi(t) \tag{5.32}$$

$$\phi(t) = 90^\circ - 2\arctan\left(\sqrt{-\frac{\pi a^2 B_3 B_1}{\pi a^2 B_3 B_2 + 2\sqrt{l(t)/\pi}}}\right) \tag{5.33}$$

$$c(t) = -K_{IC}\left(2\pi a^2 B_3 B_1\right)^{-1} \tan\left(45^\circ - \phi(t)/2\right) \tag{5.34}$$

where the shear failure angle $\psi(t) = 45^\circ + \phi(t)/2$.

5.2 Crack Angle Effect on Progressive Fracture

5.2.1 *Effect of Crack Angle Stress–Strain Relation*

Effect of crack angle on stress–strain relation under different friction coefficients is shown in Fig. 5.3a–c. Comparing Fig. 5.3a–c, it is observed that the number of curve for stress–strain decreases with increment of friction coefficient, which shows that the range of crack angle causing rock failure is decreasing. The most disadvantage crack angle (i.e., corresponding to the smallest peak stress) is 55°, 45°, 35° in Fig. 5.3a–c respectively, which shows that the most disadvantage crack angle decreases with the increasing friction coefficient. Furthermore, the most disadvantaged crack angle at $\mu = 0$ is 55° or so, it agrees well with the experimental result of Ashby and Hallam (1986) and numerical result of Zhou et al. (2008) in which the minor open flaw is considered.

Comparing Figs. 5.3 and 5.4, it is observed that peak stress decreases with increment of initial crack size at the given crack angle. Comparing Figs. 5.3c and 5.5 peak stress increases with increment of confining pressure at the given crack angle. Figure 5.6 shows the relation between crack length and axial stress which is corresponding to the stress–strain relation of Fig. 5.5. The study results of effect of model parameters on the relation between crack length and axial stress is same to the results of effect of model parameters on stress–strain relation.

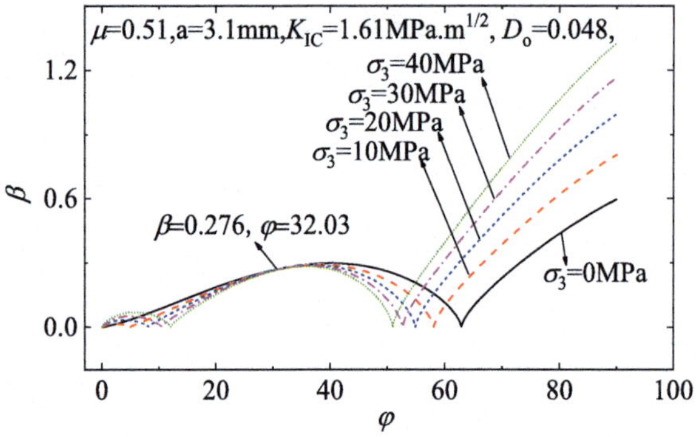

Fig. 5.2 Illustration for correction factor β and microcrack angle φ

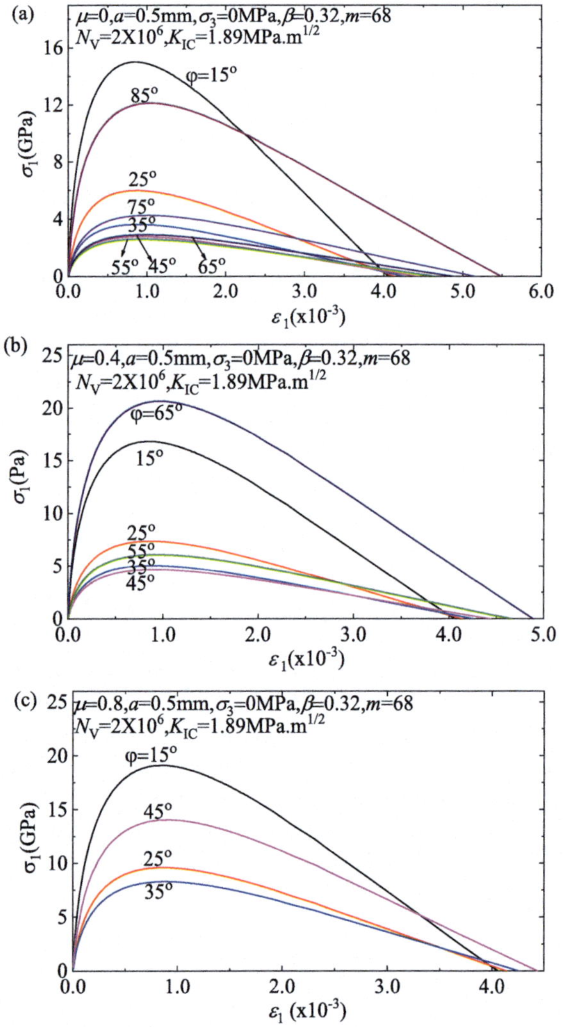

Fig. 5.3 Effects of crack angle on relationship between axial stress and strain under different friction coefficients **a** $\mu = 0$, **b** $\mu = 0.4$, **c** $\mu = 0.8$, ($a = 0.5$ mm, $\sigma_3 = 0$ MPa)

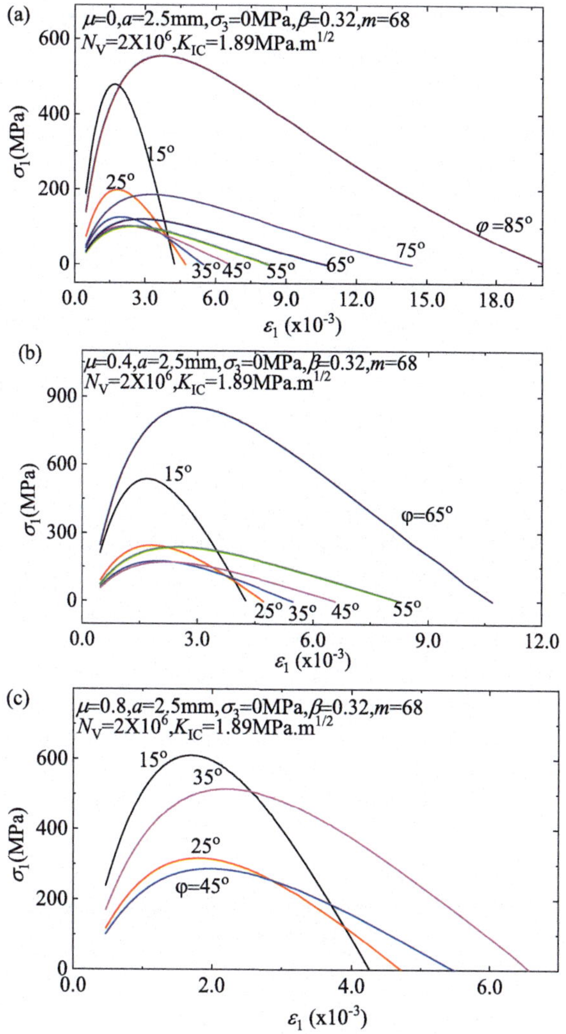

Fig. 5.4 Effects of crack angle on relationship between axial stress and strain under different friction coefficients **a** $\mu = 0$, **b** $\mu = 0.4$, **c** $\mu = 0.8$, ($a = 2.5$ mm, $\sigma_3 = 0$ MPa)

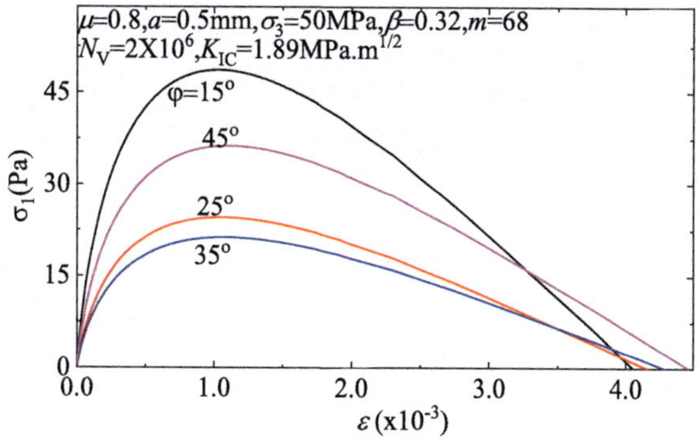

Fig. 5.5 Effects of crack angle on relationship between axial stress and strain ($a = 0.5$ mm, $\sigma_3 = 50$ MPa)

5.2.2 *Effects of Crack Angle on Strength and Crack Initiation Stress*

Based on the above illustration of stress–strain relation and stress-crack relation in different crack angles, the effects of crack angle on crack initiation stress and peak stress will be studied below. Due to the random distribution of initial microcracks in brittle rocks, the effects of crack angle on mechanical properties of brittle rock almost cannot be investigated directly. However, the effects of the minor pre-existing flaw angle on mechanical behaviors of brittle rocks were widely investigated in compressive loadings (Lajtai 1974; Ashby and Hallam 1986; Sagong and Bobet 2002; Zhou et al. 2008; Park and Bobet 2010; Lee and Jeon 2011; Zhang and Wong 2012, 2013; Cao et al. 2016; Vergara et al. 2016). Thus, the analytical results will approximately be compared with the these experimental result below.

As shown Figs. 5.7, 5.8 and 5.9, peak stress (i.e., rock strength) decreases and accesses a minimum value, and increases with increment of crack angle. This analytical phenomenon is good agreement with the published study results in Fig. 5.10 (Lee and Jeon 2011; Zhang and Wong 2012, 2013; Vergara et al. 2016). Furthermore, crack initiation stress also decreases and accesses a minimum value, and increases with increment of crack angle, which is similar to the published study results of relation between crack initiation stress and flaw inclination angle under compression in Fig. 5.10 (Lee and Jeon, 2011; Zhang and Wong, 2012, 2013). The range of crack angle (which crack growth can occur in compression) decreases with increment of friction coefficient. The crack initiation stress and peak stress both increase with increment of friction coefficient for given crack angle, it agrees well with the experiment results in which friction between closed flaw interfaces is considered (Park and Bobet 2010).

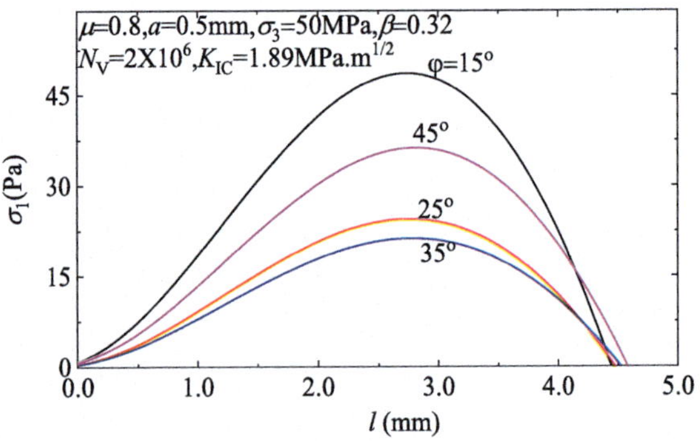

Fig. 5.6 Effects of crack angle on relationship between axial stress and crack growth

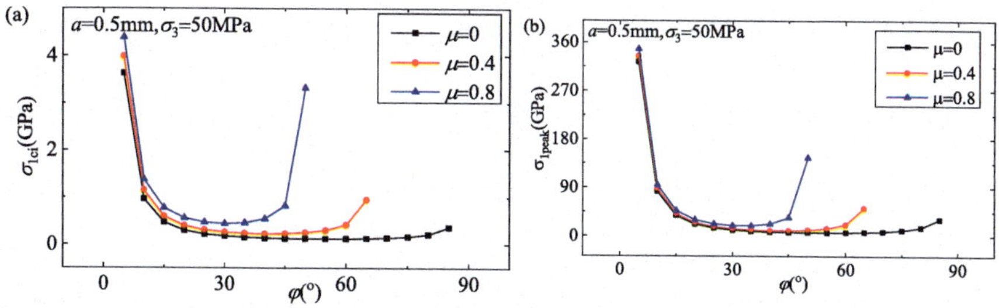

Fig. 5.7 Effects of friction coefficient on relationships between **a** crack initiation stress σ_{ci}, **b** peak stress σ_{peak} and crack angle φ

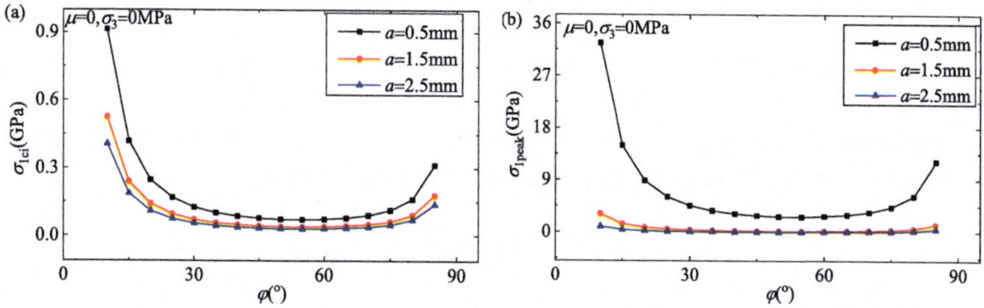

Fig. 5.8 Effects of initial crack size on relationships between **a** crack initiation stress, **b** peak stress and crack angle

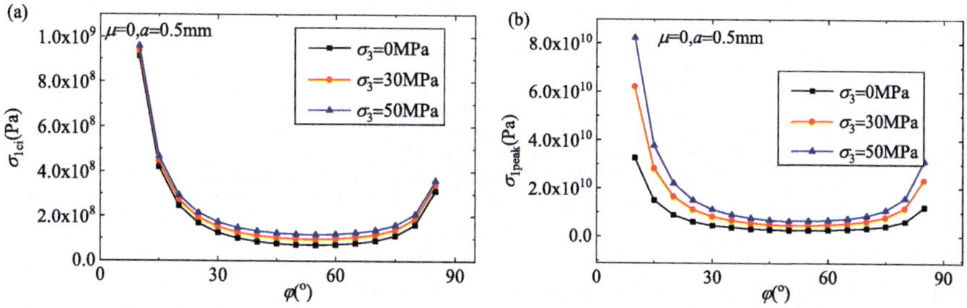

Fig. 5.9 Effects of confining pressure on relationships between **a** crack initiation stress, **b** peak stress and crack angle

As shown in Fig. 5.8, the crack initiation stress and peak stress (i.e., rock strength) both decrease with increment of initial microcrack size for given crack angle. As shown in Fig. 5.9, crack initiation stress and peak stress both increase with increment of confining pressure for given crack angle, which agrees well with the study results of Vergara et al. (2016).

5.2.3 *Effect of Crack Angle on Shear Strength*

Effects of initial crack size on shear properties under different crack inclination angles are shown in Figs. 5.11, 5.12, 5.13 and 5.14. Internal friction angle, cohesion, shear failure plane angle and shear strength decrease firstly, and then increase with increment of crack inclination angle. When crack angle accesses to 0° or 90°, the cohesion in Fig. 5.12 and the shear strength in Fig. 5.14 are very larger, and the internal friction angle in Fig. 5.11 is close to 90°, which shows that the appearance of rock failure is difficult. Furthermore, from the relation between macroscopic failure plane angle and microcrack angle shown in Fig. 5.13, it is observed that the shear failure angle is close to 90° when crack angle access to 0° or 90°, which is similar to the experimental results of Lee and Jeon (2011) and Wong and Li (2013). A minimum of internal friction angle, cohesion, shear failure angle and shear strength are founded, which is corresponding to a disadvantage crack angle related to axial peak stress in Fig. 5.7.

These study results provide a theoretical assistance in prediction for effects of crack geometry on shear properties under compressive loading.

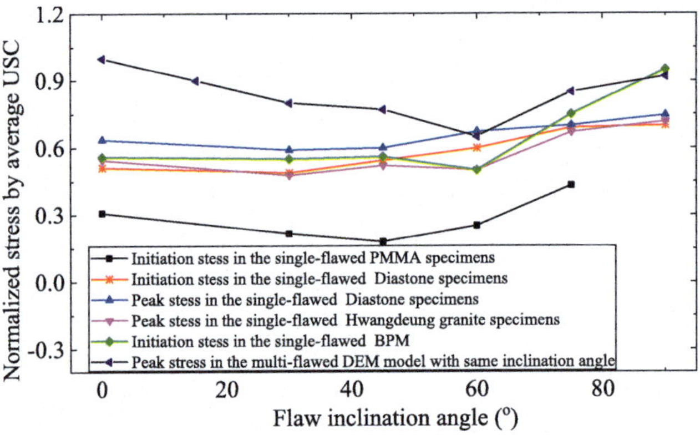

Fig. 5.10 Relationships between normalized stress by average UCS (uniaxial compressive strength) and flaw inclination angle from the published experimental data of PMMA, Diastine and Hwangdeung granite specimens (Lee and Jeon 2011), numerical data of BPM (i.e., Bonded-particle model) (Zhang and Wong 2012, 2013) and DEM (i.e., discrete element method) model (Vergara et al. 2016). *Note* This flaw inclination angle is the angle between inclination of flaw and horizontal direction

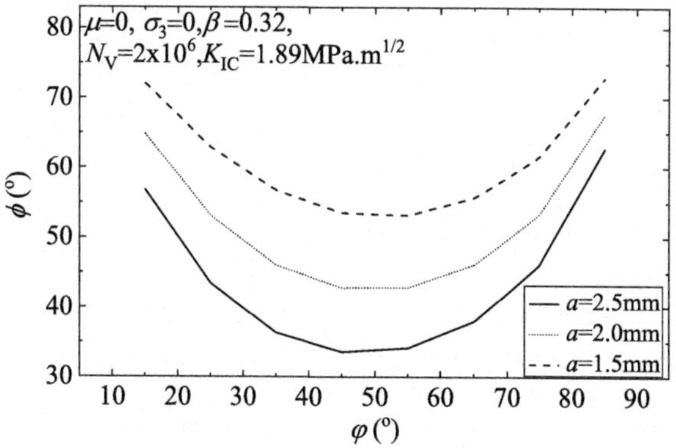

Fig. 5.11 Effect of initial crack size on relationship between internal friction angle and crack inclination angle

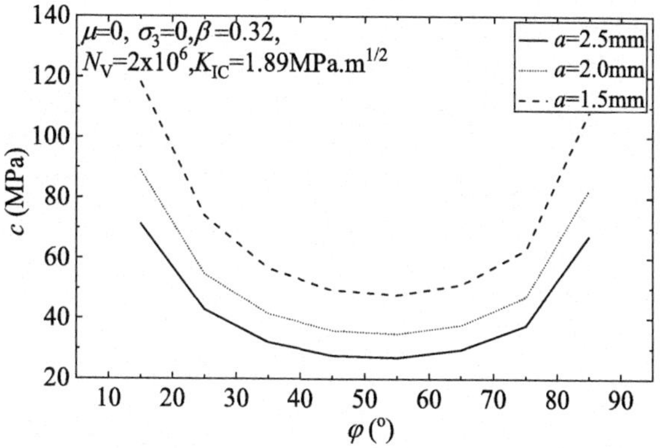

Fig. 5.12 Effect of initial crack size on relationship between cohesion and crack inclination angle

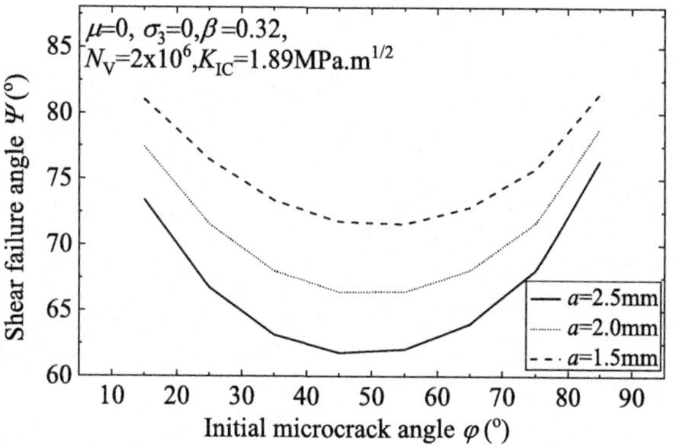

Fig. 5.13 Effect of initial crack size on relationship between shear failure angle and crack inclination angle

5.2.4 Effects of Crack Angle on Crack Growth- and Axial Strain- Dependent Shear Strength

Effect of crack angle on crack growth-dependent cohesion is shown in Fig. 5.15. The crack growth-dependent cohesion of Fig. 5.15 is corresponding to the post-peak phase of stress–strain curve. Cohesion is decreasing with increment of wing crack growth during the post-peak phase of stress–strain curve. Crack growth is corresponding to damage evolution. Thus, cohesion is also decreasing with damage evolution. This phenomenon was measured by Martin and Chandler (1994). Furthermore, the locus of initiation value of cohesion is plotted in Fig. 5.15. It is noted that this locus of initiation value of cohesion is corresponding to the locus of peak value of axial stress in the stress–strain curve. The macroscopic description of relation between cohesion and axial strain is shown in Fig. 5.16, which is consistent with the relation between cohesion and wing crack growth of Fig. 5.15.

Effect of crack angle on crack growth-dependent internal friction angle is shown in Fig. 5.17. It is noted that the crack growth-dependent internal friction angle of

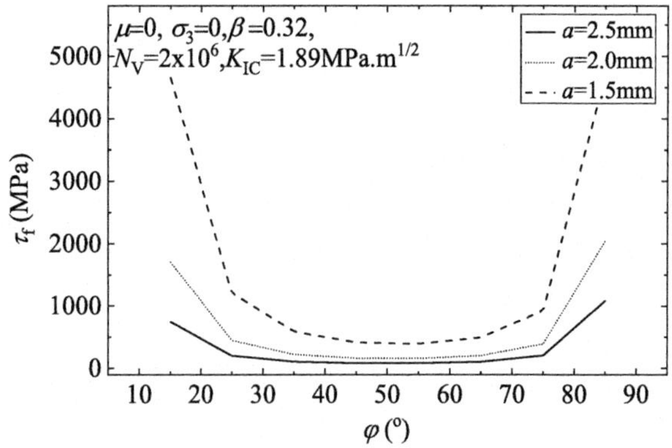

Fig. 5.14 Effect of initial crack size on relationship between shear strength and crack inclination angle

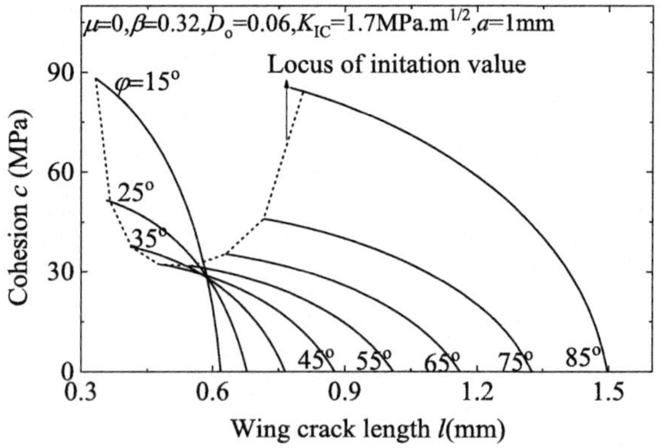

Fig. 5.15 Relationship between cohesion and crack growth under different crack angles

Fig. 5.17 is corresponding to the post-peak phase of stress–strain curve. Internal friction angle increases firstly, and decreases with increment of wing crack growth during the post-peak phase of stress–strain curve. Crack growth is corresponding to damage evolution. It shows that the relationship between internal friction angle and damage exists a similar variation trend as Fig. 5.17. Internal friction angle increases firstly, and decreases with damage evolution, which was also obtained by experiment of Martin and Chandler (1994). The locus of initiation value of internal friction angle is plotted in Fig. 5.17. The locus of initiation value of internal friction angle is corresponding to the locus of peak value of axial stress of stress–strain curve. Furthermore, the locus of peak value of internal friction angle is also plotted in Fig. 5.17. It can be concluded that crack length at peak value of internal friction angle is larger than crack length at peak value of axial stress of stress–strain curve. It shows that the internal friction angle dose not reach to the maximum value when the axial stress reaches to the peak value in post-peak failure tests. It provides a meaning implication that rock failure often occurs at the post-peak phase of stress–strain curve in the triaxial compressive tests.

Peak internal friction angle under different crack angles is shown in Fig. 5.19. Peak internal friction angle decreases firstly, and increases with increment of crack angle. The macroscopic description of relationship between internal friction angle and axial strain is shown in Fig. 5.18, which is consistent with the relationship between internal friction angle and wing crack growth of Fig. 5.17.

Effect of crack angle on crack growth-dependent shear strength is shown in Fig. 5.20. It is noted that the crack growth-dependent shear strength of Fig. 5.20 is corresponding to the post-peak phase of stress–strain curve. Shear strength decreases with increment of wing crack growth during the post-peak phase of stress–strain curve. Crack growth is corresponding to damage evolution. It shows that the relationship between shear strength and damage exists a similar variation trend as Fig. 5.20. Shear strength decreases with damage evolution, which was also obtained by experiment of Martin and Chandler (1994). The locus of initiation value of shear strength

Fig. 5.16 Relationship between cohesion and axial strain under different crack angles

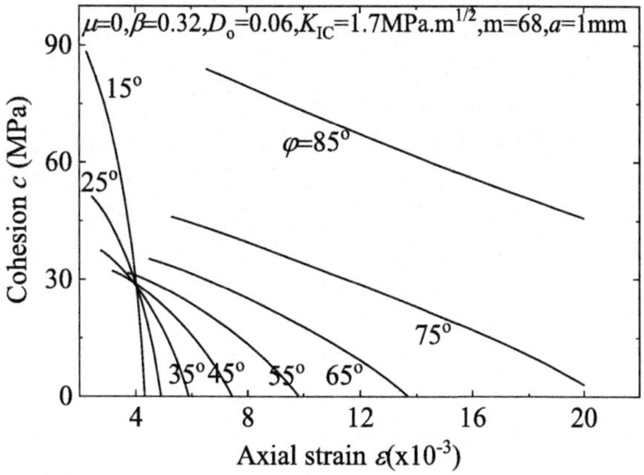

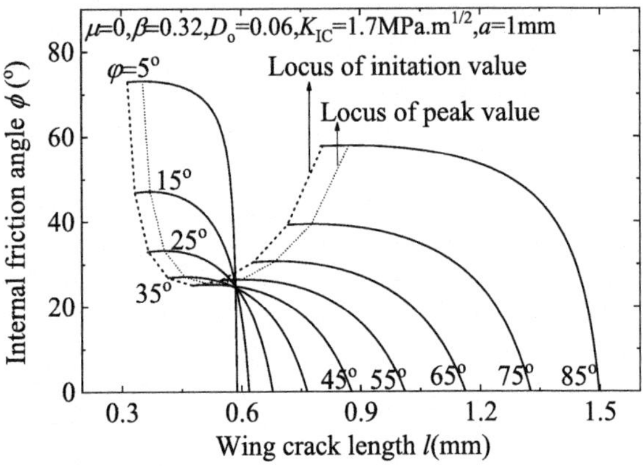

Fig. 5.17 Relationship between internal friction angle and crack growth under different crack angles

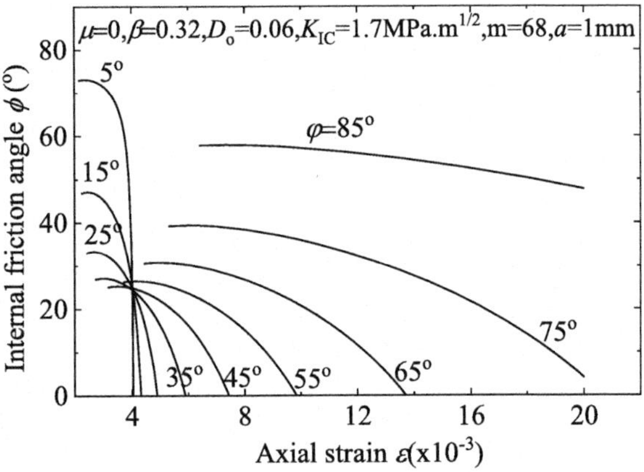

Fig. 5.18 Relationship between internal friction angle and axial strain under different crack angles

is plotted in Fig. 5.20. The locus of initiation value of shear strength is corresponding to the locus of peak value of axial stress in stress–strain curve.

Peak shear strength under different crack angles is shown in Fig. 5.22. Peak shear strength decreases firstly, and increases with increment of crack angle. The macroscopic description of relationship between shear strength and axial strain is shown in Fig. 5.21, which is consistent with the relationship between shear strength and wing crack growth of Fig. 5.20. Furthermore, this phenomenological behavior is similar to Ohnaka's (1997) law governing shear failure of intact rocks in the brittle-plastic transition regime under the triaxial compressive test, which is corresponding to the whole process of axial stress–strain curve in the triaxial compressive tests. However, it is the deficiency of our theoretical results that only can provide the strain- dependent shear strength during the post-peak phase of stress–strain curve. Moreover, it is noted that shear strength at the post-peak phase of stress–strain curve is equivalent to the applied shear stress (Zhao and Cai, 2010).

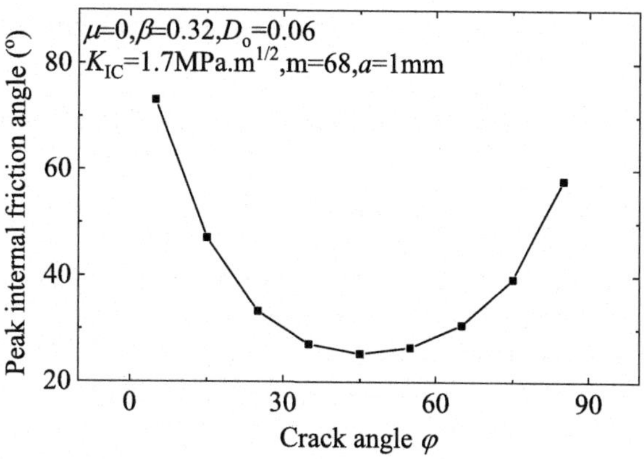

Fig. 5.19 Relationship between peak internal friction and angle crack angle

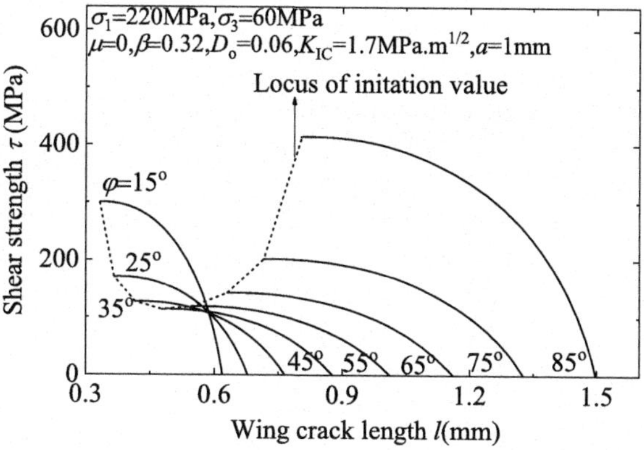

Fig. 5.20 Relationship between shear strength and crack growth under different crack angles

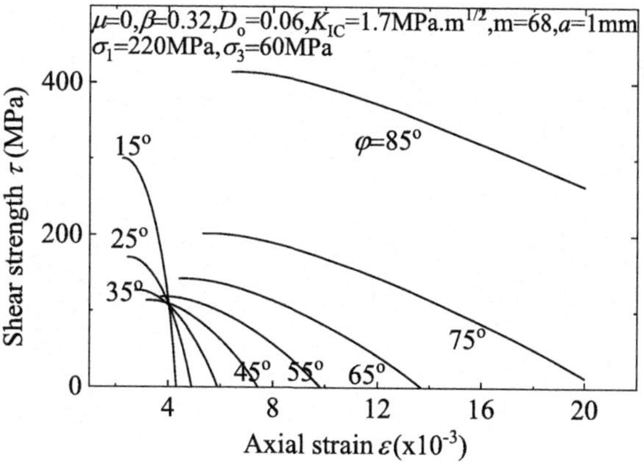

Fig. 5.21 Relationship between shear strength and axial strain under different crack angles

5.3 Crack Angle Effect on Creep Fracture

5.3.1 *Effects of Crack Angle on Creep Behaviors*

Effect of crack angle on evolution of wing crack length during creep is shown in Fig. 5.22. The creep failure crack length increases with increment of crack angle. For the given crack number N_V per unit volume in rocks, this phenomenon can be clearly explained by the geometry of crack in Fig. 2.1. The creep failure time decreases firstly, and increases with increment of crack angle from 25° to 65°. The shortest creep failure time occurs at crack angle 45°. This phenomenon is also shown by the macroscopic evolution of axial strain during creep in Fig. 5.23. Furthermore, it is also predicted that the creep failure strain increases with increment of crack angle in Fig. 5.23. The reason is that the crack growth is monotonically increasing with increment of the axial strain.

Influences of crack inclination on the creeping strain under the axial stepping stress for the distinct confining pressures (i.e., $\sigma_3 = 0$ MPa and $\sigma_3 = 2$ MPa) are shown in Fig. 5.24a, b, respectively. For the function of the axial stepping stress under the distinct crack inclination angles, the cycle time T is 3600 s, and the step value $\Delta\sigma$ is 5 MPa. In Fig. 5.24a, under the condition of unconfined pressure, the accelerated creep failure appears at the third step for crack inclination $\varphi = 35°$ and 40°, the fourth step for crack inclination $\varphi = 30°$ and 45°, the twelfth step for crack inclination $\varphi = 50°$, the eighth step for crack inclination $\varphi = 25°$, and the twentieth step for crack inclination $\varphi = 55°$.

In Fig. 5.24b, under confining pressure $\sigma_3 = 2$ MPa, the accelerated creep failure appears at the fifth step for crack inclination $\varphi = 35°$ and 40°, the seventh step for crack inclination $\varphi = 30°$ and 45°, the ninth step for crack inclination $\varphi = 50°$, the eleventh step for crack inclination $\varphi = 25°$, and the twenty-fourth step for crack inclination $\varphi = 55°$.

Furthermore, the step number and the axial stress happening at creep failure firstly decreases, and then increases with the increasing crack inclination. Comparing

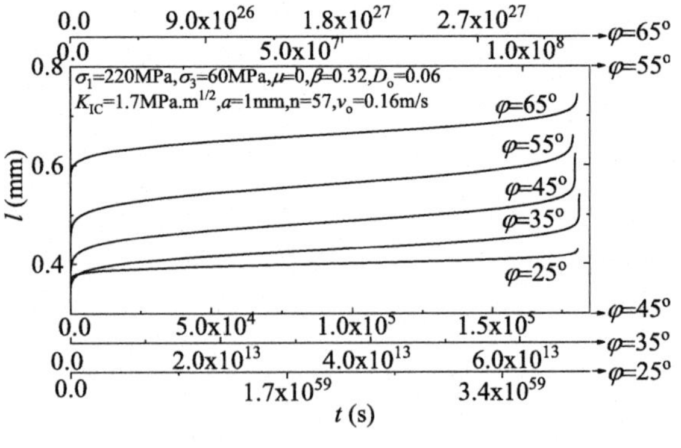

Fig. 5.22 Effect of crack angle on evolution of wing crack length during creep

Fig. 5.23 Effect of crack angle on evolution of axial strain during creep

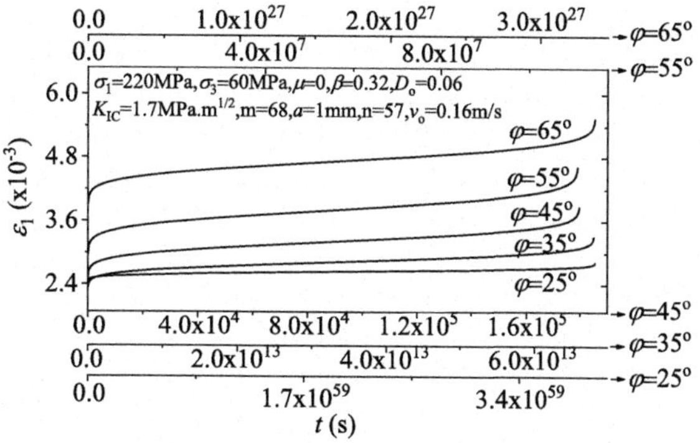

Fig. 5.24a, b, the increment of confining pressure increases the number of loading step and the time of creep failure for the given crack inclination.

Influences of crack inclination on the time happening at creep failure under the distinct friction coefficients are shown in Table 5.1. The time happening at creep failure t_f is firstly decreasing, and then is increasing with increment of crack inclination for given friction coefficient, initial crack size and stress state (σ_1, σ_3). The shortest failure time of rock appears nearby the worst crack inclination (e.g., 55° at $\mu = 0$, 45° at $\mu = 0.4$ and 35° at $\mu = 0.8$). The time happening at creep failure is increasing with increment of friction coefficient for given crack inclination, initial crack size and stress state. Furthermore, it is noted that even some small changes for crack inclination angle could result in an increase or decrease by several orders of magnitude of the creep failure time, which shows that the extreme dependence of creep failure time on crack inclination angle at given stress state. Due to the simplification of same initial crack inclination in model, it can be predicted that the creep failure time of theoretical model at the worst inclination angle is not larger than the creep failure time of natural rock containing randomly distributed microcracks, which provides a significant reference to judge the long-term mechanical properties of brittle rock.

Table 5.1 Influences of crack inclination on the time happening at creep failure t_f for $a = 0.5$ mm, $\sigma_3 = 0$ and $\sigma_1 = 1.79$ GPa

φ(°)	$\mu = 0$	$\mu = 0.4$	$\mu = 0.8$
	t_f (s)		
15	4.48e+42	8.24e+44	2.96e+47
25	2.82e+24	3.75e+28	5.98e+33
35	2.73e+14	1.00e+21	7.00e+30
45	1.69e+09	2.70e+19	2.40e+41
55	6.06e+07	5.03e+24	–
65	1.32e+10	1.23e+49	–
75	4.64e+17	–	–
85	3.01e+38	–	–

Table 5.2 Influences of crack inclination on the time happening at creep failure for $\mu = 0.4$, $\sigma_3 = 0$ MPa and $\sigma_1 = 1.79$GPa

$\varphi(^{o})$	$a = 0.5$ mm	$a = 0.6$ mm
		t_f (s)
15	8.24e+44	2.72e+37
25	3.75e+28	1.31e+21
35	1.00e+21	3.83e+13
45	2.70e+19	1.15e+12
55	5.03e+24	2.41e+17
65	1.23e+49	6.70e+41

As shown in Table 5.1, the value of creep failure time is null at the larger or smaller crack inclination angles, which shows that crack cannot grow because the applied axial stress cannot reach crack initiation stress (e.g., 0°–15° in $\mu = 0$, 0.4 and 0.8) or compressive stress never lead to crack growth, i.e., stress locking (e.g., 85°–90° in $\mu = 0$, 75°–90° in $\mu = 0.4$, and 55°–90° in $\mu = 0.8$). Table 5.2 shows the influence of crack inclination on the time happening at creep failure in the distinct initial crack sizes. The time happening at creep failure increases with decrement of initial crack size.

5.3.2 *Effects of Crack Angle on Shear Properties During Creep*

Evolutions of cohesion, internal friction angle and shear strength under different crack angle are shown in Figs. 5.25, 5.26 and 5.27, respectively. As shown in Fig. 5.25, cohesion experiences a long-time steady-state weakening process, and an accelerated drop finally. In Fig. 5.26, internal friction angle experiences a long-time steady-state strengthening process, and an accelerated drop finally. In Fig. 5.27, shear strength experiences a long-time steady-state strengthening process, and an accelerated drop finally. Cohesion, internal friction angle and shear strength at the given time increase firstly, and decrease with increment of crack angle. This phenomenon of time-dependent shear strength is similar to the study results of the Rate- and state-variable friction law in Dieterich (1979a, b). Evolution of shear stress between faults was studied during the stable and unstable sliding (Dieterich, 1979a, b). In the phase of stable sliding, shear stress approximately remains constant with the long-term creep sliding, which is associated with evolution of shear properties during aseismic fault creep. During the unstable sliding, shear stress has a relatively slow increment, and a sudden drop with the short-term sliding. The process of laboratory unstable sliding is closely related to the instable process of earthquakes on faults. Thus, the proposed theoretical results provide a meaning insight into evaluation for earthquake mechanism triggered by microcrack growth in brittle rocks.

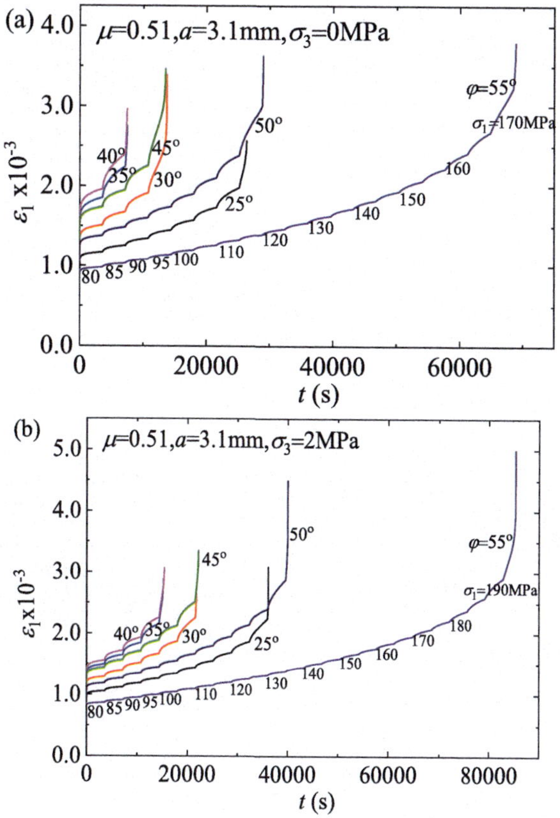

Fig. 5.24 The stepping creep curves influenced by crack inclination under distinct confining pressures **a** $\sigma_3 = 0$ MPa, **b** $\sigma_3 = 2$ MPa

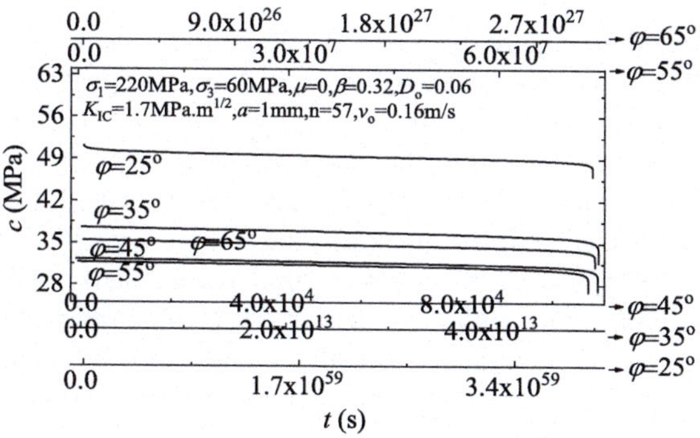

Fig. 5.25 Evolution of cohesion under different crack angles during creep

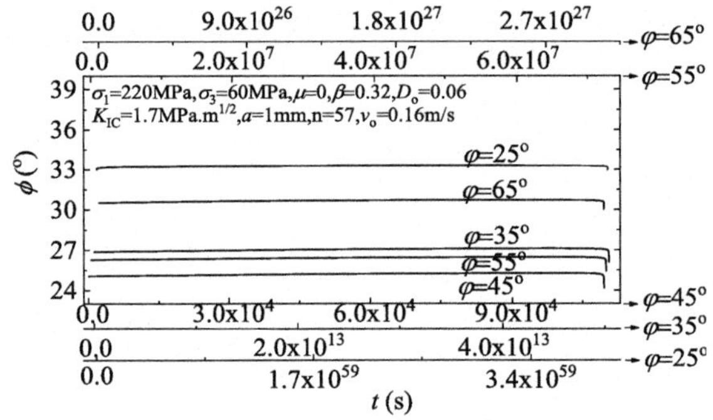

Fig. 5.26 Evolution of internal friction angle under different crack angles during creep

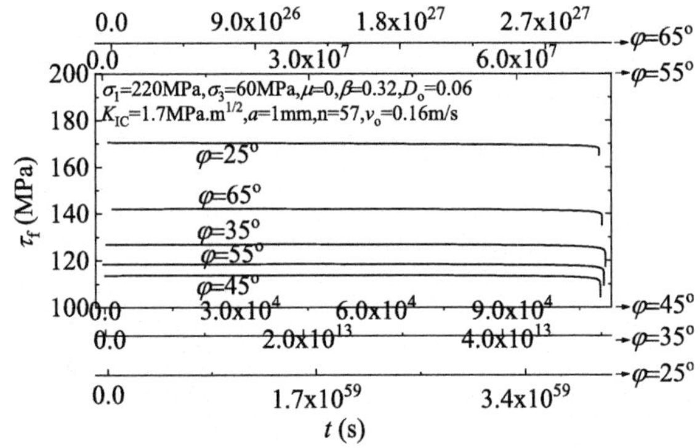

Fig. 5.27 Evolution of shear strength under different crack angles during creep

5.4 Conclusions

A micro–macro method is proposed to evaluate the microcrack angle effect on the mechanical properties of intact rocks during progressive and brittle creep. Effects of crack angle on the stress-crack relation, the stress–strain relation, the compressive and shear strengths, the crack initiation stress, the evolutions of wing crack length and axial strain, and the time- dependent shear strength are discussed.

The crack initiation stress, the compressive and shear strengths, the internal friction angle and the cohesion all decrease firstly and reach a minimum value, and then increase with increment of crack angle from 0° to 90° under compression. The range of crack angle in which could result in crack initiation and rock failure of brittle rock decreases with increment of friction coefficient in compression. When the crack angle is beyond this range, the initial microcrack could not grow. The creep failure time decrease firstly, and increase with increment of crack angle. Wing crack length and axial strain at creep failure increase with increment of crack angle.

References

Ashby MF, Hallam SD. The failure of brittle solids containing small cracks under compressive stress states. Acta Metall. 1986;34(3):497–510.

Ashby MF, Sammis CG. The damage mechanics of brittle solids in compression. Pure and Appl Geophys. 1990;133(3):489–521.

Bhat HS, Sammis CG, Rosakis AJ. The micromechanics of Westerly granite at large compressive loads. Pure Appl Geophys. 2011;168(12):1–18.

Budiansky B, O'Connel RJ. Elastic moduli of a cracked solid. Int J Solids Struct. 1976;12:81–97.

Cao RH, Cao P, Lin H, Pu CZ, Ou K. Mechanical behavior of brittle rock-like specimens with pre-existing fissures under uniaxial loading: experimental studies and particle mechanics approach. Rock Mech Rock Eng. 2016;49(3):763–83.

Dieterich JH. Modeling of rock friction 1. Experimental results and constitutive equations. J Geophys Res. 1979a;84(B5): 2161−2168.

Dieterich JH. Modeling of rock friction 2. Simulation of preseismic slip. J Geophys Res. 1979b;84(B5): 2169−2145.

Lajtai EZ. Brittle fracture in compression. Int J Fract. 1974;10(4):525–36.

Lee H, Jeon S. An experimental and numerical study of fracture coalescence in pre-cracked specimens under uniaxial compression. Int J Solids Struct. 2011;48:979–99.

Martin CD, Chandler NA. The progressive fracture of Lac du Bonnet granite. Int J Rock Mech Mining Sci Geomech Abstracts. 1994;31(6):643–59.

Ohnaka M, Akatsu M, Mochizuki H, Odedra A, Tagashira F, Yamamoto Y. A constitutive law for the shear failure of rock under lithospheric conditions. Tectonophys. 1997;277:1–27.

Park CH, Bobet A. Crack initiation, propagation and coalescence from frictional flaws in uniaxial compression. Eng Fract Mech. 2010;77:2727–48.

Sagong M, Bobet A. Coalescence of multiple flaws in a rock-model material in uniaxial compression. Int J Rock Mech Min Sci. 2002;39(2):229–41.

Vergara MR, Jan MVS, Lorig L. Numerical model for the study of the strength and failure modes of rock containing non-persistent joints. Rock Mech Rock Eng. 2016;49(4):1211–26.

Wong LNY, Li HQ. Numerical study on coalescence of two pre-existing coplanar flaws in rock. Int J Solids Struct. 2013;50:3685–706.

Zhang XP, Wong LNY. Cracking processes in rock-like material containing a single flaw under uniaxial compression- a numerical study based on parallel bonded-particle model approach. Rock Mech Rock Eng. 2012;45:711–37.

Zhang XP, Wong LNY. Crack initiation, propagation and coalescence in rock-like material containing two flaws: a numerical study based on bonded-particle model approach. Rock Mech Rock Eng. 2013;46(5):1001–21.

Zhao XG, Cai M. A mobilized dilation angle model for rocks. Int J Rock Mech Min Sci. 2010;47(3):368–84.

Zhou XP, Zhang YX, Ha QL, Zhu KS. Micromechanical modelling of the complete stress- strain relationship for crack weakened rock subjected to compressive loading. Rock Mech Rock Eng. 2008;41(5):747–69.

Chapter 6
Crack Nucleation During Creep Fracture

Some stress drops appear during the process of deformation under triaxial compressive tests in intact brittle rocks in Fig. 6.1, and each stress drop is associated with an individual shear band caused by strain localization from the accumulation and nucleation of distributed microcracks (Brantut et al. 2011). The presence of an individual shear band induced by microcrack nucleation causes a larger increment of initial damage of rocks. Thus, the localized microcrack nucleation have great effect on the mechanical properties during progressive and creep failure. This chapter will propose a micro–macro method to explore the microcrack nucleation effect on the stress–strain curve during progressive failure and the strain–time or stress-time curve during creep failure.

6.1 Theory

For realizing the experimental phenomenon of stress drops in intact brittle rocks (Fig. 6.1), the process of the presence of shear bands can be approximately understood by corresponding to the step initial damage with the increasing time. Because the initial crack size a is assumed as a constant value, the increment of step initial damage D_0 causes the increment of crack number N_V. The step damage caused by the increment of crack number N_V can be approximately understood as the increment of damage caused by the shear band formation from the localized wing crack growth and coalescence. This global variation of initial damage in the model can be defined by Fig. 2.1. Inserting this function of stepping initial damage with deformation or time (see Fig. 6.2) into the equation derived by a proposed micro–macro model in Chaps. 2 or 5, an analytical result explaining the experimental phenomenon of stress drops at the stress–strain curve during progressive failure or the shear stress-time curves during long-term creeping can be realized.

X. Li et al., *Static Creep Micro-Macro Fracture Mechanics of Brittle Solids*,
https://doi.org/10.1007/978-981-99-8203-5_6

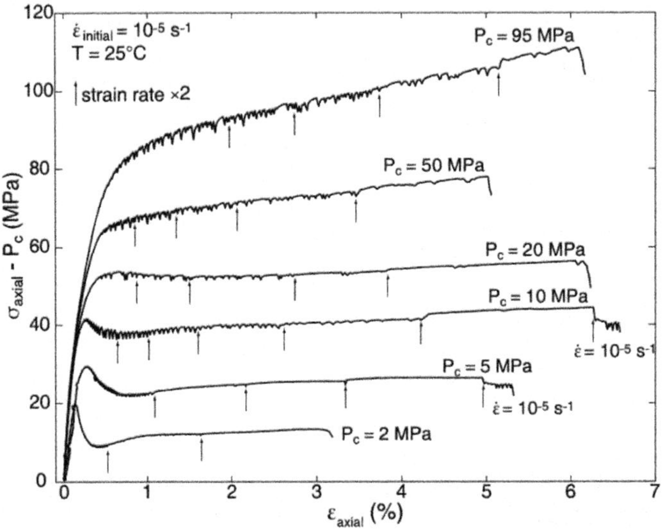

(a)

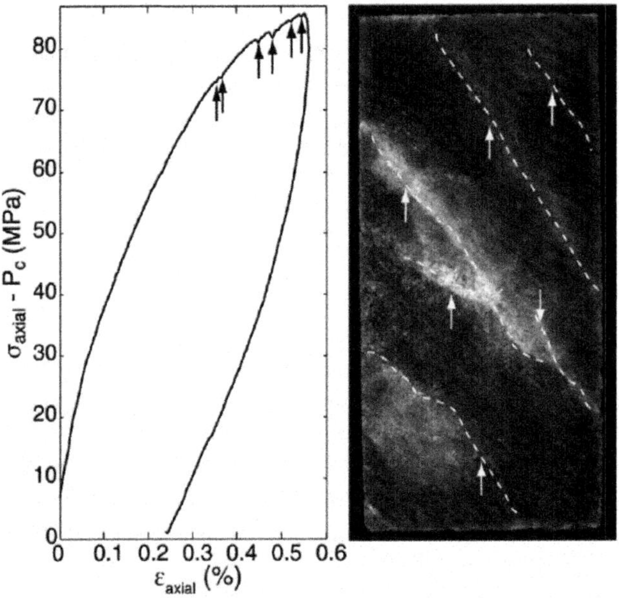

(b)

Fig. 6.1 **a** The whole stress–strain curve with multi-stress drops under different confining pressures, **b** stress–strain curve and the corresponding photograph in brittle materials. Each stress drop in the stress–strain curve is corresponding to an individual presence of shear band (Brantut et al. 2011)

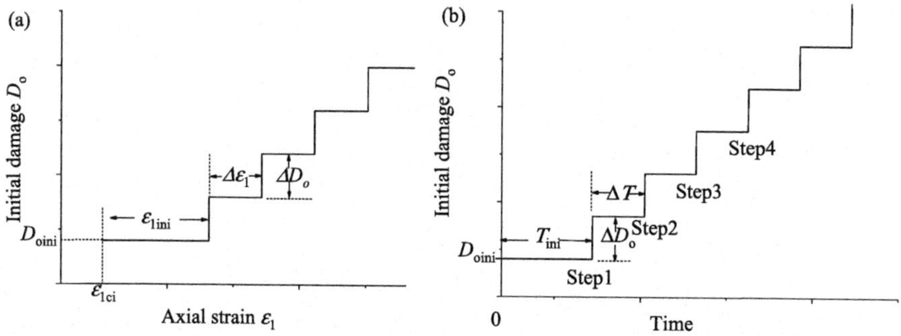

Fig. 6.2 Variation of stepping initial damage with **a** axial strain or **b** time which causes the stress drops in intact rocks, which can be described by a function. Note: Assuming that the parameters of initial damage difference ΔD_o, strain difference $\Delta\varepsilon_1$ and time difference ΔT at each step is same in this study, the stress drops caused by stepping initial damages are studied qualitatively. Actually, the value of parameters of ΔD_o and ΔT should be random in the natural rocks. D_{oini} is initial value

Figure 6.3 also shows approximately a schematic of this proposed damage rate- and state- fracture model (DRSFM). The number (from 1 to 7) represents the order of appearance for the individual shear band, and the macroscopic shear band is caused by the coalescence of the second, third, fourth, sixth and seventh shear bands. This suggested sequence of shear bands is to explain the mechanism of appearance of localized shear bands. Actually, due to randomly distributed microcracks in natural rocks, the sequence of the small shear bands is also random in natural rocks. If making the model be in the scale of earth crusts, the appearance of small shear bands causes the smaller earthquakes, and the macroscopic shear fracture causes the bigger earthquake. The presence of the individual shear band is corresponding to an initial damage variable ΔD_o in Fig. 6.2, and the deformation or time differences in presence between two consecutive shear bands is corresponding to the strain variable $\Delta\varepsilon_1$ or the time variable ΔT in Fig. 6.2. The detailed analyses see the results and discussions below.

Furthermore, this chapter adopts another equation of strain-defined damage containing confining pressure effect to replace the damage definition of Eq. (2.13), and the same theoretical derivation process will not be shown in this section.

Based on the Weibull probability distribution, the damage can be expressed as (Chen et al. 1997b):

$$D = 1 - \exp\left[-\frac{\varepsilon_1 - \frac{2\gamma|\sigma_3|}{E}}{\varepsilon_o}\right]^m \tag{6.1}$$

where γ is passion rate, E is elastic modulus, ε_1 is axial strain, ε_o, m is material constant, | | represents for the absolute value. Based on thermodynamics theory, a similar exponential damage expression relating to strain in brittle solids was proposed Mazars (1986).

Assuming the value of damage obtained by Eq. (6.1) is equal to that of Eq. (5.10),

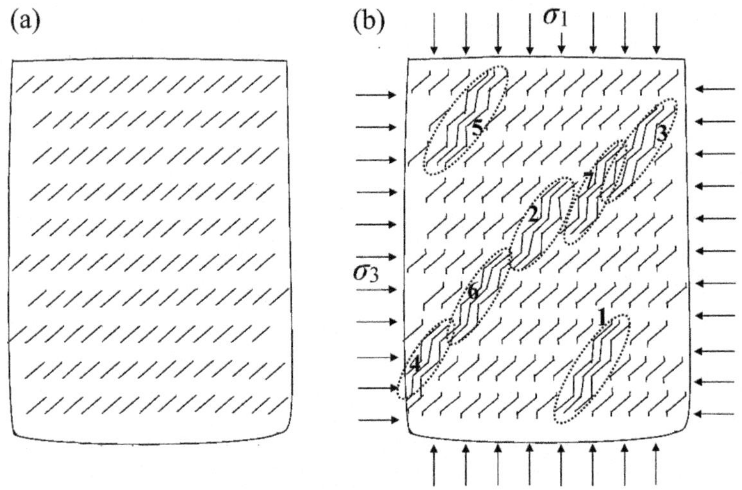

Fig. 6.3 A schematic diagram of the process of formation from the small individual shear bands to the macroscopic shear fracture band in intact rocks. **a** The rock model containing initial microcracks before initial crack sliding and wing crack growth; **b** rock model containing seven individual small shear bands caused by local initial crack sliding and wing microcrack growth, accumulation and coalescence

the relationship between crack growth and axial strain can be approximately derived as:

$$\varepsilon_1 = 2E^{-1}\gamma|\sigma_3| + \varepsilon_o\left(-\ln\left[1 - D_o\left(l/a + 1\right)^3\right]\right)^{\frac{1}{m}} \tag{6.2}$$

The slip displacement D_{sp} along shear fracture plane can be expressed as:

$$D_{sp} = \frac{\Delta L}{\sin\psi} = \frac{\varepsilon_1 L}{\sin\psi} \tag{6.3}$$

where ΔL is the axial displacement of rocks sample, L is the axial length of the rocks sample.

Combining quasi-static variation of stress with wing crack growth and static variation of wing crack growth over time in Eqs. (5.8) and (5.16) respectively, Eq. (6.2) will provide significant assistance for evaluating the micro–macro mechanism.

Inserting time-dependent cracking l(t) solved by Eq. (5.16) into Eq. (6.2), the time-dependent axial strain caused by subcritical cracking can be expressed as:

$$\varepsilon_1(t) = 2E^{-1}\gamma|\sigma_3| + \varepsilon_o\left(-\ln\left[1 - D_o\left(l(t)/a + 1\right)^3\right]\right)^{\frac{1}{m}} \tag{6.4}$$

Substituting time-dependent axial strain of Eq. (6.4) into Eq. (6.3), the time-dependent slip displacement caused by subcritical cracking can be expressed as:

$$D_{sp}(t) = \left[2E^{-1}\gamma|\sigma_3| + \varepsilon_o\left(-\ln\left[1 - D_o\left(l(t)/a + 1\right)^3\right]\right)^{\frac{1}{m}}\right]\frac{L}{\sin\psi} \tag{6.5}$$

Furthermore, the shear stress acting on the shear failure plane in rocks can be expressed as:

$$\tau_{sf} = \frac{\sigma_1 - \sigma_3}{2}\sin 2\psi \tag{6.6}$$

where ψ is the shear fracture plane, $\psi = 45° + \phi/2$, ϕ is internal friction angle.

Inserting function of axial stress relating to wing crack growth (i.e., $\sigma_1(l)$) of Eq. (5.8) into Eq. (6.6), the expression of shear stress acting on shear failure plane with wing crack growth can be obtained as:

$$\tau_{sf}(l) = \left[\frac{K_{IC} - \sigma_3\left(\pi a^2 B_3 B_2 + 2\sqrt{l/\pi}\right)}{2\pi a^2 B_3 B_1} - \frac{\sigma_3}{2}\right]\sin 2\psi \tag{6.7}$$

In Eq. (6.7), the wing crack growth is caused by the remoted axial stress. If the wing crack growth is caused by time-dependent stress corrosion, the time-dependent shear stress can be obtained, which will provide an important theoretical implication for the earthquake evolution process.

Substituting the time-dependent crack growth relating to shear displacement of Eq. (6.5) into the expression of variation of shear stress with wing crack growth of Eq. (6.7), the relationship between shear stress and time subjected to the remoted constant stress state (σ_1, σ_3) during creep can be expressed as:

$$\tau_{sf}(t) = \left[\frac{K_{IC} - \sigma_3\left(\pi a^2 B_{3t} B_2 + 2\sqrt{J_t/\pi}\right)}{2\pi a^2 B_{3t} B_1} - \frac{\sigma_3}{2}\right]\sin 2\psi \tag{6.8}$$

where

$$B_{3t} = \frac{1}{[\pi(J_t + \beta a)]^{\frac{3}{2}}} + \frac{2\sqrt{J_t/\pi}}{\pi^{1/3}[3/(4N_V)]^{2/3} - \pi(J_t + \alpha a)^2} \tag{6.9}$$

$$J_t = a\left(D_o^{-\frac{1}{3}}\left(1 - \exp\left[-\varepsilon_o^{-m}\left(\frac{D_{sp}(t)\sin\psi}{L} - \frac{2\gamma|\sigma_3|}{E}\right)^m\right]\right)^{\frac{1}{3}} - 1\right) \tag{6.10}$$

The final failure values of axial strain and shear displacement in the intact brittle rocks can be approximately predicted by inserting the length of crack coalescence l_{lim} between adjacent cracks into Eqs. (6.2) and (6.3).

6.2 Crack Nucleation Effect on Progressive Failure

Model parameters of Jinping marble in China are used in the proposed model. Other model parameters $\varphi = 32.03$, $\beta = 0.276$, $E = 30\text{GPa}$, $\gamma = 0.2$, $L = 25$ cm, $\phi = 35$.

6.2.1 Multi-stress Drops in Stress–Strain Curves Under Different Confining Pressures

Figure 6.4 shows the stress–strain curves under different confining pressures. Axial stress increases linearly before crack initiation, and then increases nonlinearly after crack initiation, and decreases after peak stress with the increment of axial strain. Peak stress and crack initiation stress increase both with an increment of confining pressure, and the corresponding axial strain of the two stress states increase also with an increment of confining pressure. Furthermore, the stress–strain curve at the transformation phase of crack initiation is not smooth, and there is a phenomenon of stiffness enhancement. The reasons may be that (1) the numerous initial crack size and angle in brittle materials are both assumed as an average value, (2) the rock fracture from strain localization is equivalently studied by the fracture from the global crack growth in the model, (3) the error of model parameters. However, the results cannot explain the phenomenon of multi-stress drops.

Figure 6.5 shows the stress–strain curves containing multi-stress drops under different confining pressures. A stress drop is caused by a stepping increment of initial damage, which is corresponding to the presence of an individual shear band. Furthermore, the value of stress drop increases with the increment of confining pressure, and the value of stress drop decreases with the stress drop sequence in Fig. 6.6. Figure 6.7 shows another stress–strain curve containing more multi-stress

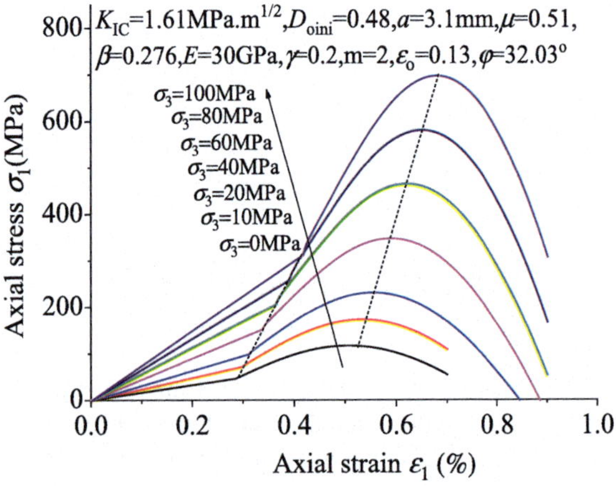

Fig. 6.4 Stress–strain curves under different confining pressures

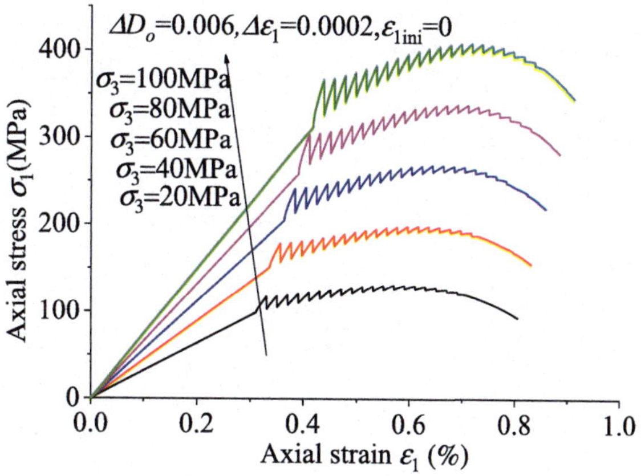

Fig. 6.5 Stress–strain curves containing multi-stress drops under different confining pressures

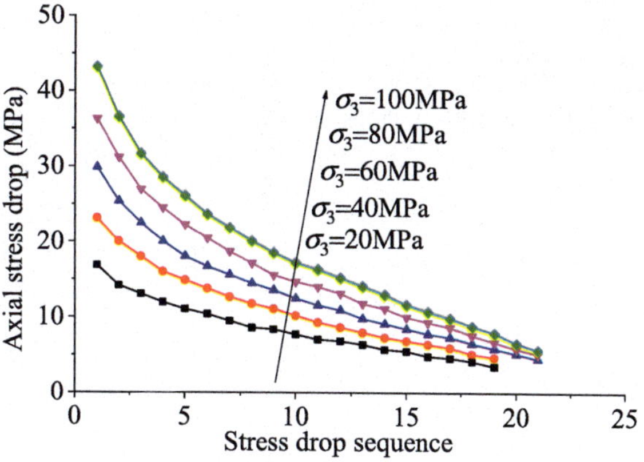

Fig. 6.6 Relationship between axial stress drop and stress drop sequence in Fig. 6.5

drops after crack initiation. The magnified area can observe clearly the variation of stress–strain curves.

Because the natural spatial random distribution of numerous microcracks in rocks is hardly realized in this model, the real occurrence sequence and position of multi-shear bands cannot be studied. Thus, the proposed theoretical phenomena of multi-stress drops in the stress–strain curve cannot be totally identical to the experimental data. But, the variation tendency of theoretical multi-stress drops in normalized stress–strain curves of intact rocks is similar to that of the experimental results (Brantut et al. 2011; Li et al. 2015; Xia et al. 2017) in Fig. 6.8, which approximately provides verification of rationality of the proposed analytical model. The experimental phenomena of multi-stress drops are qualitatively analyzed by the theoretical model. The differences between theoretical and experimental results in Fig. 6.8 can be clearly explained by analyzing the effects of parameters ΔD_{o}, $\varepsilon_{1\text{ini}}$, and $\Delta\varepsilon_1$ on stress drops in the stress–strain curve below.

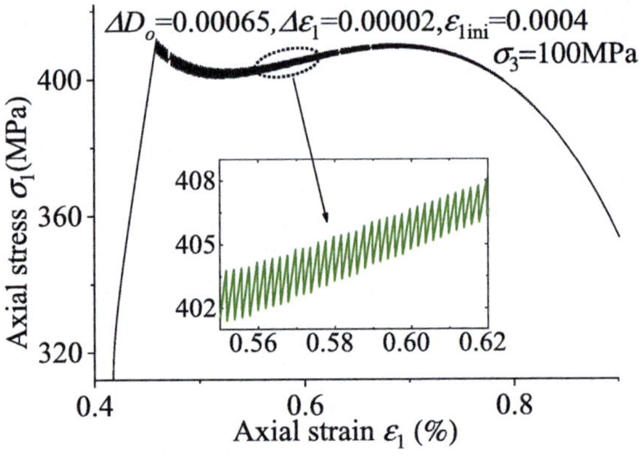

Fig. 6.7 Stress–strain curve containing more multi-stress drops at $\Delta D_o = 0.00065$, $\Delta\varepsilon_1 = 0.00002$ and $\varepsilon_{1ini} = 0.0004$ after crack initiation

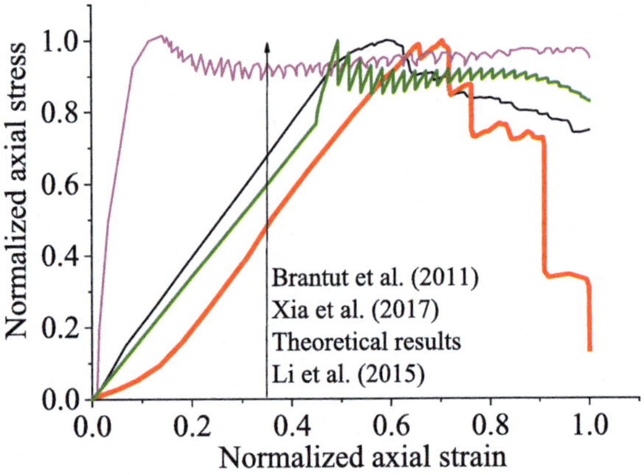

Fig. 6.8 Comparison of theoretical and experimental results in normalized stress–strain curves containing multi-stress drops

6.2.2 *Sensitivities of Model Parameters on Multi-stress Drops in Stress–Strain Curves*

As shown in Fig. 6.9a–c, the sensitivities of model parameters ΔD_o, ε_{1ini}, and $\Delta\varepsilon_1$ on the phenomena of multi-stress drops in the stress–strain curve are discussed in details. With the decrement of ΔD_o and the increment of ε_{1Fir}, and $\Delta\varepsilon_1$, the axial stress corresponding to the certain axial strain is increasing.

In Fig. 6.9a, c, from the view of a global variation of the stress–strain curve at $\Delta D_o = 0.005$ and $\Delta\varepsilon_1 = 0.0005$, the maximum or minimum of axial stress at each stepping damage phase is increasing to a peak value, and then is decreasing with an increment of axial strain. The peak strength happens at the axial strain $\varepsilon_1 = 0.007$ or so. In Fig. 6.9b, from the view of a global variation of the stress–strain curve at $\varepsilon_{1ini} = 0.0006$, the maximum or minimum of axial stress at each stepping damage phase is decreasing with an increment of axial strain. The peak strength happens at $\varepsilon_1 = 0.005$ or so. In Fig. 6.9a–c, from the view of global variation of stress–strain

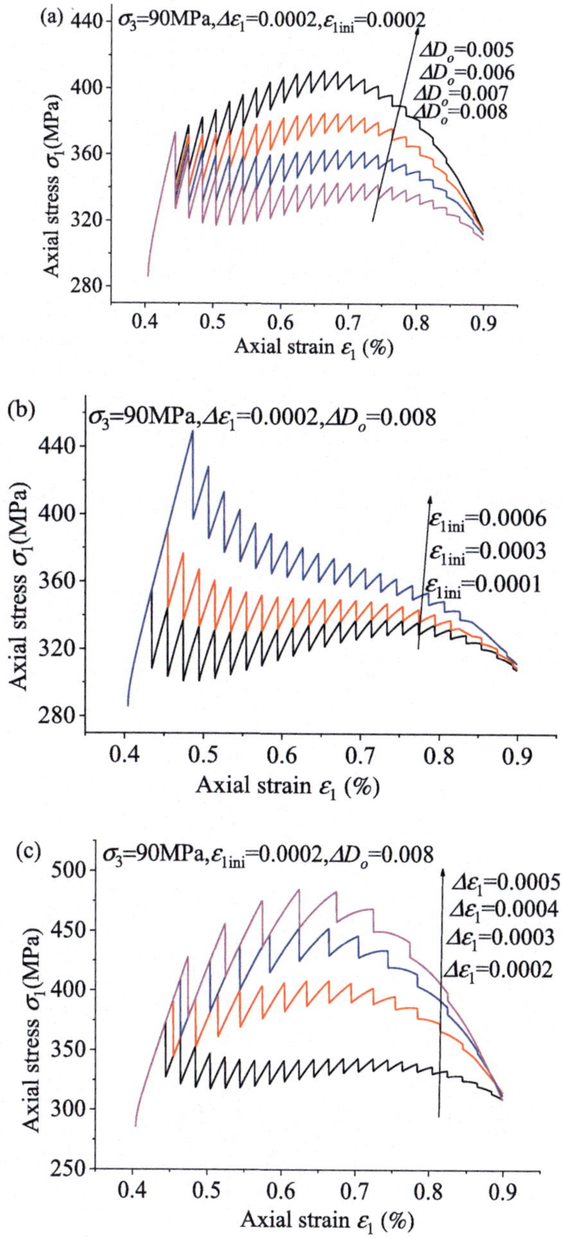

Fig. 6.9 Sensitivity of parameter **a** ΔD_o, **b** ε_{1ini} and **c** $\Delta\varepsilon_1$ on stress–strain curves containing stress drops after crack initiation

curve at $\Delta D_o = 0.008$, $\varepsilon_{1ini} = 0.0001$, and $\Delta\varepsilon_1 = 0.0002$, the maximum or minimum of axial stress at each stepping damage phase is decreasing to a bottom value, and then is increasing to a peak value, and finally is decreasing with increment of axial strain. Furthermore, two peak values happen at the global variation of the stress–strain curve. The first peak value happens at $\varepsilon_1 = 0.0045$ or so, and the second peak value happens at $\varepsilon_1 = 0.007$ or so. The peak strength is corresponding to the first peak value.

According to Fig. 6.9a, c, the relationships between axial stress drop and parameters ΔD_o or $\Delta\varepsilon_1$ at given strain are shown in Fig. 6.10a, b. The drop of axial stress is increasing with the increment of parameters ΔD_o and $\Delta\varepsilon_1$.

The three parameters ΔD_o, ε_{1ini}, and $\Delta\varepsilon_1$ reflect the damage size in the individual minor localized shear band, the initial strain appearing at the first individual minor localized shear band, and strain difference between two minor localized shear bands, respectively. These above theoretical phenomena show qualitatively that parameters ΔD_o, ε_{1ini}, and $\Delta\varepsilon_1$ all have a great influence on stress–strain curves. Furthermore, for natural rocks, the development process of multi-stress drops in stress–strain curves

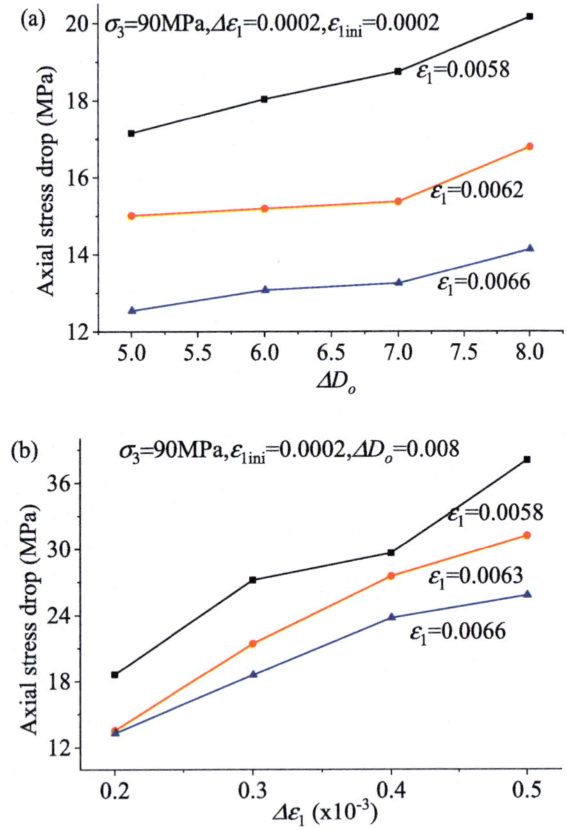

Fig. 6.10 Relationships between the axial stress drop and parameters **a** ΔD_o and **b** $\Delta\varepsilon_1$ at the given strain

can be explained by the compositive effects of the three parameters ΔD_o, ε_{1ini}, and $\Delta\varepsilon_1$.

Figure 6.11a shows the effects of parameter ΔD_o on the stress–strain curve containing multi-stress drops, and the parameter ΔD_o is varying from 0.006 to three different values (i.e., 0.003, 0.006, and 0.009) after axial strain $\varepsilon_1 = 0.0062$. It is observed that the axial stress is decreasing and the drop of axial stress is increasing at given axial strain with the increment of parameter ΔD_o.

Figure 6.11b shows the effects of parameter $\Delta\varepsilon_1$ on the stress–strain curve containing multi-stress drops, and the parameter $\Delta\varepsilon_1$ is varying from 0.0002 to three different values (i.e., 0.0001, 0.0002, and 0.0003) after axial strain $\varepsilon_1 = 0.0062$. The axial stress and the drop of axial stress are decreasing at given strain, and the number of stress drops is increasing with the decrement of parameter $\Delta\varepsilon_1$.

Furthermore, the proposed multi-stress drops are caused by the stepping initial damage, and the damage proverbially is relative to the microcrack growth and strain increment. Thus, the proposed model can be defined as a model of crack, strain or damage variation-dependent stress drops. The above phenomena of the effects of

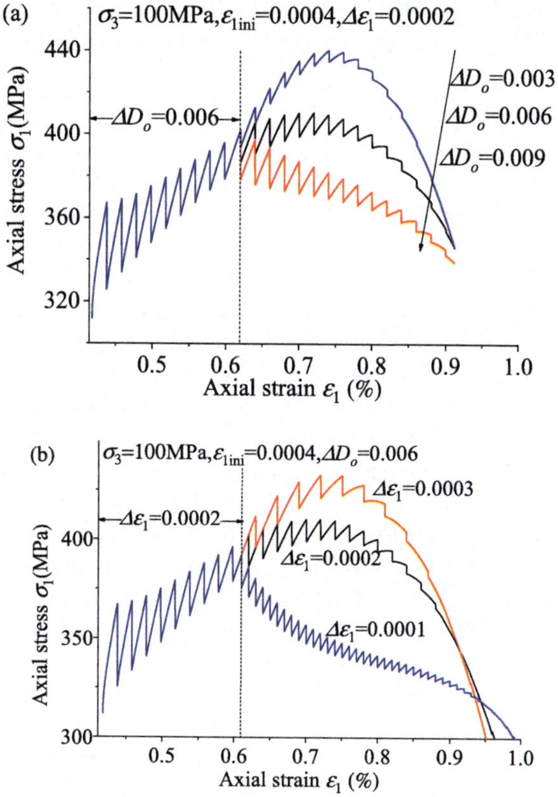

Fig. 6.11 Stress–strain curves with multi-stress drops in which **a** the parameter ΔD_o is varying from 0.006 to three different values, **b** the parameter $\Delta\varepsilon_1$ is varying from 0.0002 to three different values after axial strain $\varepsilon_1 = 0.0062$

continuous changing parameters ΔD_o and $\Delta \varepsilon_1$ on the multi-stress drops in stress–strain curves of intact rocks qualitatively provide a certain understanding for evaluation of earthquake mechanisms during geological movement and the safety of deep underground engineering during excavation. The stepping initial damage is corresponding to the presence of an individual shear fracture band caused by the accumulation and coalescence of many microcracks, which causes a larger stress drop in intact rocks. This stress drop causes the rapid weakening of mechanical properties in brittle rocks under compression. For the rocks subjected to the larger remoted constant loadings in deep earth crusts, the rapid weakening of the rock mechanical properties could lead to the sudden larger deformation in the local surrounding rocks of deep underground engineering, or lead to the sudden larger slip causing the local earthquakes in earth crusts.

6.3 Crack Nucleation Effect on Creep Fracture

6.3.1 Time-Dependent Shear Stress and Deformations During Brittle Creep Without Step Damages

A correspondence of time-dependent shear stress τ_{sf}, shear stress rate τ_{sf}', wing crack growth l, axial strain ε_1, shear displacement D_{sp}, damage D, and damage rate D' during static creep is shown in Fig. 6.12. The damage is relative to wing crack growth, axial strain and shear displacement. The damage, wing crack growth, axial strain and shear displacement all experience three phases: the decelerated, stationary and accelerated increment variations. Figure 6.12 only shows the variation curve of damage rate. The variation tendency of other curves in the rates of wing crack growth, axial strain, and shear displacement is similar to the curve of damage rate.

Based on the time-dependent variation curves of shear stress, shear stress rate and the damage rate in Fig. 6.12, the relationship between shear stress and damage rate is drawn in Fig. 6.13. Shear stress experiences three phases: (1) the strengthening phase, (2) the weakening phase I, and (3) the weakening phase II. In the strengthening phase, shear stress is increasing with the decrement of damage rate. In the weakening phase I, shear stress is decreasing with the decrement of damage rate. In the weakening phase II, shear stress is decreasing with the increment of damage rate (Note: the damage rate is corresponding to shear displacement rate). Furthermore, shear stress is approximately linearly varying with damage rate at larger damage rate, and shear stress is nonlinearly varying with damage rate at smaller damage rate, which shows the simpler mechanical relationship of shear tress and damage rate at the larger damage rate. A turning point in the curve of shear stress and damage rate appears at damage rate 2×10^{-15} or so. The sensitivity of damage rate on shear stress at smaller damage rate is higher than that of larger damage, and the turning point at damage rate 2×10^{-15} is the boundary between the smaller and larger damage rates.

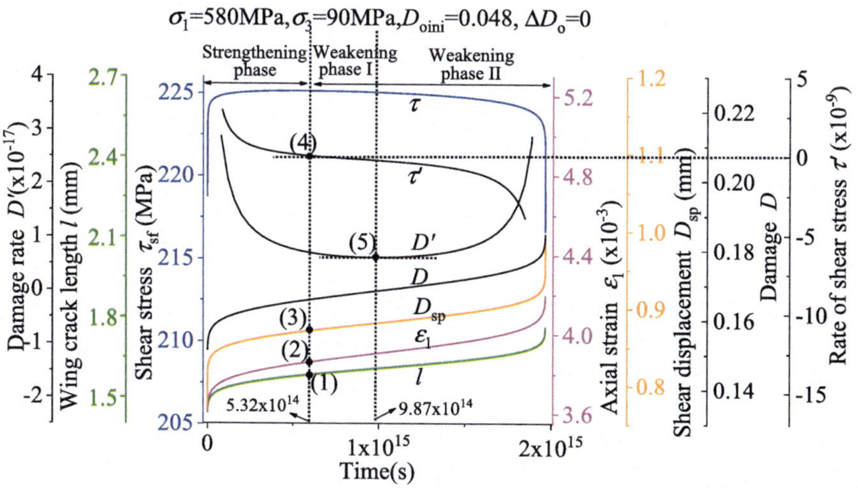

Fig. 6.12 Correspondence of time-dependent shear stress, shear stress rate, wing crack growth, axial strain, shear displacement, damage and damage rate during static creep. The solid points (1–4) respectively represent for the wing crack length, axial strain, shear displacement, and shear stress rate appearing at the transformation moment from the strengthening to weakening phase of shear stress, and the solid point (5) presents for minimum damage rate

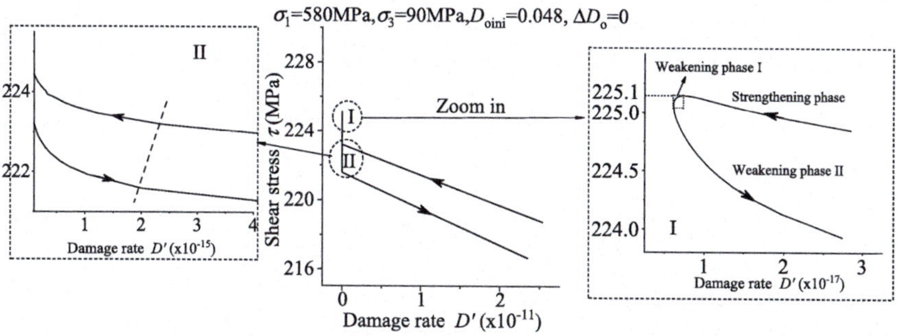

Fig. 6.13 Variation of shear stress with damage rate in Fig. 6.12. The larger arrow represents for the flow tendency corresponding to rock creep (the same below)

This process of time-dependent shear fracture in intact rocks is similar to the process of time-dependent shear friction from static to slip friction in faults. In the strengthening phase, this variation tendency of the creep strengthening phase in intact rocks is identical to that of static friction in faults, which both have a time-dependent slowly increase of shear stress (Dieterich et al. 1979a, b; Scholz 1998). In the weakening phase, the variation tendency of creep weakening phase in intact rocks is similar to that of the slip friction in faults (Dieterich et al. 1979a, b; Scholz 1998). However, it is noteworthy that Fig. 6.3 only shows an evolution process of rocks from hardening to softening phases corresponding to the process of faults from static to slip friction, which cannot explain the phenomenon corresponding to the process of faults from slow stable slip friction to rapid unstable slip friction. The well-known rate-and state- friction model proposed by Dieterich (1979a, b) and Ruina (1983) explained the variation process of shear friction in faults from slow stable to rapid unstable slip friction. In the well-known rate-and state- friction model, the equation of friction at steady state is:

$$\tau_{\text{ss}} = \sigma_n[\mu_o + (\text{a} - \text{b})ln[V/V_o]] \tag{6.11}$$

where V is slip velocity, V_o a reference velocity, μ_o the steady-state friction at $V = V_o$, and a and b are material properties. If a-b > 0, the material is velocity-strengthening and will always be stable, and shear stress is increasing with an increment of slip velocity. If a-b < 0, the material is in the velocity-weakening field, there is a Hopf bifurcation between unstable and stable phases. Due to slip velocity at static friction is zero, the process from static to slip friction may also be understood approximately as a process from slow sable to rapid unstable slip friction. Thus, the strengthening phase of shear stress in Fig. 6.12 can be corresponding to the strengthening phase at a-b > 0 in Eq. (6.11); the weakening phase of shear stress in Fig. 6.12 can be corresponding to the weakening phase at a-b < 0 in Eq. (6.11). However, in the strengthening phase, Eq. (6.11) shows that shear stress is increasing with an increment of slip velocity, but, Fig. 6.12 shows that shear stress is increasing with decrement of damage rate (i.e., shear displacement velocity). In the weakening phase, Eq. (6.11) shows that shear stress is decreasing with increment of slip velocity, which is identical to the damage rate-dependent shear stress during weakening phase II in Fig. 6.12. The weakening phase I in Fig. 6.12 has no correspondence to Eq. (6.11).

Based on a simple spring–slider model with fixed stiffness k, the bifurcation occurs at a critical value of effective normal stress (Scholz 1998):

$$\sigma_{nc} = k\,L_c/(\text{b} - \text{a}) \tag{6.12}$$

where L_c is the critical slip distance. If applied normal stress $\sigma_n > \sigma_{nc}$, sliding is unstable under quasi-static loading. If applied normal stress $\sigma_n < \sigma_{nc}$, sliding is stable under quasi-static loading but can become unstable under dynamic loading if subjected to a velocity jump exceeding ΔV. The velocity jump is a sudden variation of slip velocity, which will have an important effect on the weakening of shear friction (i.e., shear stress drop). It is noted that the velocity jump is caused by dynamic loading. However, under lithospheric conditions, the stress conditions in rocks may keep a constant state, i.e., creep. The sudden appearance of a small shear fracture band from the damage variable ΔD_o induced by long-term subcritical crack accumulations and coalescences may cause a velocity jump, which will cause a weakening of shear stress (i.e., shear stress drop). The effect of the stepping initial damage ΔD_o on the shear stress drop will be studied and discussed in detail below. Some experimental studies found a conclusion that the onset of stress drop caused by shear fracture was similar to the onset of stress drop caused by frictional stick–slip, which both have the same underlying physical process with the onset of earthquakes (Brace and Byerlee

1966; Ohnaka 1995). Thus, our proposed model has a theoretical verification of this conclusion.

Based on the time-dependent damage, shear displacement and damage rate in Fig. 6.12, the variations of shear displacement and damage with damage rate are drawn in Fig. 6.14. The increment of damage and shear displacement are increasing with the decreasing damage rate firstly, and then are increasing with increasing damage rate. Based on the time-dependent shear stress, wing crack growth, axial strain and shear displacement in Fig. 6.12, the variations of shear stress with wing crack length, axial strain and shear displacement under constant stress state ($\sigma_1 =$ 580 MPa and $\sigma_3 = 90$ MPa) during creep are drawn in Fig. 6.15, which does not show the deformations before crack initiation. The solid points (1), (2), (3) and (4) in Fig. 6.12 happen at the identical time (i.e., $5.32 \times 10^{14\,\mathrm{s}}$), and the solid point (4) represents the transformation from positive to negative of the rate of shear stress. Moreover, the values of wing crack length, axial strain and shear displacement at solid points (1), (2) and (3) in Fig. 6.12 are corresponding to that of the solid points (1), (2) and (3) in Fig. 6.15. Thus, the hardening and softening phases of wing crack length, shear displacement and axial strain in Fig. 6.15 are corresponding to the strengthening and weakening phases of shear stress in Fig. 6.12. The solid points (1), (2), and (3) in Fig. 6.12 happen prior to the minimum creep strain rate (i.e., solid point (5) in Fig. 6.12), which also shows that the wing crack length, axial strain and shear displacement appearing at the minimum creep strain rate (i.e., solid point (5) in Fig. 6.12) happen behind the wing crack length, axial strain and shear displacement at peak shear stress in Fig. 6.15.

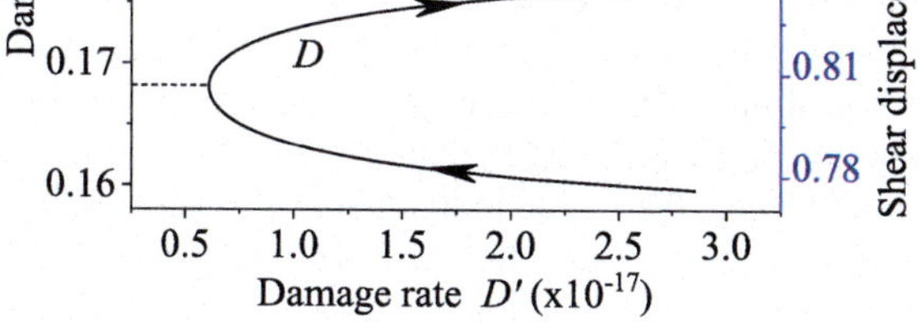

Fig. 6.14 Variations of damage and shear displacement with damage in Fig. 6.12

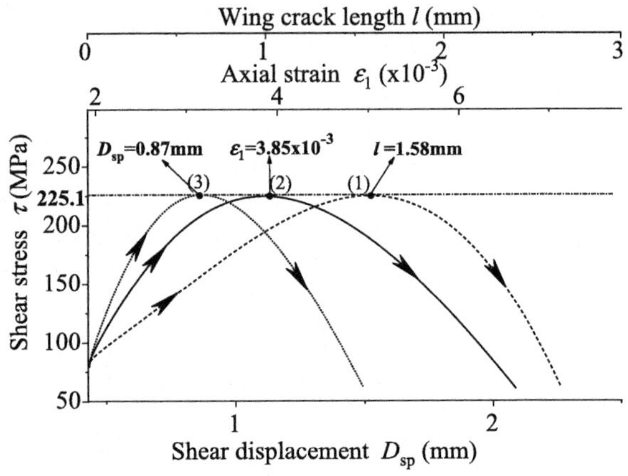

Fig. 6.15 Variations of shear stress with wing crack length, axial strain and shear displacement, which is plotted by drawing the corresponding shear stress, shear displacement, axial strain and wing crack length under different time in Fig. 6.12

6.3.2 Time-Dependent Shear Stress and Deformations During Brittle Creep with Step Damages

In the function of stepping initial damage over time in Fig. 6.2b, the parameters of initial damage difference ΔD_0 and time difference ΔT have an important influence on evolution processes of shear stress, shear displacement, axial strain, and wing crack growth during static creep. The value of parameter ΔD_0 is corresponding to the damage variable caused by the presence of a shear band. The value of parameter ΔT is corresponding to the time difference between the presence of two shear bands. Introducing the effect of time-dependent stepping initial damage into the static creep model, the time-dependent shear stress, shear displacement, axial strain, wing crack growth, damage, and damage rate are shown in Figs. 6.16 and 6.25. Some stress drops appear at the variation curve of shear stress over time in Figs. 6.16a and 6.25a. A stepping increment of initial damage causes a drop of shear stress, which is corresponding to the presence of an individual shear band in rocks induced by microcrack accumulation and coalescence. The last big drop of shear stress is corresponding to the shear fracture of intact rocks. In Figs. 6.16b–g and 6.25b–g the stepping increment of initial damage causes the stepping increment of shear displacement, axial strain, wing crack length, damage, and damage rate. Between the two stress drops during creep, there is an individual process containing the shear stress strengthening and weakening phases. In Figs. 6.16f and 6.16g, due to the huge difference between the minimum and maximum damage rates, the minimum and maximum damage rates are obtained respectively, which provides a clear illustration for the value of damage rate and the variation of damage rate under different parameters ΔD_0.

Figures 6.16a–g show the effect of parameter ΔD_0 on the time-dependent shear stress, shear displacement, axial strain, wing crack growth, damage and damage rate, respectively. The final failure time is descending with the increment of parameter $\Delta D_{0.}$ As shown in Fig. 6.16a, with the increment of parameter ΔD_0, the shear stress is descending after the first stress drop, and the stress drop at the position of the stepping

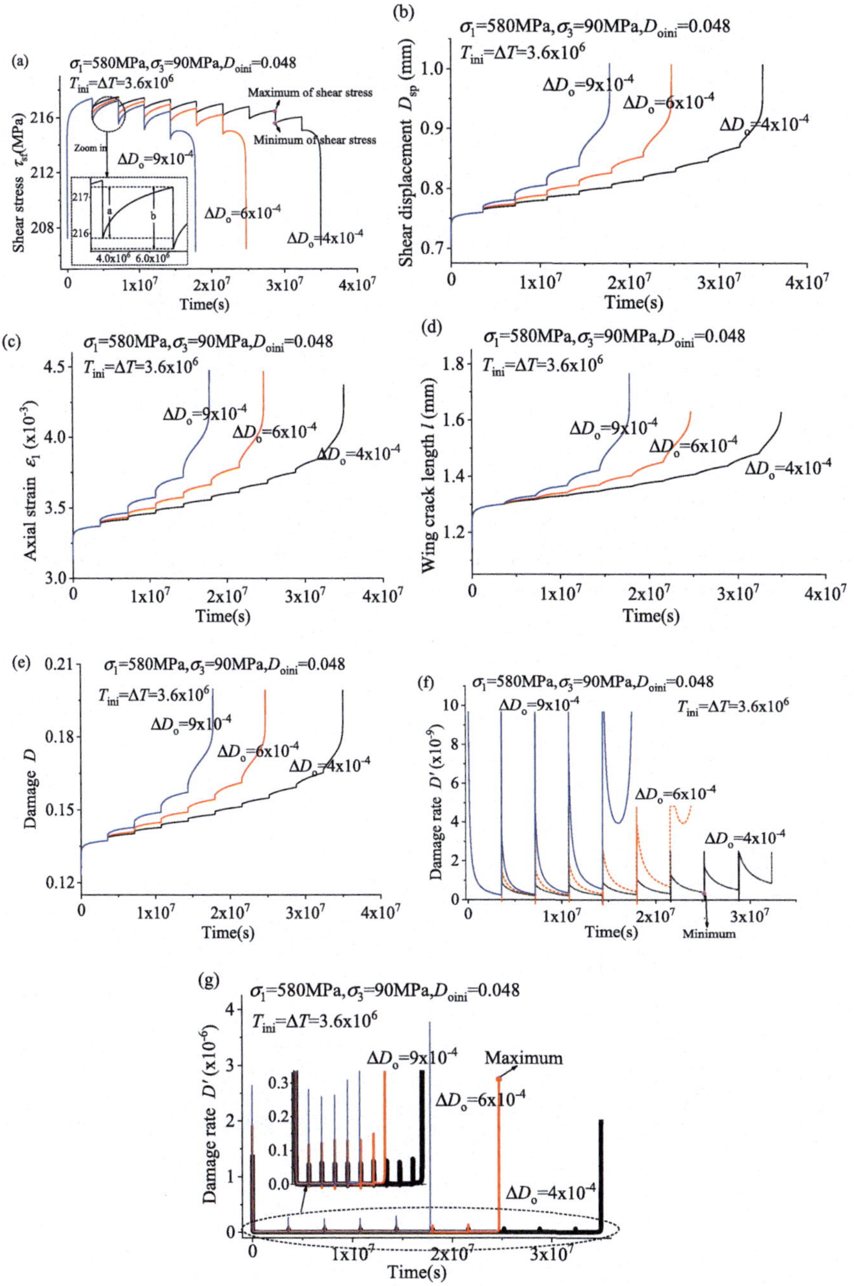

Fig. 6.16 Time-dependent **a** shear stress, **b** shear displacement, **c** axial strain, **d** wing crack growth, **e** damage, **f** damage rate near minimum, and **g** damage rate near maximum subjected to the stepping initial damage under different parameter ΔD_o during static creep. Note that the positions of the maximum and minimum of shear stress or damage rate are marked by the circular solid point in Fig. 6.16a, f and g

initial damage is ascending at a given time, which is also clearly shown in Fig. 6.17a–c. Figure 6.17a–c is obtained by drawing the data of maximum and minimum of shear stress at each position of the stepping initial damage in Fig. 6.16a. In the magnified area in Fig. 6.16a, two parameters a and b are defined in an individual strengthening and weakening phase. Parameter b is larger than parameter a, which shows that rock globally is in a weakening process during creep. This global weakening process in Fig. 6.16a is corresponding to the weakening phase II in Fig. 6.12. Both parameters have a similar function to that of Eqs. (6.11) and (6.13) in the rate-and state- friction model (Dieterich 1979a, b; Ruina 1983).

As shown in Fig. 6.16b–g, with the increment of parameter ΔD_0, the values of shear displacement, axial strain, wing crack growth, damage and damage rate are ascending, and the increments of shear displacement, axial strain, wing crack growth, damage and damage rate at the position of the stepping initial damage are also ascending at given time. The variations of damage rate and increment of damage rate with the increment of parameter ΔD_0 are clearly shown in Fig. 6.18a–c by drawing the data of maximum and minimum of damage rate at the position of stepping initial damage in Fig. 6.16f, g. The variation of increment of shear displacement with the increment of parameter ΔD_0 is also clearly shown in Fig. 6.17d by drawing the data of maximum and minimum of shear displacement at the position of the stepping initial damage in Fig. 6.16b.

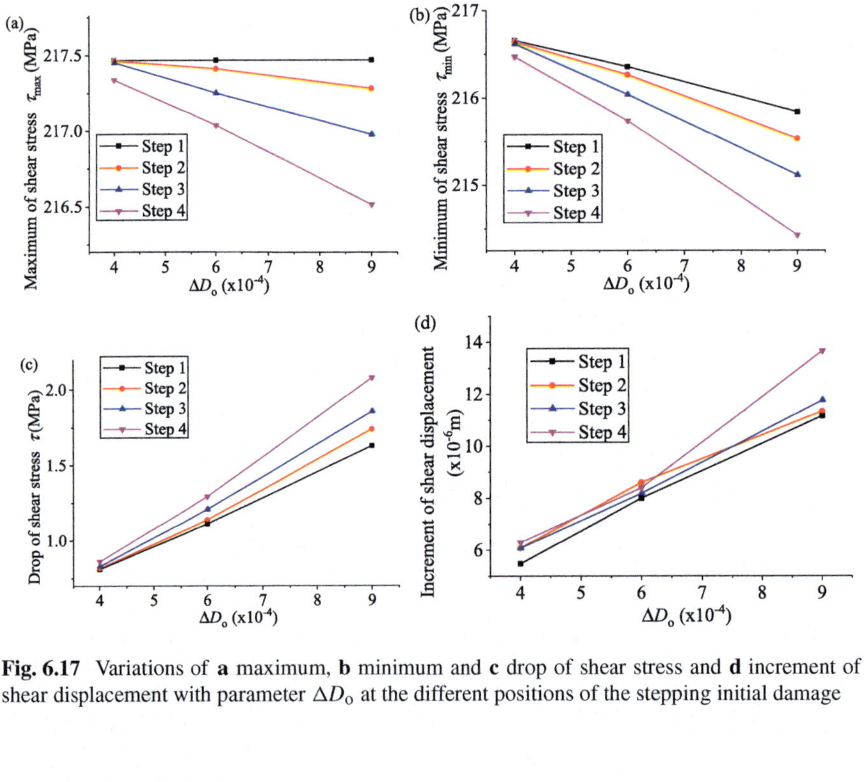

Fig. 6.17 Variations of **a** maximum, **b** minimum and **c** drop of shear stress and **d** increment of shear displacement with parameter ΔD_0 at the different positions of the stepping initial damage

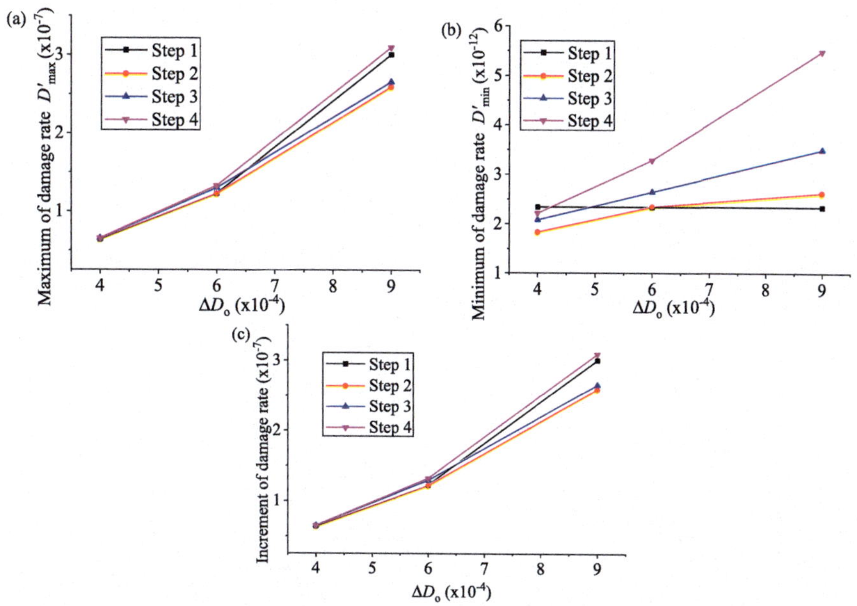

Fig. 6.18 Variations of **a** maximum, **b** minimum and **c** increment of damage rate with parameter ΔD_0 at the different positions of the stepping initial damage

Based on these above phenomena in Figs. 6.17 and 6.18, it is concluded that the larger damage rate leads to the smaller shear stress, and the larger increment of damage rate leads to the larger drop of shear stress and the larger increment of shear displacement, which is clearly illustrated in Fig. 6.19. In Fig. 6.19, the maximum shear stress is descending with the increment of the maximum damage rate, the minimum shear stress is descending with the increment of the minimum damage rate, and the drop of shear stress is ascending with the ascending increment of damage rate at each position of the stepping initial damage. The theoretical results have a good agreement with the experimental results of shear failure in intact rocks under lithospheric conditions (i.e., triaxial compressive loadings) (Ohnaka et al. 1997). The experimental results showed that the breakdown stress drop and the critical slip displacement are increasing with the increment of strain rate. However, this experimental result was only obtained by the many independent experiments in different rock samples, which cannot reflect the continuous constitutive relationship of stress drop caused by many smaller shear faults in the individual rock sample. Furthermore, the experimental conditions were the quasi-static conditions, which cannot reflect the real static conditions liking lithospheric conditions during long-term creep (i.e., constant stress state (σ_1, σ_3)) of the suggested model in this study.

The variation tendency of damage in Fig. 6.16b is corresponding to that of the shear displacement in Fig. 6.16e. Thus, the increment of shear displacement rate caused by damage leads to the decrement of shear stress, and the increment of variation in shear displacement rate leads to the increment of the drop of shear stress. The appearance

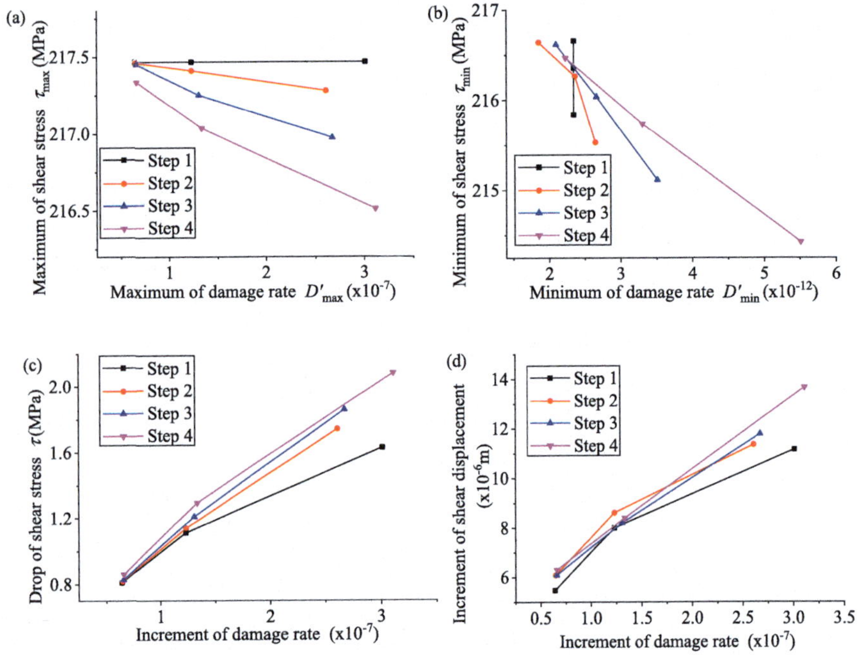

Fig. 6.19 Variations of **a** maximum and **b** minimum of shear stress with maximum and minimum of damage rate; **c** drop of shear stress and **d** increment of shear displacement with an increment of damage rate respectively at the different positions of presence in stepping initial damage

of stepping initial damage in rocks causes a sudden increase and then a decrease in the rate of shear displacement. This conclusion of drops of shear stress during creep in intact rock is also very similar to the well-known rate-and state- friction model between faults proposed by Dieterich (1979a, b) and Ruina (1983) and many experiment results of slip friction between faults (Blanpied et al. 1987; Goldsby and Tullis 2011; Spagnuolo et al. 2016; Scholz 1998). The well-known rate-and state-friction model can be expressed as:

$$\tau = \sigma_n[\mu_o + \mathrm{a}ln[V/V_o] + \mathrm{b}ln[(\theta V_o)/L_c]] \tag{6.13}$$

$$d\theta \big/ dt = 1 - \theta V \big/ L_c \tag{6.14}$$

where θ is the state variable. An experimental-based phenomenon of suddenly imposed bigger increase and then decrease in sliding velocity was proposed by this rate-and state- friction model. It can be observed that the suggested stepping initial damage relating to damage rate in our proposed model has an identical role with this state variable θ relating to sliding velocity in the well-known rate-and state-friction model. Moreover, the theoretical results of time-dependent shear stress and displacement in Fig. 6.16a, b have an identical variation tendency with the long-term static creep slip friction results (Sleep and Blanpied 1992) and the quasi-static slip friction experimental results (Leeman et al. 2016, 2018).

The bigger stepping increment of initial damage causes the bigger drop of shear stress and the bigger increment of shear displacement, axial strain, and wing crack growth. In the scale of earth crusts, the presence of bigger shear fault will cause the bigger rapid drop of shear stress and the bigger rapid increment of shear slip displacement, which is corresponding to a bigger earthquake. And the presence of smaller shear fault will cause a smaller rapid drop of shear stress and the smaller rapid increment of shear slip displacement, which is corresponding to a smaller earthquake (i.e., slow earthquake which is formed between stable creep and destructive earthquakes (Linde et al. 1996; Ide et al. 2007; Ikari et al. 2013; Leeman et al. 2016, 2018). So, based on the above analyses of the relationship between shear stress and damage rate during long-term creep in intact brittle rocks under triaxial compressive loadings (Figs. 6.12, 6.13, 6.14, 6.15, 6.16, 6.17, 6.18 and 6.19), the proposed model can be called as the Damage Rate-and State- Fracture Model (DRSFM) for evaluating earthquake mechanisms.

For the above analysis, the theoretical results in the suggested damage rate-and state- fracture model (DRSFM) are concluded by the comparing the evolution curves of shear stress, shear displacement, axial strain, wing crack growth, damage, and damage rate under different parameters ΔD_0. However, for a given parameter ΔD_0, based on the comparisons of shear stress, shear displacement, axial strain, wing crack growth, damage and damage rate over time, it is observed that the increment of damage rate leads to the decrement of shear stress (i.e., the larger damage rate is corresponding to smaller shear stress), and the increment of variation in damage rate leads to the increment of drop of shear stress (i.e., the larger increment of damage rate is corresponding to larger drop of shear stress). This phenomenon can be illustrated in Figs. 6.20, 6.21 and 6.22. Figures 6.20 and 6.21 are obtained by drawn the data in Fig. 6.16a, f and g. In Fig. 6.20, the maximum damage rate, the minimum damage rate, and the increment of damage are ascending with an increment of step sequence at given parameter ΔD_0. In Fig. 6.21, the maximum shear stress and the minimum shear stress are descending, and the drop of shear stress is ascending with increment of step sequence at given parameter ΔD_0. Figure 6.22 is obtained by combining Figs. 6.20 and 6.21. In Fig. 6.22, the maximum shear stress is descending with the increment of the maximum damage rate, the minimum shear stress is descending with the increment of the minimum damage rate, and the drop of shear stress is ascending with increment of damage rate at given parameter ΔD_0.

It can be observed that the variation tendencies of shear stress with damage rate and drop of shear stress with an increment of damage rate shown in Figs. 6.19 and 6.22 are identical. Thus, whether the damage rate is compared in the same time under different curves of time-dependent shear stress or the damage rate is compared in the different time under one curve of time-dependent shear stress, the theoretical results are good agreement with this proposed the damage rate-dependent shear stress during the weakening phase II in damage rate-and state- fracture model (DRSFM).

The above micro–macro model of stress drops during brittle creep in rocks only provides a qualitative explanation for stress drops caused by localized shear bands

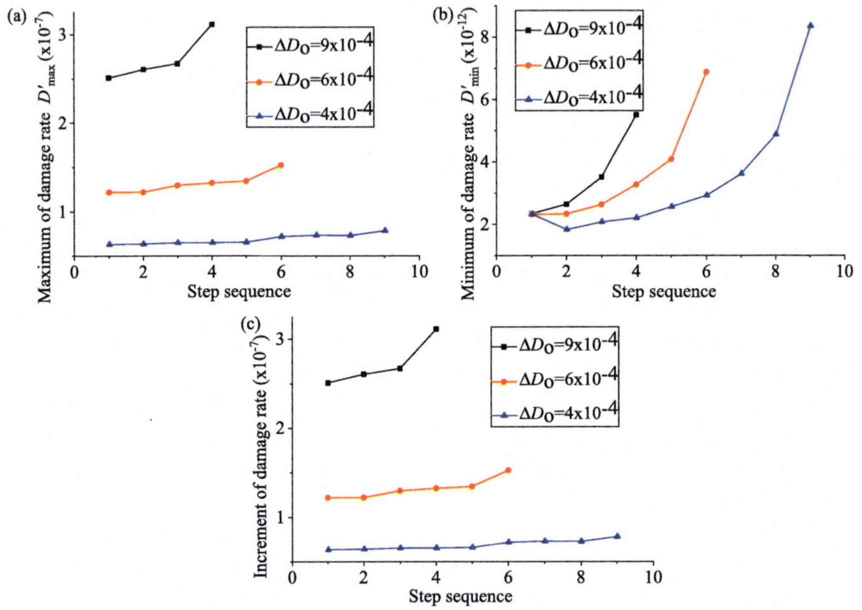

Fig. 6.20 Variations of **a** maximum, **b** minimum and **c** increment of damage rate with step sequence at the different parameters ΔD_o

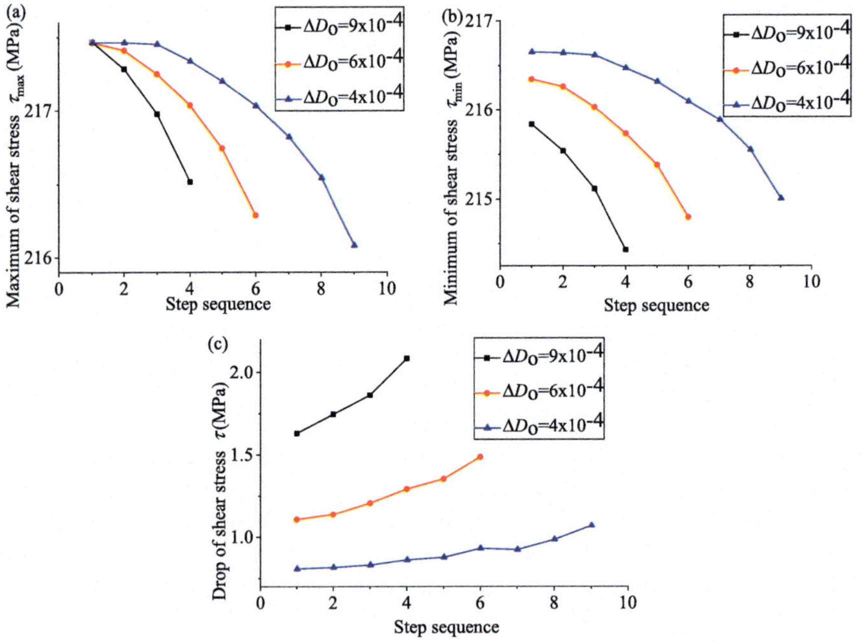

Fig. 6.21 Variations of **a** maximum, **b** minimum and **c** drop of shear stress with step sequence at the different parameters ΔD_o

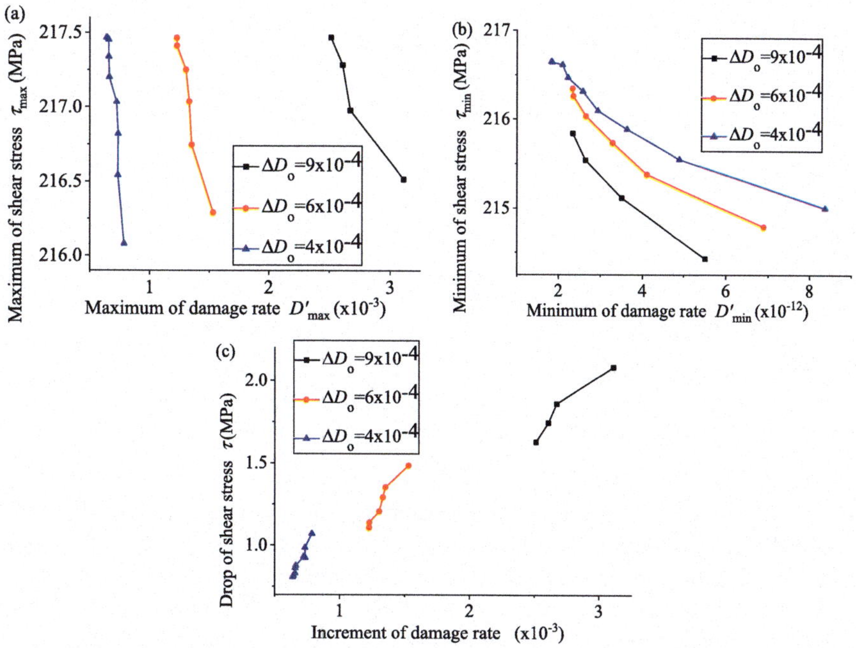

Fig. 6.22 Variations of **a** maximum, **b** minimum and **c** drop of shear stress with maximum, minimum and increment of damage rate respectively at the different parameters ΔD_o

from the localized microcracks nucleation and coalescence. This suggested sequence of shear bands in Fig. 6.3 is to explain the mechanism of the appearance of localized shear bands. Actually, due to randomly distributed microcracks in natural rocks, the sequence of the small shear bands is also random in natural rocks. At present, the relationship between the analytical and experimental sequences of stress drops is hardly achieved. The proposed analytical results of stress drops are difficult to keep identical with the experimental results in natural rocks. Thus, the approximate comparisons of the analytical and experimental results in the normalized deformation (Zhu et al.2012) or the normalized crack damage (Sun et al. 2013) over the normalized time during creep of Jinping marble are shown in Fig. 6.23a and b. The variation tendency of the theoretical results is similar to that of the experimental results. This corresponding relationship between the analytical and experimental stress drops is very important to the engineering applications, which will be further studied in the future.

Based on the data in Fig. 6.16a and b, the relationships between shear stress and shear displacement subjecting to constant stress state during creep under different parameters ΔD_o are shown in Fig. 6.24. The static theoretical results of stress drops are very similar to the experimental results of stress drops in loading-displacement or strain-strain curves under quasi-static triaxial compressive loadings (Brace and Byerlee 1966; Brantut et al. 2011). However, in the laboratory experimental results,

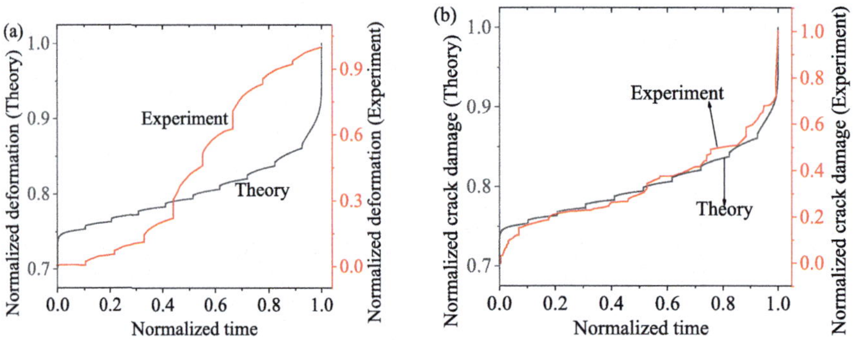

Fig. 6.23 Comparisons of the theoretical and experimental results in **a** the normalized deformation (Zhu et al. 2012) or **b** the normalized crack damage (Sun et al. 2013) over the normalized time due to the appearance of localized shear bands during static creep of Jinping marble

the real constant stress conditions acting on the rock during long-term creep are hardly implemented in a short time. Thus, our theoretical model provides well important help for evaluating rock mechanical properties in earth-crust during long-term creep.

Figure 6.25a–g show the effect of parameter ΔT on the time-dependent shear stress, shear displacement, axial strain, wing crack growth, damage, and damage rate. The final failure time is descending with the decrement of parameter ΔT. The shear stress is ascending with increment of the parameter ΔT at a given time (Fig. 6.25a). The damage rate and the increment of damage rate are both ascending with increment of the parameter ΔT at a given time (Fig. 6.25f–g). The phenomena of relationship between shear stress and damage rate also agree with the suggested damage rate-and state- fracture model (DRSFM). These results show that the time difference between two stepping initial damage (i.e., two presences of the smaller shear fracture bands) also has a great influence on the evolution process of shear stress and deformation (i.e., the evolution process of earthquake).

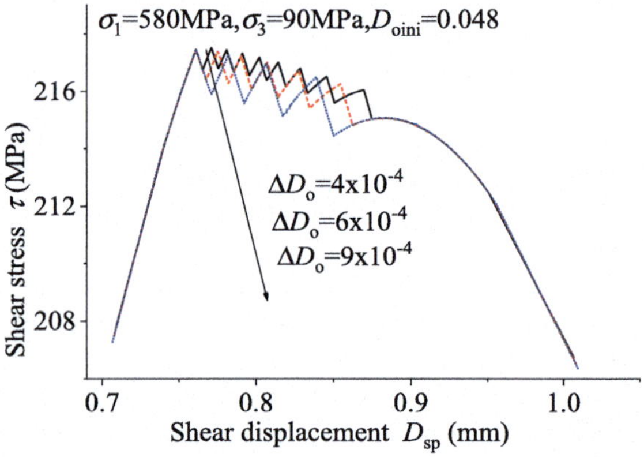

Fig. 6.24 Relationship between shear stress and shear displacement along with the shear fracture plane subjecting to constant stress state during creep under different parameters ΔD_o, which is obtained by drawing the data from Fig. 6.16a, b

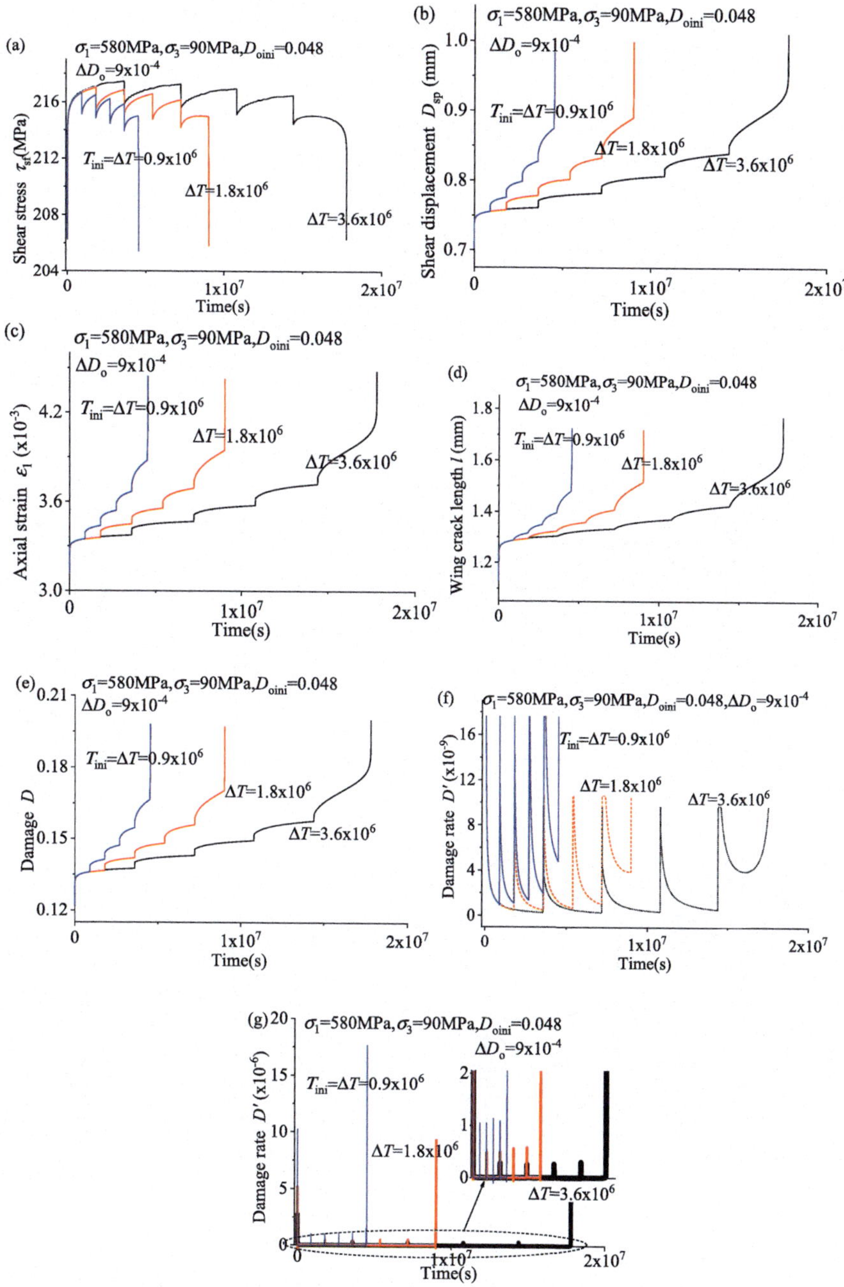

Fig. 6.25 Time-dependent **a** shear stress, **b** shear displacement, **c** axial strain, **d** wing crack growth, **e** damage, **f** damage rate near minimum, **g** damage rate near maximum subjected to the stepping initial damage under different parameter ΔT during static creep

6.3.3 *Coupling Effect of Multi-step Damage Variable or Multi-time Difference on Creep Behaviors*

From the above analytical results, the parameters ΔT and ΔD_o are both keeping a constant in the whole time-dependent curves of shear stress, shear displacement, axial strain, and wing crack growth during creep of intact brittle rocks. In Figs. 6.26 and 6.27, the whole time-dependent curves of shear stress, shear displacement, axial strain, wing crack growth damage, and damage rate under different parameters ΔD_o and ΔT are studied.

In Fig. 6.26a, the parameter ΔD_o experiences a variation from high value (i.e., $\Delta D_o = 9 \times 10^{-4}$) to low value (i.e., $\Delta D_o = 1 \times 10^{-4}$) over time. This variation includes three phases: (1) the first phase in the variation curve at three stages of

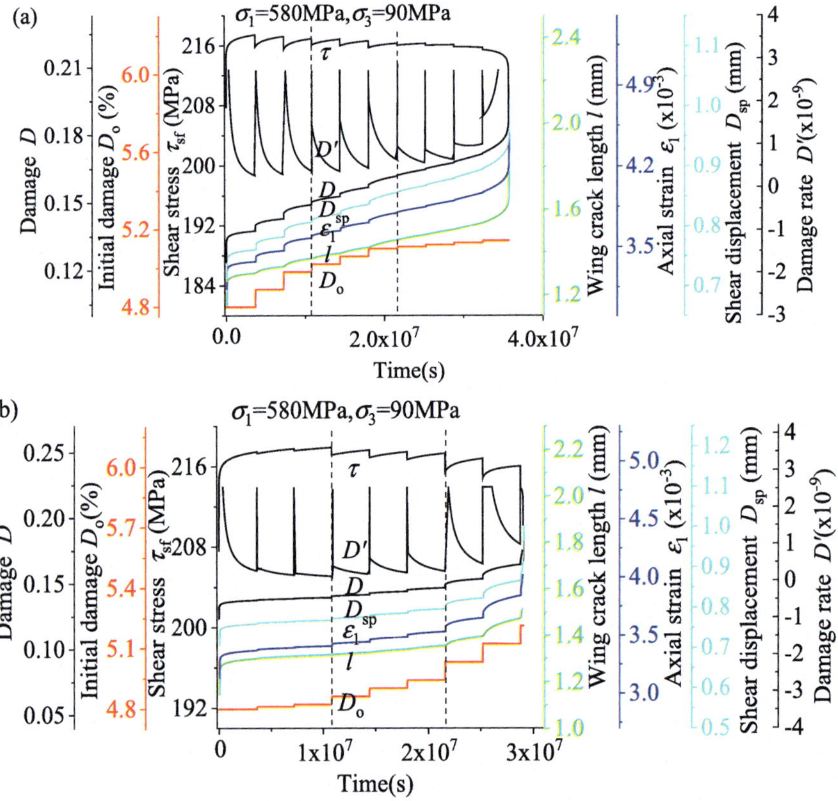

Fig. 6.26 Correspondence of time-dependent shear stress, shear displacement, axial strain, wing crack growth, damage, and damage rate subjected to the stepping initial damage (**a** $\Delta D_o = 9 \times 10^{-4}$, 4×10^{-4}, and 1×10^{-4} over time, **b** $\Delta D_o = 1 \times 10^{-4}$, 4×10^{-4}, and 9×10^{-4} over time, and $\Delta T = 3.6 \times 10^6$) during static creep. Note that the dashed line represents the different phases of different ΔD_o, the same below

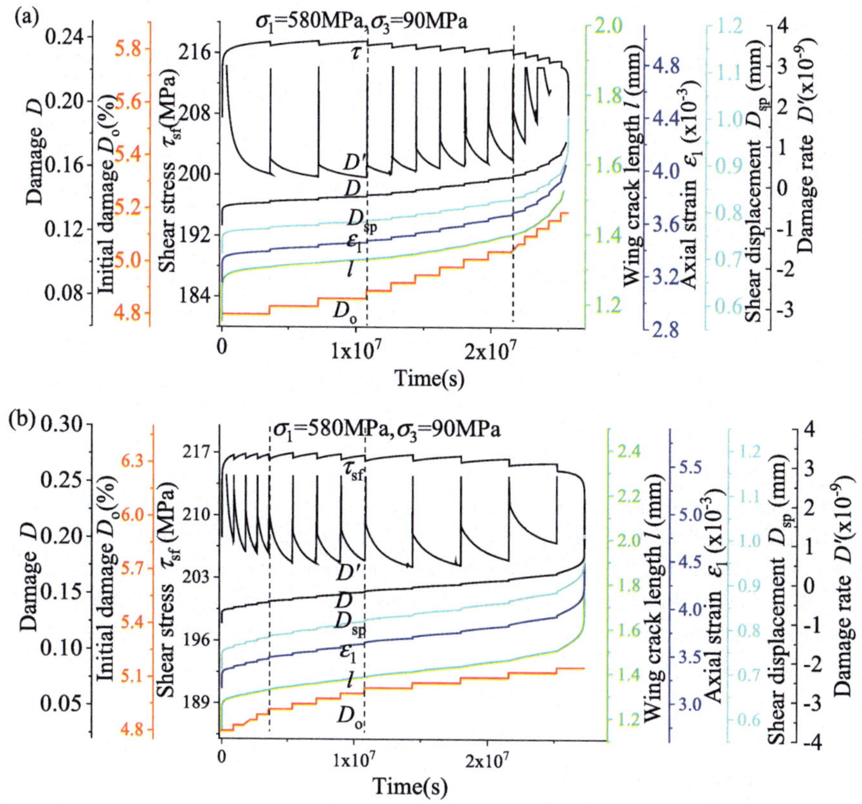

Fig. 6.27 Correspondence of time-dependent shear stress, shear displacement, axial strain, wing crack growth damage and damage rate subjected to the stepping initial damage (**a** $\Delta T = 3.6 \times 10^6$, 1.8×10^6, and 0.9×10^6 over time, **b** $\Delta T = 0.9 \times 10^6$, 1.8×10^6, and 3.6×10^6 over time, and $\Delta D_o = 3 \times 10^{-4}$) during static creep

$\Delta D_o = 9 \times 10^{-4}$, (2) the second phase in variation curve at three stages of $\Delta D_o = 4 \times 10^{-4}$, (3) the third phase in variation curve at the final stages of $\Delta D_o = 1 \times 10^{-4}$. The drop of shear stress is obviously decreasing with the decrement of parameter ΔD_o at given parameter ΔT (i.e., the drop of shear stress is decreasing with the decrement of variation in damage rate) from the first phase to the third phase. The shear displacement, axial strain, and wing crack growth also experience three phases at the parameter $\Delta D_o = 9 \times 10^{-4}, 4 \times 10^{-4}$, and 1×10^{-4} over time, respectively. The increment of shear displacement, axial strain, and wing crack growth are decreasing with the decrement of parameter ΔD_o at given parameter ΔT.

In Fig. 6.26b, the parameter ΔD_o experiences a variation from low value (i.e., $\Delta D_o = 1 \times 10^{-4}$) to high value (i.e., $\Delta D_o = 9 \times 10^{-4}$) over time. This variation also includes three phases: (1) the first phase in the variation curve at three stages of $\Delta D_o = 1 \times 10^{-4}$, (2) the second phase in variation curve at three stages of $\Delta D_o = 4 \times 10^{-4}$, (3) the third phase in variation curve at the final stages of $\Delta D_o = 9 \times 10^{-4}$.

The drop of shear stress is obviously increasing with the increment of parameter ΔD_o at given parameter ΔT (i.e., the drop of shear stress is increasing with the increment of variation in damage rate) from the first phase to the third phase. The shear displacement, axial strain and wing crack growth also experience three phases at the parameter $\Delta D_o = 1 \times 10^{-4}, 4 \times 10^{-4}$, and 9×10^{-4} over time, respectively. The increment of shear displacement, axial strain, and wing crack growth are increasing with the increment of parameter ΔD_o at given parameter ΔT.

In Fig. 6.27a and b, the correspondence of the whole time-dependent shear stress, shear displacement, axial strain, wing crack growth, damage, and damage rate at different parameters ΔT over time during static creep is shown. Figure 6.18a shows the time-dependent shear stress, shear displacement, axial strain and wing crack growth at parameter $\Delta T = 3.6 \times 10^6, 1.8 \times 10^6$, and 0.9×10^6 over time, respectively. Figure 6.27b shows the time-dependent shear stress, shear displacement, axial strain and wing crack growth at parameter $\Delta T = 0.9 \times 10^6$, 1.8×10^6, and 3.6×10^6 over time, respectively.

In general, the smaller stress drops during creep are corresponding to some smaller earthquakes, and the final larger stress drop is corresponding to the bigger earthquake. These theoretical results provide qualitatively an important understanding for evaluation of the evolution process of larger earthquakes during long-term crustal creep under the different time differences ΔT between two smaller earthquakes and the different damage differences ΔD_o caused by the smaller earthquake.

6.4 Conclusions

A micromechanics-based Damage Rate- and State- Fracture Model (DRSFM) is proposed to explain the mechanism of stress drops in intact rocks during long-term creep. The formulation of this DRSFM is based on the wing crack growth model, the subcritical cracking law, the correlations of crack, strain and shear displacement and the function of time-dependent step initial damage. This time-dependent step initial damage has a great significance for evaluating the stress drops caused by the presence of shear bands from localized numerous microcracks accumulation and coalescence in rocks during brittle creep, which provides an implication for evaluation of earthquake mechanism. The rationality of this DRSFM is approximately verified by comparing the published theoretical and experimental data. Some conclusions are as follows:

(1) The effect of confining pressure on the stress–strain curve with multi-stress drops is studied. With the increment of confining pressure, the values of axial stress and stress drop are increasing at the given axial strain. Corresponding to the function of stepping initial damage relating to axial strain, the value of stress drop in the stress–strain curve decreases with the stress drop sequence at given confining pressure. Parameters ΔD_o, $\varepsilon_{1\mathrm{ini}}$, and $\Delta\varepsilon_1$ in the stepping damage function reflect the damage size in individual minor localized shear fracture

band, the initial axial strain at the first individual minor localized shear fracture band, and the strain difference between two minor localized shear fracture bands, respectively. For natural rocks, the phenomena of multi-stress drops in stress–strain curves can be explained by compositive effects of the three parameters ΔD_o, ε_{1ini}, and $\Delta\varepsilon_1$. The axial stress at a given axial strain is increasing with the decrement of ΔD_o and the increment of ε_{1ini} and $\Delta\varepsilon_1$. The drops of axial stress are increasing with the increment of parameters ΔD_o and $\Delta\varepsilon_1$.

(2) During long-term brittle creep without step initial damage, shear stress experiences a strengthening phase, a weakening phase I and another weakening phase II in the intact rocks. In the strengthening phase, shear stress is increasing with the decrement of damage rate. In the weakening phase I, shear stress is decreasing with the decrement of damage rate. In the weakening phase II, shear stress is decreasing with the increment of damage rate. This evolution process of shear stress in intact rocks is similar to that of shear friction from static to slip friction in faults. The weakening phase II has a good agreement with the slip velocity-dependent friction in rate-and state- friction model proposed by Dieterich and Ruina, which also provides a certain help for the understanding of a conclusion that the onset of stress drop caused by shear fracture is similar to the onset of stress drop caused by frictional stick–slip.

(3) During the long-term brittle creep with step initial damage, some stress drops appear at the variation curve of shear stress over time. A stepping increment of initial damage causes a drop of shear stress, which is corresponding to the presence of an individual shear band in rocks induced by microcrack accumulation and coalescence. The last big drop of shear stress is corresponding to the shear fracture of intact rocks. Between the two stress drops during creep, shear stress experiences an individual process of strengthening and weakening phases. When the stress drops all appear at the weakening phase II, the shear stress is decreasing with increment of damage rate, and drop of shear stress is increasing with the increasing increment of damage rate. The stepping increment of initial damage causes the stepping increment of shear displacement, axial strain, wing crack length, damage, and damage rate. Increment of shear displacement is increasing with the increasing increment of damage rate. The time variable has an important meaning for evaluating the time difference between two earthquakes. This stepping initial damage has an identical effect on shear stress liking the state variable relating to slip velocity in rate- and state- friction model proposed by Dieterich and Ruina, which provides a certain understanding for evaluation of mechanism of the numerous smaller (or slow) earthquakes and the larger earthquake.

(4) It is noteworthy that the above all studies provide an understanding qualitatively for the step damage variable-induced stress drops in intact rocks during long-term brittle creep. However, how to establish the relationship of time-dependent step damage between in natural rocks and the DRSFM will be a more meaningful study for accurately evaluating the earthquake mechanism. The relationship between Acoustic Emissions signals and microcracks growth-induced damage could provide important help for evaluation of the real time-dependent step

damage, which will be further studied in the future. Furthermore, the formation of stress drops induced by the small individual shear bands, which is caused by strain localization from the local microcracks accumulation and coalescence, is approximately simulated by the global variation of time-dependent initial damage in the proposed DRSFM. The macroscopic shear fracture of intact rocks is studied approximately by the global failure caused by the global microcrack coalescence in the model. The suggested model may propose a certain theoretical understanding of the earthquake mechanism.

References

Blanpied ML, Tullis TE, Weeks JD. Frictional behavior of granite at low and high sliding velocities. Geophys Res Lett. 1987;14(5):554–7.

Brace WF, Byerlee JD. Stick-slip as a mechanism for earthquakes. Science. 1966;153(3739):990–2.

Brantut N, Schubnel A, Guéguen Y (2011) Damage and rupture dynamics at the brittle-ductile transition: the case of gypsum. J Geophys Res. 116(B1)

Chen ZH, Fu YF, Tang CA. Confining pressure effect on acoustic emissions in rock failure. Chinese J Rock Mech Eng. 1997;16(1):65–70.

Dieterich JH (1979a) Modeling of rock friction 1. Experimental results and constitutive equations. J Geophys Res. 84(B5):2161–2168

Dieterich JH (1979b) Modeling of rock friction 2. Simulation of preseismic slip. J Geophys Res. 84(B5):2169–2145

Goldsby DL, Tullis TE. Flash heating leads to low frictional strength of crustal rocks at earthquake slip rates. Science. 2011;334(6053):216–8.

Ide S, Beroza GC, Shelly DR, Uchide T. A scaling law for slow earthquakes. Nature. 2007;447(7140):76.

Ikari MJ, Marone C, Saffer DM, Kopf AJ. Slip weakening as a mechanism for slow earthquakes. Nat Geosc. 2013;6(6):468.

Leeman JR, Saffer DM, Scuderi MM, Marone C. Laboratory observations of slow earthquakes and the spectrum of tectonic fault slip modes. Nat Commun. 2016;7:11104.

Leeman JR, Marone C, Saffer DM. Frictional mechanics of slow earthquakes. J Geophys Res. 2018;123:7931–49.

Li H, Li H, Gao B, Jiang D, Feng J (2015) Study of acoustic emission and mechanical characteristics of coal samples under different loading rates. Shock and Vibrat 458519

Linde AT, Gladwin MT, Johnston MJ, Gwyther RL, Bilham RG. A slow earthquake sequence on the San Andreas fault. Nature. 1996;383(6595):65.

Mazars J. A description of micro- and macroscale damage of concrete structures. Eng Fract Mech. 1986;25(5–6):729–37.

Ohnaka M. A shear failure strength law of rock in the brittle-plastic transition regime. Geophys Res Lett. 1995;22(1):25–8.

Ohnaka M, Akatsu M, Mochizuki H, Odedra A, Tagashira F, Yamamoto Y. A constitutive law for the shear failure of rock under lithospheric conditions. Tectonophysics. 1997;277:1–27.

Ruina A. Slip instability and state variable friction laws. J Geophys Res. 1983;88(B12):10359–70.

Scholz CH. Earthquakes and friction laws. Nature. 1998;391(391):37–42.

Sleep NH, Blanpied ML. Creep, compaction and the weak rheology of major faults. Nature. 1992;359(6397):687.

Spagnuolo E, Nielsen S, Violay M, Di Toro G. An empirically based steady state friction law and implications for fault stability. Geophys Res Lett. 2016;43:3263–71.

Sun JS, Chen M, Jiang QH, Lu WB, Zhou CB. Numerical simulation of mesomechanical characteristics of creep demage evolution for Jingping marble. Chinese J Rock Soil Mech. 2013;34(12):3601–8.
Xia YJ, Li LC, Tang CA, Li XY, Ma S, Li M. A new method to evaluate rock mass brittleness based on stress-strain curves of class I. Rock Mech Rock Eng. 2017;50(5):1123–39.
Zhu WS, Qi YP, Guo YH, Yang WM. 3D damage rheology analysis of deformation and fracture of surrounding rocks in Jinping I hydropower station underground powerhouse. Chinese J Rock Mech Eng. 2012;31(5):865–72.

Chapter 7
Crack Recovery During Creep Fracture

Fatigue failure under cyclic static loading and unloading (i.e., cyclic/fatigue creep failure) in brittle rocks has a great significance for the evaluation of lifetime in underground engineering. The visco-elastic rebound strain is an important mechanical phenomenon of brittle rocks during the unloading. However, the unloading-induced plastic rebound is rarely studied. This chapter will propose a method to explore the mechanism of plastic rebound caused by microcrack recovery. The time-dependent visco-elastic–plastic deformation during creep failure will be analyzed under constant confining pressure and cyclic static loading and unloading of axial stress.

7.1 Theory

A micro–macro model of fatigue creep failure under triaxial compression (i.e., $\sigma_1 > \sigma_3 = \sigma_2$) is proposed in Fig. 7.1, which is consisted of the function of cyclic static loading and unloading path, the Hooke-Kelvin viscoelastic model and the established micro–macro failure model in Chap. 5. The microcrack recovery-induced rebound of plastic deformation is found. The relationship between rebounds of the elastic and viscoelastic deformation and the plastic deformation is explained. Figure 7.1a shows the mechanical schematic of microcrack interactions during wing crack growth or close in brittle solids. Figure 7.1b shows the path of cyclic static loading and unloading of axial stress over time.

X. Li et al., *Static Creep Micro-Macro Fracture Mechanics of Brittle Solids*,
https://doi.org/10.1007/978-981-99-8203-5_7

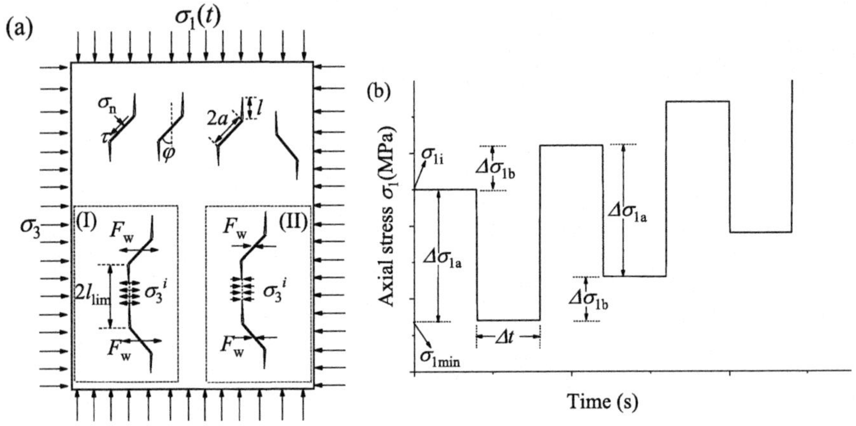

Fig. 7.1 Illustration of the microcrack-triggered model under confined static cyclic loading and unloading. **a** Brittle solid model showing mechanical schematic of microcrack interactions in wing crack growth and close, **b** the path of cyclic static loading and unloading of axial stress over time

7.1.1 Progressive Fracture Considering Viscoelastic Deformation

The wedge force F_w of Eq. (5.1) applied to the initial crack plane is related to the remoted axial stress and confining pressure acting on the model, τ and σ_n are shear stress and normal stress acting on the initial crack plane. It is noted that the compressive and tensile stresses are denoted as negative and positive in the theoretical calculations respectively. Thus, if wedge force is positive (i.e., the area of I in Fig. 7.1a), the friction induced by normal stress is smaller than the shear stress on initial crack plane, and the open and growth of wing crack could be caused; if wedge force is negative (i.e., the area of II in Fig. 7.2a), the friction induced by normal stress is larger than the shear stress on initial crack plane, and the close and recovery of wing crack could be caused. Furthermore, for convenient observation, the compressive and tensile stresses are denoted positive and negative in the graphic descriptions respectively below.

Thus, a critical axial stress $\sigma 1w$ at wedge force $Fw=0$ can be solved as:

$$\sigma_{1w} = -\frac{\sigma_3(\sin 2\varphi + \mu + \mu \cos 2\varphi)}{\mu - \mu \cos 2\varphi - \sin 2\varphi} \tag{7.1}$$

This critical axial stress σ_{1w} is equal to zero at confining pressure $\sigma_3 = 0$, which shows the wedge force F_w always keeps positive at any applied compressive axial stresses and the close of wing crack cannot be caused. However, this critical axial stress σ_{1w} only overcomes the friction on initial crack plane and can be regarded as the maximum axial stress causing linear elastic deformation. But the wing crack could not open and grow at the applied axial stress $|\sigma_1| > |\sigma_{1w}|$ (| | represents for

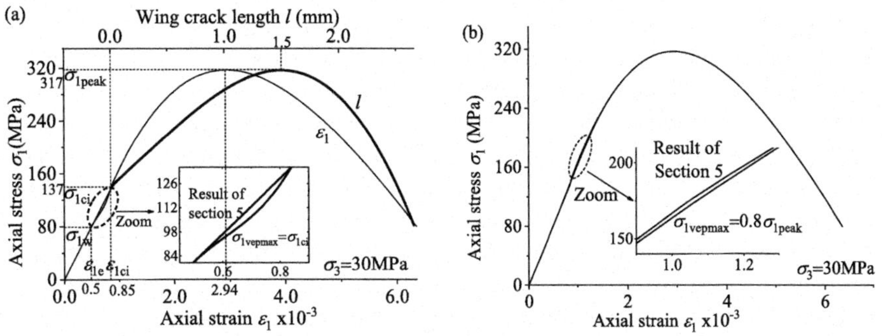

Fig. 7.2 **a** Correspondence of axial stress with wing crack growth and strain increment at $\sigma_{1\text{vepmax}} = \sigma_{1\text{ci}}$, **b** stress–strain curve at $\sigma_{1\text{vepmax}} = 0.8\,\sigma_{1\text{peak}}$

the absolute value), which is because that the applied stress does not overcome the fracture toughness of crack in brittle solids. The critical stress of wing crack initiation will be studied below.

The open and growth of wing crack cause the increasing axial plastic deformation, and the close and recovery of wing crack cause the decreasing axial plastic deformation (i.e., a dramatic phenomenon of the rebound of plastic deformation). However, this rebound of plastic strain caused by close and recovery of wing crack is attributed to the presence of the confining pressures.

Furthermore, a cyclic static loading and unloading path of axial stress over time (Fig. 7.2c) is introduced in this above proposed model, which provides an important evaluation for the phenomenon of rebound plastic strain caused by wing crack recovery under cyclic fatigue creep in brittle solids.

The quasi-static stress–strain constitutive relationship describing the whole strain-hardening and -softening phases during wing crack growth was obtained in Eq. (5.13). However, the viscoelastic stress–strain constitutive relationship before crack initiation (i.e., $|\sigma_1|<|\sigma_{1\text{ci}}|$) cannot be explained. Based on Eq. (5.8) and Eq. (7.1), the relationship between this critical axial stress $\sigma_{1\text{w}}$ and the crack initiation stress $\sigma_{1\text{ci}}$ can be expressed as:

$$\sigma_{1\text{ci}} = \frac{\beta^{3/2}\sqrt{\pi}K_{IC}}{B_1\sqrt{a}} + \sigma_{1\text{w}} \tag{7.2}$$

Crack initiation stress $\sigma_{1\text{ci}}$ is larger than this critical stress $\sigma_{1\text{w}}$ (i.e., $|\sigma_{1\text{ci}}| > |\sigma_{1\text{w}}|$). When the applied axial stress is smaller than the critical axial stress $\sigma_{1\text{w}}$ (i.e., $|\sigma_1| < |\sigma_{1\text{w}}|$), the close and recovery of the extended wing crack could happen. But, the open and growth of wing crack could not happen at $|\sigma_1| > |\sigma_{1\text{w}}|$, and the open of growth of wing crack will happen at $|\sigma_1| > |\sigma_{1\text{ci}}|$.

Furthermore, it is known that the total quasi-static stress–strain curves consist of elastic, viscoelastic, plastic and viscoplastic deformations. The elastic deformation is recoverable and time-independent, the viscoelastic deformation is recoverable and time-dependent, the plastic deformation is irrecoverable and time-independent, and the viscoplastic deformation is irrecoverable and time-dependent. In this study, the plastic and viscoplastic deformations are collectively called as plastic deformation below. According to the proposed two critical axial stress values σ_{1w} and σ_{1ci}, the elastic deformation happens at $|\sigma_1| < |\sigma_{1w}|$, the viscoelastic deformation happens at $|\sigma_{1w}| < |\sigma_1| < |\sigma_{1vepmax}|$, and the plastic and viscoplastic deformations happen at $|\sigma_1| > |\sigma_{1ci}|$. Furthermore, it is noteworthy that the viscoelastic and viscoplastic deformations could simultaneously happen at $|\sigma_{1ci}| < |\sigma_1| < |\sigma_{1vepmax}|$ ($\sigma_{1vepmax}$ represents for a maximum of applied axial stress causing a maximum of accumulated viscoelastic deformation). However, the proportion between viscoelastic and plastic deformations at $|\sigma_{1ci}| < |\sigma_1| < |\sigma_{1vepmax}|$, and this critical applied axial stress $\sigma_{1vepmax}$ are difficult to be determined accurately. Furthermore, a critical value of axial stress from stable to unstable wing crack growth is $0.8\sigma_{1peak}$ in brittle rocks (Martin and Chandler 1994). The new viscoelastic deformation could less generate when applied axial stress is larger than this critical axial stress. Thus, the viscoelastic deformation at $|\sigma_{1ci}| < |\sigma_1| < |0.8\sigma_{1peak}|$ will be compared and discussed graphically below. The proportion of viscoelastic deformation at $|\sigma_{1ci}| < |\sigma_1| < |\sigma_{1vepmax}|$ contributing to final accumulated pure viscoelastic deformation can be obtained approximately by the time-dependent viscoelastic deformation under constant loadings below.

Furthermore, the elastic strain is approximately obtained by combining Eqs. (5.8), (5.11) and (7.1) as:

$$\varepsilon_{1e} = \sigma_{1w}\frac{\varepsilon_{1ci}}{\sigma_{1ci}} = \frac{\sigma_3 B_2 \varepsilon_o(-\ln[1-D_o])^{1/m}}{\sigma_3 B_2 - K_{IC}\beta^{3/2}\sqrt{\pi/a}} \tag{7.3}$$

The elastic stress–strain constitutive relationship can be expressed as:

$$\sigma_1(\varepsilon_1) = E_w\varepsilon_1 = \frac{\sigma_{1w}}{\varepsilon_{1e}}\varepsilon_1 = \left(\frac{K_{IC}\beta^{3/2}\sqrt{\pi/a} - \sigma_3 B_2}{\varepsilon_o B_1(-\ln[1-D_o])^{1/m}}\right)\varepsilon_1, \quad (0 < \varepsilon_1 < \varepsilon_{1e}) \tag{7.4}$$

where Ew is elastic modulus.

For the phase of viscoelastic deformation (i.e., $\varepsilon_{1e} < \varepsilon_1 < \varepsilon_{1vepmax}$, where $\varepsilon_{1vepmax}$ is axial strain at applied stress $\sigma_{1vepmax}$ in stress–strain constitutive curve of Eq. (5.13)), the transient rate of axial stress and strain is E at ε_{1e}, and the transient rate (i.e., E_{vepmax}) of axial stress and strain at $\varepsilon_{1vepmax}$ can be obtained by solving strain derivate of Eq. (5.13). Due to the complexity and lengthiness of this equation of rate E_{vepmax}, this equation is not given.

The variable in the rate of axial stress and strain from E to E_{vepmax} (i.e., visco-elastic–plastic modulus E_{vep}) can be defined by a function relating to axial strain ε_1 as:

$$E_{vep}(\varepsilon_1) = E_w\left(1 + Z_1 + Z_2\varepsilon_1 + Z_3\varepsilon_1^2\right) \tag{7.5}$$

The known conditions are as follows:

$$\left.\begin{array}{l} E_{\text{vep}}(\varepsilon_{1e}) = E_w \\ E_{\text{vep}}\left(\varepsilon_{1\text{vepmax}}\right) = E_{\text{vepmax}} \\ \sigma_{1\text{vepmax}} - \sigma_{1w} = \int\limits_{\varepsilon_{1e}}^{\varepsilon_{1\text{vepmax}}} E_{\text{vep}}(\varepsilon_1)\,d\varepsilon_1 \end{array}\right\} \tag{7.6}$$

Combining Eqs. (7.6) and (7.6), the parameters Z_1, Z_2 and Z_3 can be solved.

The viscoelastic constitutive relation of axial stress and strain can be expressed as:

$$\sigma_1(\varepsilon_1) = \sigma_{1\text{w}} + \int\limits_{\varepsilon_{1\text{e}}}^{\varepsilon_1} E_{\text{w}}\left(1 + Z_1 + Z_2\varepsilon_1 + Z_3\varepsilon_1^2\right) d\varepsilon_1 \quad \left(\varepsilon_{1\text{e}} < \varepsilon_1 < \varepsilon_{1\text{vepmax}}\right) \tag{7.7}$$

Thus, combining Eqs. (5.13), (7.4) and (7.7), the total stress–strain constitutive relationship describing the elastic–viscoelastic-plastic deformation can be expressed as:

$$\sigma_1(\varepsilon_1) = \begin{cases} \left(\dfrac{K_{IC}\beta^{3/2}\sqrt{\pi/a} - \sigma_3 B_2}{\varepsilon_o B_1(-\ln[1 - D_o])^{1/m}}\right)\varepsilon_1, \\ (0 < \varepsilon_1 \le \varepsilon_{1e}, 0 < |\sigma_1| \le |\sigma_{1w}|) \\ -\dfrac{\sigma_3 B_2}{B_1} + \displaystyle\int\limits_{\varepsilon_{1e}}^{\varepsilon_1} \left(\dfrac{K_{IC}\beta^{3/2}\sqrt{\pi/a} - \sigma_3 B_2}{\varepsilon_o B_1(-\ln[1 - D_o])^{1/m}}\right)\left(1 + Z_1 + Z_2\varepsilon_1 + Z_3\varepsilon_1^2\right) \\ d\varepsilon_1 \quad \left(\varepsilon_{1e} < \varepsilon_1 \le \varepsilon_{1\text{vepmax}}, |\sigma_{1w}| < |\sigma_1| \le \left|\sigma_{1\text{vepmax}}\right|\right) \\ \dfrac{K_{IC} - \sigma_3\left(\pi a^2 B_3 B_2 + 2\sqrt{l/\pi}\right)}{\pi a^2 B_3 B_1} \\ \left(\begin{array}{l} \varepsilon_{1\text{vepmax}} < \varepsilon_1 < \varepsilon_{1f}, \left|\sigma_{1\text{vepmax}}\right| \\ < |\sigma_1| < \left|\sigma_{1\text{peak}}\right| or \left|\sigma_{1f}\right| < |\sigma_1| < \left|\sigma_{1\text{peak}}\right| \end{array}\right) \end{cases} \tag{7.8}$$

In the third part of Eq. (7.8) showing the constitutive relationship of plastic deformation, the applied axial stress causing plastic deformation prior to reaching peak stress is $\left|\sigma_{1\text{vepmax}}\right| < |\sigma_1| < \left|\sigma_{1\text{peak}}\right|$, , and the applied axial stress after peak stress is smaller than peak stress and is larger than failure stress (i.e., $|\sigma_{1\text{f}}| < |\sigma_1| < \left|\sigma_{1\text{peak}}\right|$). Furthermore, it is noted that the elastic strain happens at the total deformation process (i.e., $0 < \varepsilon_1 < \varepsilon_{1\text{f}}$), the viscoelastic strain happens during $\varepsilon_{1\text{w}} < \varepsilon_1 < \varepsilon_{1\text{f}}$, and the plastic strain happens during $\varepsilon_{1\text{ci}} < \varepsilon_1 < \varepsilon_{1\text{f}}$.

7.1.2 Confined Fatigue Creep Fracture Considering Viscoelastic and Plastic Rebound Deformations

Based on the proposed stress–strain constitutive relationship above, for time-dependent creep under cyclic static loading and unloading, the elastic strain at loading and the rebound elastic strain at the unloading of axial stress can be obtained by Eq. (7.8).

Furthermore, Hooke-Kelvin viscoelastic model can be expressed as:

$$|\sigma_1|+\frac{\eta}{E_{\text{vet}}+E_{\text{w}}}|\dot{\sigma}_1|=\frac{E_{\text{vet}}E_{\text{w}}}{E_{\text{vet}}+E_{\text{w}}}\varepsilon_1+\frac{\eta E_{\text{w}}}{E_{\text{vet}}+E_{\text{w}}}\dot{\varepsilon}_1 \tag{7.9}$$

where E_{vet} is viscoelastic modulus, η is viscosity.

Based on the known conditions $|\dot{\sigma}_1| = 0$ and $\varepsilon_1(t = 0) = |\sigma_{1\text{L}}|/E_{\text{w}}$, the time-dependent axial viscoelastic strain under constant loading can be solved by Hooke-Kelvin viscoelastic model as:

$$\varepsilon_{1\text{veL}}(t) = \frac{|\sigma_{1\text{L}}|}{E_{\text{w}}} + |\sigma_{1\text{L}}|\left(1 - e^{-tE_{\text{vet}}/\eta}\right)/E_{\text{vet}} \tag{7.10}$$

where $\sigma_{1\text{L}}$ represents the value of static loading of axial stress. Furthermore, $\varepsilon_{1\text{veL}}$ is decelerated increasing to a maximum $\varepsilon_{1\text{vemax}}$ over time. This maximum of accumulated viscoelastic axial strain $\varepsilon_{1\text{vemax}}$ is larger than axial strain $\varepsilon_{1\text{ci}}$ at crack initiation in the stress–strain constitutive curve of Eq. (5.13), which shows that the accumulated pure viscoelastic deformation increases by $\Delta\varepsilon_{1\text{ve}}$ (i.e., $\varepsilon_{1\text{vemax}}$-$\varepsilon_{1\text{ci}}$) after wing crack initiation.

Based on the known conditions $|\dot{\sigma}_1| = 0$, $|\sigma_1| = |\sigma_{1\text{UL}}|$ and $\varepsilon_{11} = \varepsilon_1(t = t_{\text{veL}}) = \varepsilon_{1\text{veL}}(t_{\text{veL}}) - (|\sigma_{1\text{L}}| - |\sigma_{1\text{UL}}|)/E_{\text{w}}$, the time-dependent rebound of axial viscoelastic strain when axial stress is unloading to a given value $\sigma_{1\text{UL}}$ can be expressed by Hooke-Kelvin viscoelastic model as:

$$\begin{aligned}\varepsilon_{1\text{veUL}}(t) =& |\sigma_{1\text{UL}}|\left(\frac{E_{\text{w}}+E_{\text{vet}}}{E_{\text{w}}E_{\text{vet}}}\right)+\left(\varepsilon_{11} - |\sigma_{1\text{UL}}|\left(\frac{E_{\text{w}}+E_{\text{vet}}}{E_{\text{w}}E_{\text{vet}}}\right)\right)\\ &\left(e^{-(t-t_{\text{veL}})E_{\text{vet}}/\eta}\right)/E_{\text{vet}} \quad (t > t_{\text{veL}})\end{aligned} \tag{7.11}$$

where $\sigma_{1\text{UL}}$ represents the value of static unloading of axial stress under the cyclic static loading and unloading, t_{veL} is the loading time of viscoelastic strain in Eq. (7.10).

When the axial stress is unloading to zero (i.e., $\sigma_{1\text{UL}} = 0$), the time-dependent rebound of axial viscoelastic strain can be simplified as:

$$\varepsilon_{1\text{veUL}}(t) = \left(\varepsilon_{1\text{veL}}(t_{\text{veL}}) - \frac{|\sigma_{1\text{L}}|}{E_{\text{w}}}\right)\left(-e^{-(t-t_{\text{veL}})E_{\text{vet}}/\eta}\right)/E_{\text{vet}} \quad (t > t_{\text{veL}}) \tag{7.12}$$

The viscoelastic deformation under each cyclic static loading and unloading can be regarded as an independent process. For each static loading phase under cyclic static loading and unloading, the initial time at start of each static loading is defined as zero, the initial axial strain is elastic strain induced by initial applied axial stress in first static loading, and another initial axial strain is the sum of the axial strain after the adjacent static unloading and the axial elastic strain induced by stress difference between loading and unloading in another static loading. For each static unloading phase under cyclic static loading and unloading, the initial time at start of each static unloading is defined by last adjacent loading time, and the initial axial strain is the minus of the axial strain after the adjacent static loading and the axial elastic strain induced by stress difference between loading and unloading. Furthermore, it is assumed that the rebounds of elastic and viscoelastic deformations with increasing cyclic number are not reduced and keep constant in this study.

Thus, after the first cyclic loading and unloading, the initial condition at next static loading can be obtained as:

$$\varepsilon_{12} = \varepsilon_1(t = 0) = \varepsilon_{1\text{veUL}}(t_{\text{veL}} + t_{\text{veUL}}) + \frac{|\sigma_{1\text{L}}| - |\sigma_{1\text{UL}}|}{E_\text{w}} \tag{7.13}$$

The next time-dependent viscoelastic strain under static loading can be expressed by Hooke-Kelvin viscoelastic model as:

$$\varepsilon_{1\text{veLN}}(t) = \left(\varepsilon_{12} - \frac{|\sigma_{1\text{LN}}|}{E_\text{w}}\right)e^{-tE_{\text{vet}}/\eta} + \frac{|\sigma_{1\text{LN}}|}{E_\text{w}} + |\sigma_{1\text{LN}}|\left(1 - e^{-tE_{\text{vet}}//\eta}\right)/E_{\text{vet}} \tag{7.14}$$

where $\sigma_{1\text{LN}}$ is axial stress of next static loading besides the case of first static loading under cyclic loading and unloading. Besides the case of time-dependent viscoelastic strain in Eq. (7.10) at first static loading, the other time-dependent viscoelastic strain at static loading during cyclic static loading and unloading all can be expressed by Eq. (7.14).

Based on the proposed creep Eq. (5.17), the time-dependent and wing crack growth-induced creep axial strain removing the crack initiation strain under constant loadings can be rewritten by:

$$\begin{aligned} \varepsilon_{1\text{p}}(t) = \varepsilon_{1\text{wcg}}(t) - \varepsilon_{1\text{ci}} = \varepsilon_\text{o}\left\{-\ln\left[1 - D_\text{o}(l(t)/a + 1)^3\right]\right\}^{1/m} \\ -\varepsilon_{1\text{ci}}, \quad \left(|\sigma_{1\text{ci}}| < |\sigma_1| < \left|\sigma_{1\text{peak}}\right|\right) \end{aligned} \tag{7.15}$$

Furthermore, Eq. (7.15) describes time-dependent visco-elastic- plastic strains during wing crack growth, but the unloading-induced rebound of elastic and viscoelastic strains cannot be explained. Combining Eqs. (7.10–7.15), the time-dependent confined fatigue creep failure containing rebound of elastic and viscoelastic deformations under cyclic loading and unloading of axial stress can be obtained by:

$$\varepsilon_1(t) = \varepsilon_{1p}(t) + \varepsilon_{1veL}(t) + \varepsilon_{1veUL}(t) + \varepsilon_{1veLN}(t) \quad \left(|\sigma_{1ci}| < |\sigma_{1i}| < |\sigma_{1peak}|\right) \tag{7.16}$$

In Eq. (7.16), the axial stress σ_1 is defined as a time-dependent function. The time-dependent axial stress is defined as cyclic static loading in Fig. 7.1b. If the maximum of cyclic static loading is smaller crack initiation stress, the time-dependent strain can be obtained by combining Eqs. (7.10–7.14). If the maximum of cyclic static loading larger than the crack initiation stress, the time-dependent strain can be obtained by Eq. (7.16). Furthermore, in this study, an interesting rebound plastic axial strain is found under cyclic static loading and unloading of axial stress and constant confining pressure. This dramatic rebound plastic axial strain is caused by close and recovery of extended wing crack subjecting to confining pressure when axial stress is unloaded to a critical value. This critical value of axial stress is σ_{1w} in Eq. (7.1).

7.2 Visco-elastic–Plastic Stress–Strain Curves

Model parameters are selected by referring to the Jinping marble in China. K_{IC} = 1.99 MPa·m$^{1/2}$, $m = 1.0$, $\varepsilon_o = 0.172$, $\varphi = 32.03°$, $\mu = 0.51$, $a = 3.1$ mm, $\beta = 0.276$, $D_o = 0.048$, $v = 0.16$, $n = 57$, $E_{vet} = 290$ GPa, $\eta = 3 \times 10^5$ GPa s.

Figure 7.2a shows that the correspondence of axial stress with wing crack growth and axial strain increment at confining pressure $\sigma_3 = 30$ MPa. The rationality of this microcrack-induced stress–strain curve has been verified in Chaps. 2 and 5. However, in Chaps. 2 and 5, the viscoelastic stress–strain curve was not proposed, and the viscoelastic stress–strain curve was assumed as a linear curve, which can be observed in the amplified area of the comparison between the assumed elastic stress–strain curve in Chap. 5. and the proposed viscoelastic stress–strain curve in this study (Fig. 7.2). Furthermore, the axial strain considering viscoelastic deformation in this study is larger than that of ignoring viscoelastic deformation in Chap. 5 at given axial stress.

Due to the difficulty of accurate determination for the critical axial stress $\sigma_{1vepmax}$, the sensitivities of axial stress $\sigma_{1vepmax}$ on visco-elastic–plastic stress–strain curves are independently analyzed in Fig. 7.3a, and the sensitivities of axial stress $\sigma_{1vepmax}$ on visco-elastic–plastic modulus E_{vep} are drawn in Fig. 7.3b. Visco-elastic–plastic modulus E_{vep} is decreasing firstly to a minimum, and then is increasing with increment of axial strain at $\sigma_{1vepmax} = \sigma_{1ci}$; visco-elastic–plastic modulus is increasing firstly to a maximum, and then is decreasing with increment of axial strain at $\sigma_{1vepmax} = 0.8\sigma_{1peak}$, $0.7\sigma_{1peak}$, $0.6\sigma_{1peak}$ or $0.5\sigma_{1peak}$, which may be because that the visco-elastic–plastic modulus E_{vep} at $\sigma_{1vepmax} = \sigma_{1ci}$ is much larger than that of at $\sigma_{1vepmax} = 0.8\sigma_{1peak}$, $0.7\sigma_{1peak}$, $0.6\sigma_{1peak}$ or $0.5\sigma_{1peak}$.

In Fig. 7.2a, the elastic axial stress σ_{1w}, crack initiation stress σ_{1ci} and peak axial stress σ_{1peak} are 80 MPa, 137 MPa, and 317 MPa respectively, the axial strain at elastic maximum, crack initiation and peak stress are 0.5×10^{-3}, 0.85×10^{-3} and 2.94×10^{-3} respectively, and the wing crack length at peak stress is 1.5 mm. The three critical stress states provide help for evaluating the fatigue creep deformation

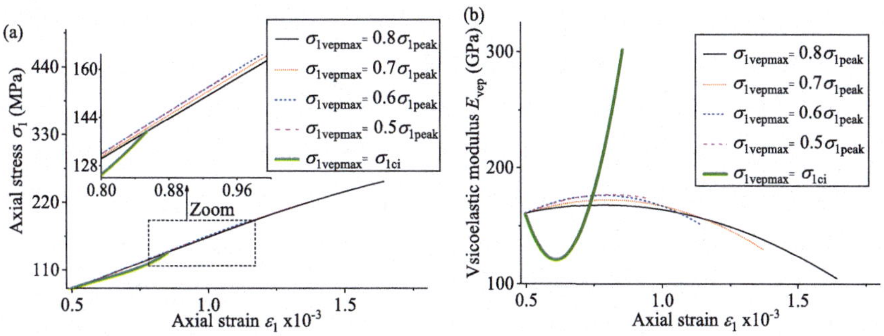

Fig. 7.3 **a** Visco-elastic–plastic stress–strain curves and **b** variations of visco-elastic–plastic modulus with axial strain under different critical axial stress $\sigma_{1vepmax}$

under cyclic static loading and unloading. The static creep plastic deformation could appear when the applied static stress is larger than crack initiation stress σ_{1ci} and is smaller than peak axial stress σ_{1peak}. Furthermore, the initial applied axial stress σ_{1i} causing fatigue creep failure in Fig. 7.1b should lie between σ_{1ci} and σ_{1peak}, i.e., $\sigma_{1ci} < \sigma_{1i} < \sigma_{1peak}$.

7.3 Fatigue Visco-elastic–Plastic Creep Fracture

7.3.1 Creep Fracture Under Cyclic Loading and Unloading

As shown in Fig. 7.4, the time-dependent uniaxial fatigue creep failure is obtained. Under cyclic static loading and unloading of uniaxial stress in Fig. 7.4a, the variations of wing crack growth, pure viscoelastic axial strain ε_{1ve}, wing crack growth-induced axial strain ε_{1wcg}, and total visco-elastic-plastic axial strain ε_1 over time are shown in Fig. 7.4b–e respectively. In Fig. 7.4a, the maximum of applied constant axial stress is 106 MPa, which is larger than the crack initiation stress σ_{1ci} and is smaller than the peak stress σ_{1peak} without confining pressure; and the minimum of applied constant axial stress is 0 MPa. The time-dependent wing crack growth in Fig. 7.4b causes the time-dependent plastic axial strain ε_{1wcg} in Fig. 7.4d, the wing crack length, and axial strain ε_{1wcg} are both rapid relatively increasing under high axial stress and keep constant under zero axial stress over time at each cyclic loading and unloading, the accumulated wing crack length and axial strain ε_{1wcg} are increasing to final failure value over time, and the wing crack growth—induced axial strain ε_{1wcg} is irreversible. In Fig. 7.4c, for each cyclic loading and unloading, the pure viscoelastic axial strain is decelerated increasing under high axial stress and is decelerated decreasing under zero axial stress over time, and the variation of pure viscoelastic axial strain over time keeps identical.

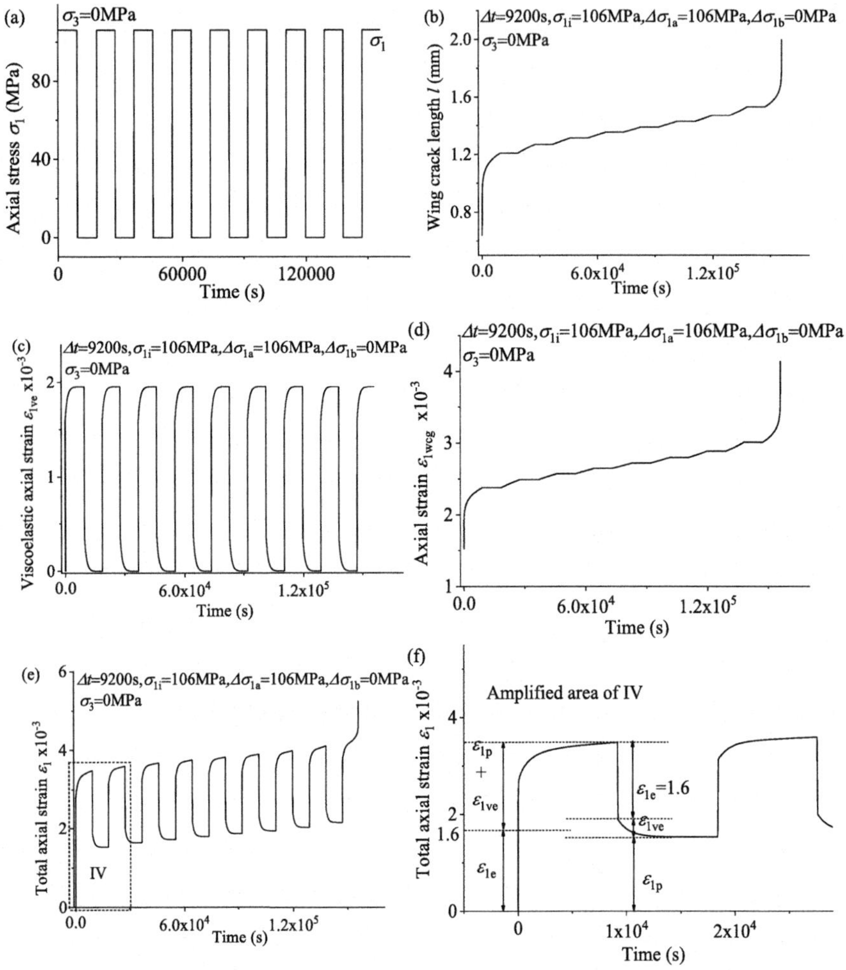

Fig. 7.4 Time-dependent **b** wing crack growth l, **c** viscoelastic axial strain ε_{1ve}, **d** wing crack growth-induced axial strain ε_{1wcg}, **e** total axial strain ε_1 under **a** cyclic static uniaxial loading and unloading. **f** The detailed illustration of the IV area in Fig. 7.4e

In Fig. 7.4e, the total axial strain containing visco-elastic–plastic deformation is obtained by combining the pure viscoelastic axial strain in Fig. 7.4c and the pure plastic axial strain in Fig. 7.4d. The total axial strain is increasing under high stress and is decreasing under zero axial stress over time for each cycle, the maximum of axial strain for each static loading increases with increasing cyclic number, the minimum of axial strain for each zero axial stress increases with increasing cyclic number, and the accumulated axial strain is increasing to final failure strain. Fig. 7.4f shows the detailed formulation of visco-elastic- plastic strain over time during each loading and unloading. Under the stress path in Fig. 7.4a, the elastic and viscoelastic strains are recoverable, and the plastic strain is irrecoverable, which agrees well with the published experimental results (Grgic and Amitrano 2009; He et al. 2010; Zivaljevic and Tomanovic 2015; Roberts et al. 2015; Chen et al. 2018).

As shown in Fig. 7.5, the time-dependent confined fatigue creep failure is obtained. Under cyclic static loading and unloading of axial stress in Fig. 7.5a, the variations of wing crack growth, pure viscoelastic axial strain $\varepsilon_{1\text{ve}}$, wing crack growth-induced axial strain $\varepsilon_{1\text{wcg}}$, and total visco- elastic- plastic axial strain ε_1 over time are shown in Figs. 7.5b–e respectively. In Fig. 7.5a, the maximum of applied constant axial stress is 279 MPa, and the minimum of applied constant axial stress is 189 MPa, which are both larger than the crack initiation stress $\sigma_{1\text{ci}}$ and are smaller than the peak stress $\sigma_{1\text{peak}}$ in Fig. 7.2. The time-dependent wing crack growth in Fig. 7.5b causes the time-dependent plastic axial strain $\varepsilon_{1\text{wcg}}$ in Fig. 7.5d, the wing crack length and axial strain $\varepsilon_{1\text{wcg}}$ are both rapid increasing relatively under high axial stress and slow increasing relatively under low axial stress over time at each cyclic loading and unloading, the accumulated wing crack length and axial strain $\varepsilon_{1\text{wcg}}$ are increasing to final failure value over time, and the wing crack growth -induced axial strain $\varepsilon_{1\text{wcg}}$ is irreversible. In Fig. 7.5c, for each cyclic loading and unloading, the pure viscoelastic axial strain is decelerated increasing under high axial stress and is decelerated decreasing under low axial stress over time, and the variation of pure viscoelastic axial strain over time presumptively keeps identical. In Fig. 7.5e, the total axial strain containing visco-elastic- plastic deformation is obtained by combining the axial strain in Fig. 7.5c, d. The total axial strain is increasing under high stress and is decreasing under low stress over time for each cycle, the maximum of axial strain for each static loading increases with increasing cyclic number, the minimum of axial strain for each static unloading increases with increasing cyclic number, and the accumulated axial strain is increasing to final failure strain.

Figure 7.5f shows the detailed formulation of visco-elastic- plastic strain over time during each loading and unloading, which agrees well with the published experimental results (Hu et al. 2019). Under the applied stress path in Fig. 7.5a, the local elastic and viscoelastic strains (i.e., $\varepsilon_{1\text{e}1}$ and $\varepsilon_{1\text{ve}1}$) are recoverable, and the other local elastic and viscoelastic strains (i.e., $\varepsilon_{1\text{e}2}$ and $\varepsilon_{1\text{ve}2}$) and the plastic strain (i.e., $\varepsilon_{1\text{p}}$) are irrecoverable. The irrecoverable elastic and viscoelastic strains are caused by the residual static axial stress after unloading, and this residual axial stress is 189 MPa in Fig. 7.5a, i.e., this irrecoverable elastic strain $\varepsilon_{1\text{e}2}$ can be solved by inserting this residual stress into the first item of Eq. (7.8), and the irrecoverable viscoelastic strain can be obtained by the use of Eqs. (7.10) and (7.11). Furthermore, loading elastic strain $\varepsilon_{1\text{e}}$ is equal to the sum of unloading elastic strain $\varepsilon_{1\text{e}1}$ and residual elastic strain $\varepsilon_{1\text{e}2}$, and loading viscoelastic strain $\varepsilon_{1\text{ve}}$ is equal to the sum of unloading viscoelastic strain $\varepsilon_{1\text{ve}1}$ and residual viscoelastic strain $\varepsilon_{1\text{ve}2}$.

Figure 7.6 shows time-dependent confined fatigue creep failure under increasing cyclic loading and unloading. Under increasing cyclic loading and unloading of axial stress in Fig. 7.6a, the variations of wing crack growth, pure viscoelastic axial strain $\varepsilon_{1\text{ve}}$, wing crack growth-induced axial strain $\varepsilon_{1\text{wcg}}$, and total visco-elastic- plastic axial strain ε_1 over time are shown in Figs. 7.6b–e respectively.

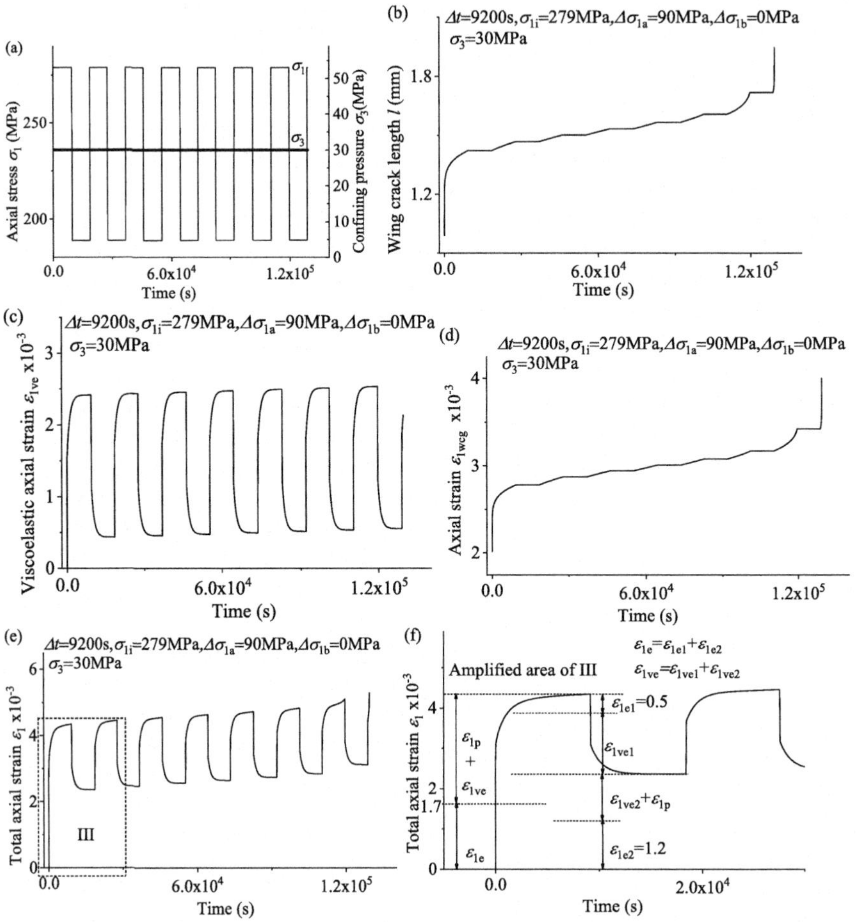

Fig. 7.5 Time-dependent **b** wing crack growth l, **c** viscoelastic axial strain $\varepsilon_{1\mathrm{ve}}$, **d** wing crack growth-induced axial strain $\varepsilon_{1\mathrm{wcg}}$, **e** total axial strain ε_1 under **a** constant confining pressure and cyclic static loading and unloading of axial stress. **f** The detailed illustration of the III area in Fig. 7.6e

In Fig. 7.6a, the maximum of applied constant axial stress at first cyclic loading is 250 MPa, and the maximum of applied constant axial stress increases by 2 MPa with increasing cyclic number; the minimum of applied constant axial stress at first cyclic unloading is 45 MPa, and the minimum of applied constant axial stress increases by 2 MPa with increasing cyclic number. The maximum of applied axial stress is larger than the crack initiation stress $\sigma_{1\mathrm{ci}}$ and is smaller than the peak stress $\sigma_{1\mathrm{peak}}$ in Fig. 7.2; the minimum of applied axial stress is smaller elastic axial stress $\sigma_{1\mathrm{w}}$, which causes the rebound of plastic axial strain in Fig. 7.6d from close and recovery of extended wing crack in Fig. 7.6b.

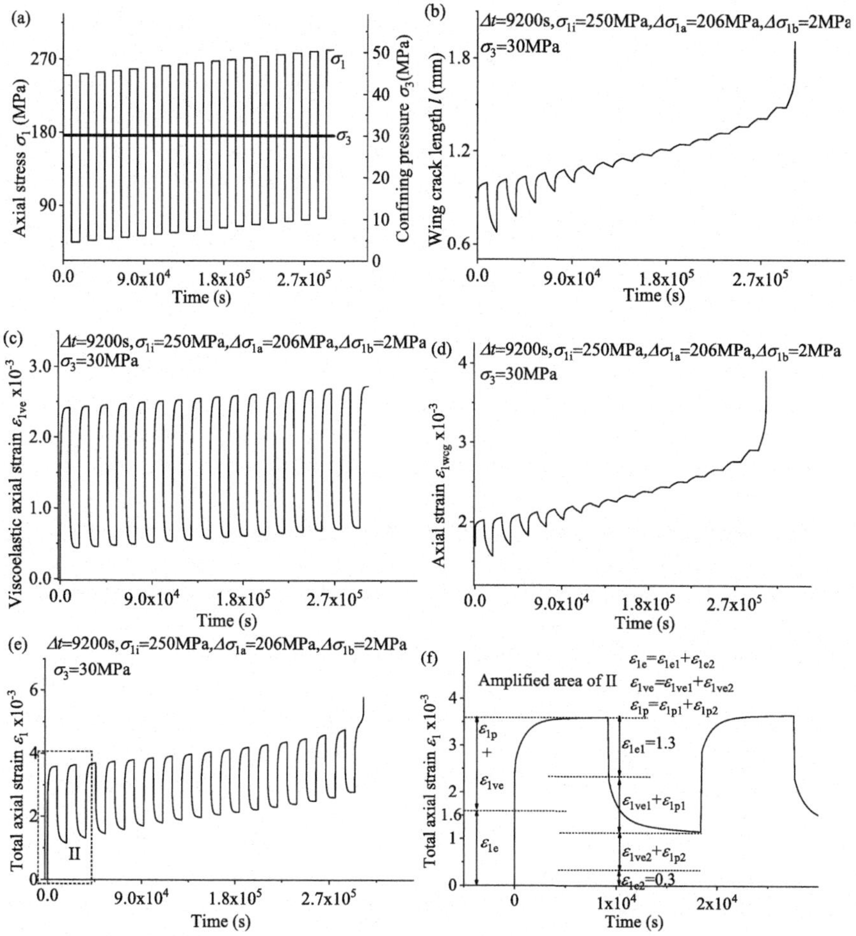

Fig. 7.6 Time-dependent **b** wing crack growth l, **c** viscoelastic axial strain ε_{1ve}, **d** wing crack growth-induced axial strain ε_{1wcg}, **e** total axial strain ε_1 under **a** constant confining pressure and cyclic increasing static loading and unloading of axial stress. **f** The detailed illustration of the II area in Fig. 7.6e

The wing crack length in Fig. 7.6b and axial strain ε_{1wcg} in Fig. 7.6d are both decelerated increasing under high axial stress and are decelerated decreasing under low axial stress over time at each cyclic loading and unloading. The accumulated wing crack length and axial strain ε_{1wcg} are increasing to final failure value over time. The decrement of wing crack length and axial strain ε_{1wcg} are decreasing with increasing cyclic number, which is because of the increasing constant axial stress of unloading with increasing cyclic number. The wing crack-induced axial strain ε_{1wcg} is reversible, which is because of the presence of confining pressure, i.e., inserting the applied constant axial stress after unloading and the confining pressure into Eq. (5.1), the wedge force is a compressive force which causes the close of extended wing crack. In other words, the reversible wing crack growth-induced axial strain ε_{1wcg} could not happen at uniaxial cyclic compressive loading and unloading.

In Fig. 7.6c, for each cyclic loading and unloading, the pure viscoelastic axial strain is decelerated increasing under high axial stress and is decelerated decreasing under low axial stress over time. The maximum of pure viscoelastic axial strain at each cyclic loading and the minimum of pure viscoelastic axial strain at each cyclic unloading are both increasing over time.

In Fig. 7.6e, the total axial strain containing visco-elastic–plastic deformation is obtained by combining the axial strain in Fig. 7.6c, d. The total axial strain is increasing under high stress and is decreasing under low stress over time for each cycle, the maximum of axial strain for each cyclic loading increases with increasing cyclic number, the minimum of axial strain for each cyclic unloading increases with increasing cyclic number, and the accumulated axial strain is increasing to final failure strain. The rebound of axial strain is decreasing with increasing cyclic number, which is because of the increment of minimum in unloading with increasing cyclic number in Fig. 7.6d.

Figure 7.6f shows the detailed formulation of visco-elastic- plastic strain over time during each loading and unloading. Under the applied stress path in Fig. 7.6a, the rebound of axial strain consists of local elastic strain ε_{1e1}, local viscoelastic strain ε_{1ve1} and local wing crack close-induced plastic strain ε_{1p1}; the irrecoverable axial strain consists of local elastic strain ε_{1e2}, local viscoelastic strain ε_{1ve2} and local wing crack growth-induced plastic strain ε_{1p2}. The irrecoverable elastic and viscoelastic strains are caused by the residual static axial stress after unloading, the recoverable plastic strain is caused by the presence of confining pressure. The loading elastic axial strain ε_{1e} is equal to the sum of the unloading elastic strain ε_{1e1} and residual elastic strain ε_{1e2}, loading viscoelastic axial strain ε_{1ve} is equal to the sum of the unloading viscoelastic strain ε_{1ve1} and residual viscoelastic strain ε_{1ve2}, and loading plastic axial strain ε_{1p} is equal to the sum of the irrecoverable plastic strain ε_{1p2} and recoverable plastic strain ε_{1p1}.

Furthermore, a comparison between the theoretical results and experimental results in the variation of normalized axial strain and normalized time during fatigue creep is shown in Fig. 7.7. The theoretical variation tendency of axial strain over time is similar to that of the experimental, which approximately verifies the rationality of the proposed theoretical fatigue creep model. However, during the constant loading phase, the predicted total strain tends to be stable, while the experimental total strain keeps increasing. The differences between theoretical and experimental curves maybe because that (1) all crack sizes and crack angles are both assumed to a constant respectively, the spatial effects of randomly distributed microcracks and the failure caused by strain localization cannot be considered, and an average and equivalent method is used to analyze qualitatively the mechanical properties in real brittle solids, (2) the different durations and sizes of constant loading in the theoretical and experimental data, (3) the different material properties in the theoretical and experimental data or (4) the errors of experiment.

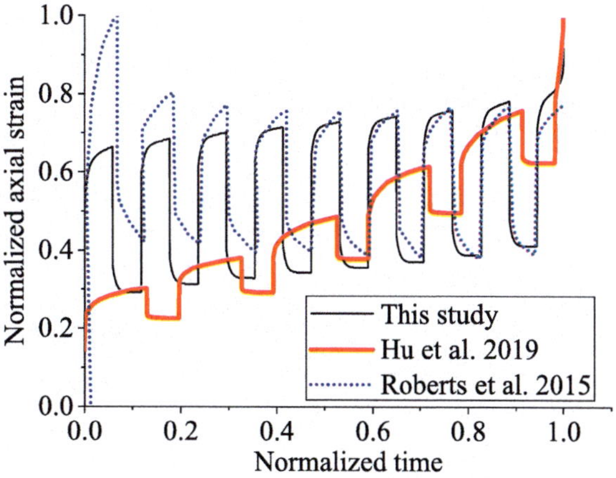

Fig. 7.7 A comparison between the theoretical and experimental (Roberts et al. 2015; Hu et al. 2019) results in the variation of normalized axial strain over normalized time during fatigue creep

7.3.2 Sensitivity Analysis of Model Parameters

From the above studies, the time-dependent fatigue creep failure explaining wing crack growth- and close-induced visco- elastic- plastic deformation under cyclic static loading and unloading is proposed. The time-dependent rebound phenomenon of viscoelastic deformation under cyclic loading and unloading was widely studied, but the time-dependent rebound phenomenon of plastic deformation under cyclic static loading and unloading was rarely studied, and the mechanism of rebound of plastic deformation was not clearly explained. Thus, ignoring the rebound phenomenon of viscoelastic deformation, the mechanical properties of brittle solids from wing crack growth- and close-induced axial strain ε_{1wcg} will be studied independently in detail below.

Figure 7.8 shows the time-dependent curves of axial strain ε_{1wcg} during fatigue creep failure under different parameter $\Delta\sigma_{1a}$. Axial strain experiences the smoothly transient, steady-state, and accelerated static deformation at $\Delta\sigma_{1a} = 0$ and $\Delta\sigma_{1b} = 0$. For the case of $\Delta\sigma_{1a} > 0$, $\Delta\sigma_{1b} = 0$, and $|\sigma_{1min}| > |\sigma_{1ci}|$ (note: $|\sigma_{1min}| = |\sigma_{1i}| - \Delta\sigma_{1a}$ in Fig. 7.1b), the axial strain has a roughly transient, steady-state, and accelerated static deformation. Furthermore, in Figs. 7.9 and 7.10, a special phenomenon of the irregularity of failure time with parameter $\Delta\sigma_{1a}$ and Δt is found, but the time to failure at the case of $\Delta\sigma_{1a} > 0$, $\Delta\sigma_{1b} = 0$ is larger than the time to failure at $\Delta\sigma_{1a} = 0$ and $\Delta\sigma_{1b} = 0$. The phenomenon shows that the cyclic static loading and unloading has an uncertain effect on fatigue creep failure time, but the failure time under cyclic static loading and unloading is always larger than that of constant static loading.

Furthermore, for some cases of $\Delta\sigma_{1a} > 0$, $\Delta\sigma_{1b} = 0$, and $|\sigma_{1min}| < |\sigma_{1ci}|$ (for instance $\Delta\sigma_{1a} = 120$ MPa and 140 MPa in Fig. 7.8), the fatigue creep failure also appears. Based on the proposed theoretical model, the fatigue creep failure could not appear when $\Delta\sigma_{1a}$ is larger than a critical value $\Delta\sigma_{1al} = 150$ MPa, i.e., $|\sigma_{1min}| < |\sigma_{1scf}| = |\sigma_{1i}| - \Delta\sigma_{1al} < |\sigma_{1ci}|$.

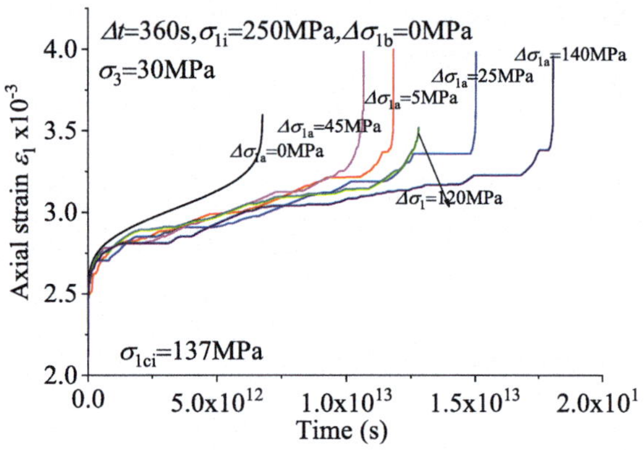

Fig. 7.8 Evolution curves of axial strain $\varepsilon_{1\text{wcg}}$ during cyclic creep failure under different parameter $\Delta\sigma_{1a}$ ($\Delta\sigma_{1b} = 0$)

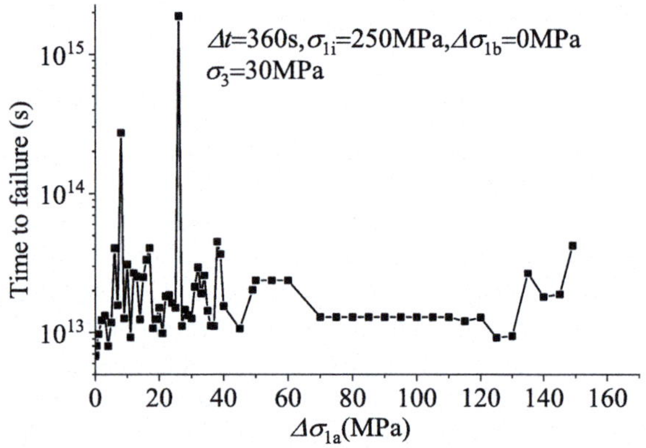

Fig. 7.9 Effect of parameter $\Delta\sigma_{1a}$ on time to failure under cyclic fatigue creep

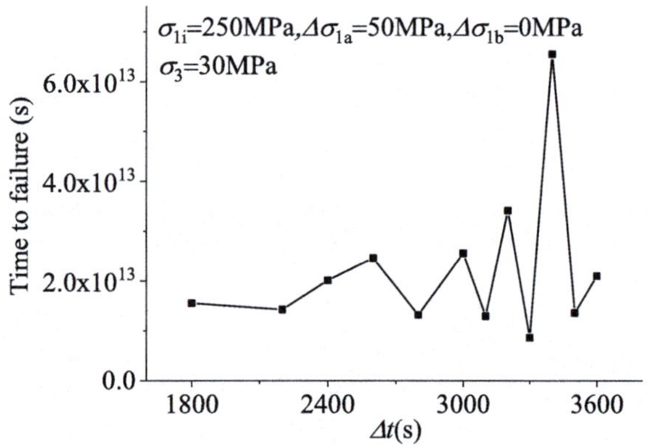

Fig. 7.10 Effect of parameter Δt on time to failure under cyclic fatigue creep

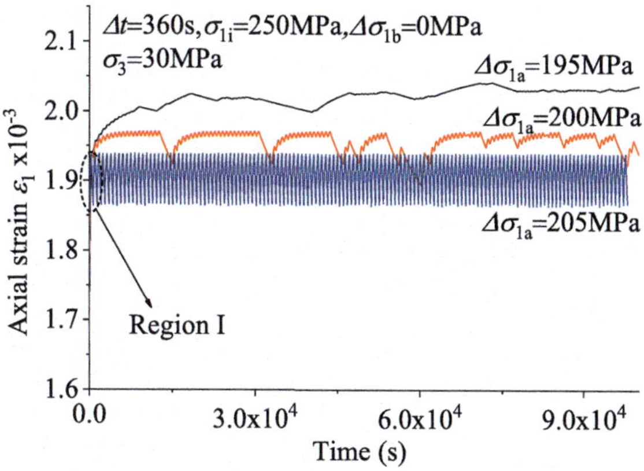

Fig. 7.11 Evolution curves of axial strain under different parameter $\Delta\sigma_{1a}$ during cyclic fatigue creep. *Note* that creep failure does not happen

However, the plastic rebound phenomenon does not appear in Fig. 7.8. In Fig. 7.11, the rebound phenomenon for axial strain curve is more regular with the increment of $\Delta\sigma_{1a}$, and the rebound strain is increasing with the increment of $\Delta\sigma_{1a}$. In other words, the rebound phenomenon appears when $\Delta\sigma_{1a}$ is larger than a critical value $\Delta\sigma_{1a2}$ (i.e., $|\sigma_{1\min}| < |\sigma_{1w}| = |\sigma_{1i}| - \Delta\sigma_{1a2}$), and the obvious and regular curve of rebound phenomenon appears when $\Delta\sigma_{1a}$ is larger than another critical value $\Delta\sigma_{1a3}$ (i.e., $|\sigma_{1\min}| < |\sigma_{1rr}| = |\sigma_{1i}| - \Delta\sigma_{1a3}$). Furthermore, the fatigue creep failure cannot be happened at $|\sigma_{1\min}| < |\sigma_{1w}|$.

Thus, a conclusion can be obtained that $\Delta\sigma_{1a1} < \Delta\sigma_{1a2} < \Delta\sigma_{1a3}$ and $|\sigma_{1ci}| > |\sigma_{1scf}| > |\sigma_{1w}| > |\sigma_{1rr}|$, where σ_{1ci} is a critical stress of wing crack initiation, σ_{1scf} is a critical axial stress of minimum applied stress $\sigma_{1\min}$ causing static cyclic failure, σ_{1w} is a critical stress causing rebound of plastic strain and recovery of wing crack, and σ_{1rr} is a critical stress causing regular rebound of plastic strain.

Figure 7.12 shows the detailed rebound phenomenon of plastic axial strain. The axial strain is decelerated increasing firstly at static loading phase and is decelerated decreasing at static unloading phase. Furthermore, the wing crack length experiences a decelerated growth firstly at static loading phase and experiences a decelerated decrement (i.e., crack recovery) at static unloading phase. The macroscopic evolution of axial strain is corresponding to the evolution of wing crack length under cyclic static loading and unloading. The analytical solution of crack recovery has a great and direct influence on the rebound plastic strain in brittle solids, which provides an interesting and important meaning for evaluating the cyclic fatigue creep failure in brittle solids containing numerous microcracks. It is noteworthy that the phenomenon of crack recovery is caused by the presence of confining pressure.

From the study above, for cyclic static deformation at the case of $\Delta\sigma_{1a} > \Delta\sigma_{1a1}$=150 MPa (i.e., $|\sigma_{1\min}| <$ 100 MPa) and $\Delta\sigma_{1b} = 0$, the solid failure could not happen. However, for the case of $\Delta\sigma_{1a} >$ 150MPa, a smaller increment of $\Delta\sigma_{1b}$ can cause the failure of solids, which will be discussed in details in Fig. 7.13.

Figures 7.13a–c show the evolution curves of axial strain during cyclic increasing static loading and unloading at $\Delta t = 360$ s. The axial strain experiences the multi-

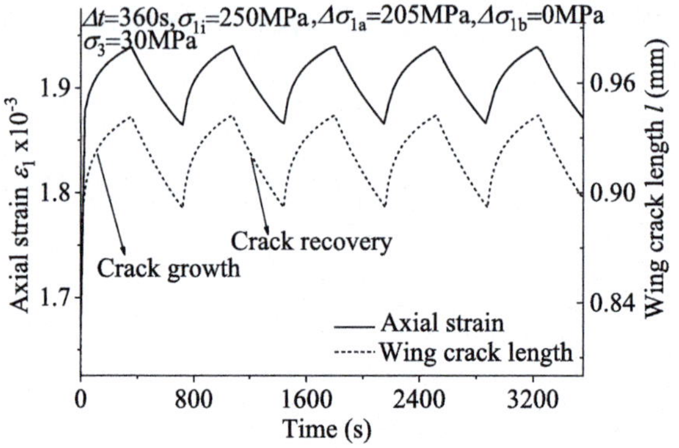

Fig. 7.12 Enlargement of rebound phenomenon of Region I in Fig. 7.12 for $\Delta\sigma_{1a} = 205$ MPa, and the corresponding relationship between axial strain and wing crack length

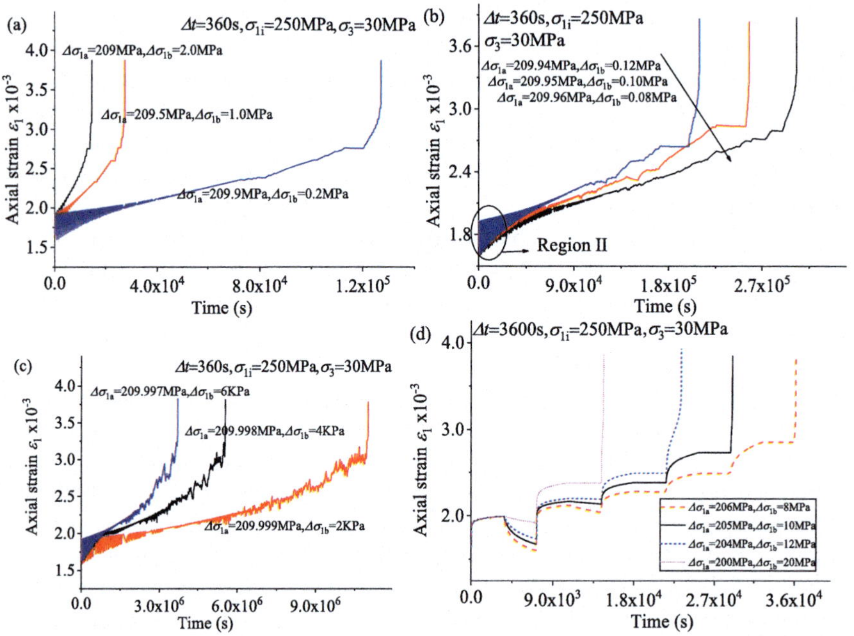

Fig. 7.13 Evolution curves of axial strain ε_{1wcg} during cyclic increasing static loading and unloading under different parameter $\Delta\sigma_{1a}$, $\Delta\sigma_{1b}$ and Δt. *Note* $\Delta\sigma_{1a} + \Delta\sigma_{1b}/2 = 210$ MPa

rebound behaviors firstly, and the accelerated creep failure happens at the last loading. Furthermore, with the increasing cyclic number, the amplitude of rebound strain is decreasing, the maximum of strain at each static loading phase is increasing, and the minimum of strain at each static unloading phase is increasing, which can be clearly observed in Fig. 7.14. Furthermore, in Fig. 7.13c, for the case that $\Delta\sigma_{1b}$ accesses

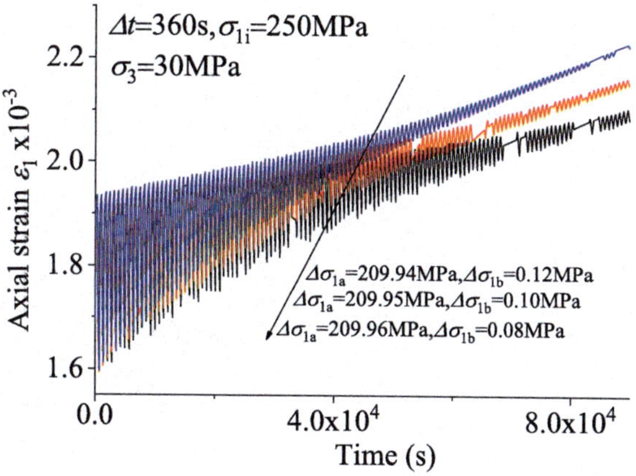

Fig. 7.14 Enlargement of Region II in Fig. 7.13b

to zero, the amplitude of rebound strain is decreasing firstly, and then is irregularly increasing at accessing to solid failure with increment of cyclic number.

Figure 7.13d also shows clearly the evolution of axial strain under cyclic increasing static loading at $\Delta t = 3600$ s. Comparing Fig. 7.13d with Fig. 7.13a–c, the lesser cyclic number causing solid failure has a more regular evolution curve before the last accelerated deformation. The irregularity of the evolution of axial strain often happens at the larger cyclic number causing solid failure. Furthermore, the value of rebound strain is decreasing with the increasing cyclic number in Fig. 7.13d.

The time to failure is acceleratedly increasing with the increment of $\Delta\sigma_{1b}$ and decrement of $\Delta\sigma_{1a}$ (Fig. 7.15), and is linearly increasing with an increment of Δt (Fig. 7.16). The time to failure tends to infinity when $\Delta\sigma_{1a}$ and $\Delta\sigma_{1b}$ access to 210 MPa and 0 MPa respectively, which shows that solid failure will hardly happen at that of stress state. Comparing with Figs. 7.9 and 7.15 or Figs. 7.10 and 7.16, it is concluded that the variations of time to failure with $\Delta\sigma_{1a}$, $\Delta\sigma_{1b}$ or Δt are more regular under the cyclic increasing static loading and unloading.

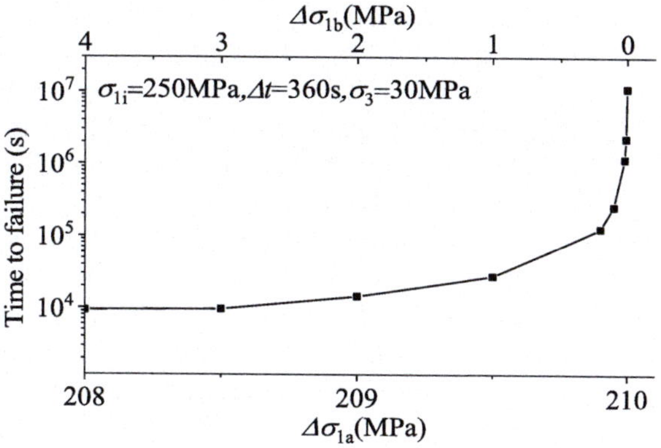

Fig. 7.15 Effect of parameter $\Delta\sigma_{1a}$ and $\Delta\sigma_{1b}$ on time to failure during cyclic increasing static loading and unloading

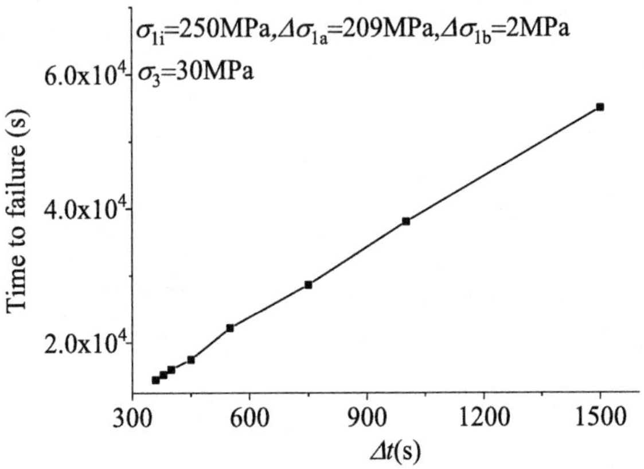

Fig. 7.16 Effect of parameter Δt on time to failure during cyclic increasing static loading and unloading

7.4 Conclusions

Based on the confined cyclic static loading and unloading path, the Hooke-Kelvin viscoelastic model, and the micro–macro mechanical model proposed by Chap. 5, an analytical confined fatigue creep failure model in brittle solids containing numerous microcracks is proposed. The total visco-elastic–plastic deformation relating to the microcrack variable is explained. Irrecoverable elastic and viscoelastic deformations caused by residual static stress after unloading are explained. A dramatic plastic rebound phenomenon induced by microcrack recovery is found. The conclusions are as follows:

(1) Some critical axial stress values (i.e., σ_{1w}, σ_{1ci}, σ_{1scf}, σ_{1rr}, and σ_{1peak}) are found. σ_{1w} is a critical stress causing the pure elastic strain, the rebound of plastic strain and the recovery of wing crack, σ_{1ci} is a critical stress of plastic strain and wing crack initiation, σ_{1scf} is a critical axial stress of minimum applied stress σ_{1min} causing failure of fatigue creep, σ_{1rr} is a critical stress causing regular rebound of plastic strain, σ_{1peak} is compressive strength of brittle solid materials. The order of size is $|\sigma_{1peak}| > |\sigma_{1ci}| > |\sigma_{1scf}| > |\sigma_{1w}| > |\sigma_{1rr}|$. The increment and rebound of plastic deformation are corresponding to the wing crack growth and recovery in brittle solids respectively. Under cyclic uniaxial loading and unloading, the critical values of σ_{1w} and σ_{1scf} are both equal to zero, there is no critical value of σ_{1rr}, there is no recovery of wing crack, and there is no rebound phenomenon of plastic deformation.
(2) Under cyclic uniaxial loading and exhausted unloading, the elastic and viscoelastic strains are recoverable, and the plastic strain is irrecoverable, the loading elastic and viscoelastic strains are equal to the unloading elastic and viscoelastic strains. Under confined cyclic loading and unloading (and there is residual stress after unloading and the residual stress is larger than critical stress σ_{1w}), loading elastic strain ε_{1e} is equal to sum of unloading elastic strain ε_{1e1} and residual elastic strain ε_{1e2}, and loading viscoelastic strain ε_{1ve} is equal to

sum of unloading viscoelastic strain ε_{1ve1} and residual viscoelastic strain ε_{1ve2}. Under confined cyclic loading and unloading (and there is residual stress after unloading and the residual stress is smaller than critical stress σ_{1w}), the loading elastic axial strain ε_{1e} is equal to the sum of the unloading elastic strain ε_{1e1} and residual elastic strain ε_{1e2}, loading viscoelastic axial strain ε_{1e} is equal to the sum of the unloading viscoelastic strain ε_{1ve1} and residual viscoelastic strain ε_{1ve2}, and loading plastic axial strain ε_{1p} is equal to the sum of the irrecoverable plastic strain ε_{1p2} and recoverable plastic strain ε_{1p1}.

(3) Under confined cyclic static loading and loading at $\Delta\sigma_{1a} > 0$, $\Delta\sigma_{1b} = 0$ and $\sigma_{1i} > \sigma_{1ci}$, the total visco-elastic–plastic axial strain experiences multiple viscoelastic increments and rebound accompanying the increasing plastic axial strain at $\sigma_{1min} > \sigma_{1w}$. The decelerated increasing total axial strain happens at the static loading phase, the decelerated decreasing total axial strain is happened at the static unloading phase, the accelerated increasing total axial strain (i.e., accelerated creep failure) happens at last static loading phase. The accumulated total axial strain is decelerated, steady-state and accelerated increasing over time. When the minimum of applied axial stress σ_{1min} under static unloading is smaller than a critical axial stress σ_{1scf}, the static cyclic failure cannot happen. When the minimum of applied axial stress σ_{1min} is smaller than a critical stress σ_{1w}, the rebound of plastic strain can happen, but the cyclic static failure cannot happen. When the minimum of applied axial stress σ_{1min} is smaller than a critical stress σ_{1rr}, the regular rebound of plastic strain can happen. The fatigue failure time under different parameter $\Delta\sigma_{1a}$ or Δt is irregular, but cyclic fatigue creep failure time is always larger than the static failure time at $\Delta\sigma_{1a} = 0$ and $\Delta\sigma_{1b} = 0$.

(4) Under confined and increasing cyclic static loading and loading at $\Delta\sigma_{1a} > 0$, $\Delta\sigma_{1b} > 0$, $\sigma_{1i} > \sigma_{1ci}$ and $\sigma_{1min} < \sigma_{1rr}$, the total visco-elastic–plastic axial strain experiences multiple viscoelastic increments and rebound accompanying with the multiple plastic increment and rebound. The decelerated increasing total axial strain happens at the static loading phase, the decelerated decreasing total axial strain is happened at static unloading phase, the accelerated increasing total axial strain (i.e., accelerated creep failure) happens at last static loading phase. The rebound of total axial strain is decreasing with the cyclic number, which is because of the increasing axial stress σ_{1min} (i.e., the minimum of applied axial stress) at static unloading. When $\Delta\sigma_{1b}$ is accessing to zero, the amplitude of rebound plastic strain is decreasing firstly and is irregularly increasing at accessing to solid failure with an increment of cyclic number. The lesser cyclic number causing solid failure has a more regular evolution curve. For the much cyclic number, an irregular curve between strain and time appears in the whole failure process. The time to failure is accelerated increasing with the increment of $\Delta\sigma_{1b}$ and decrement of $\Delta\sigma_{1a}$ and is linearly increasing with increment of Δt.

References

Chen C, Xu T, Heap MJ, Baud P. Influence of unloading and loading stress cycles on the creep behavior of Darley Dale Sandstone. Int J Rock Mech Min Sci. 2018;112:55–63.

Grgic D, Amitrano D. Creep of a porous rock and associated acoustic emission under different hydrous conditions. J Geophys Res. 2009;114(B10).

He MC, Miao JL, Feng JL. Rock burst process of limestone and its acoustic emission characteristics under true-triaxial unloading conditions. Int J Rock Mech Min Sci. 2010;47(2):286–98.

Hu B, Yang SQ, Xu P, Cheng JL. Cyclic loading–unloading creep behavior of composite layered specimens. Acta Geophys. 2019;67:449–64.

Martin CD, Chandler NA. The progressive fracture of Lac du Bonnet granite. Int J Rock Mech Min Sci Geomech Abs. 1994;31(6):643–59.

Roberts LA, Buchholz SA, Mellegard KD, Düsterloh Uwe. Cyclic loading effects on the creep and dilation of salt rock. Rock Mech Rock Eng. 2015;48(6):2581–90.

Zivaljevic S, Tomanovic Z. Experimental research of the effects of preconsolidation on the time-dependent deformations—creep of marl. Mech Time-Dependent Mater. 2015;19:43–59.

Chapter 8
Chemical Effect of Water on Creep Fracture

The safety and stability of rock engineering are closely related to water, such as roadways, dams, tunnels, and underground chambers will inevitably be affected by groundwater. The physical, chemical, and mechanical properties will change when the brittle rock encounters water chemicals. The water content of water-bearing rock will significantly influence the brittle rock's water properties, such as water absorption, frost resistance, disintegration, expansion, and water permeability. Based on Chap. 5, this chapter will reveal the evolution law of a series of mechanical characteristics of water-bearing rocks, such as peak stress intensity, crack initiation stress, elastic modulus, creep failure time, and creep rate, by studying the influence of water content ω on rock micromechanical parameters.

8.1 Theory

The friction coefficient μ (Lu et al. 2018), fracture toughness K_{IC} (Niu et al. 2021), and initial damage D_0 (Yang 2021) relations of brittle rock under the influence of water content are as follows (see Fig. 8.1),

$$\mu(\omega) = a_1\omega + a_2 \tag{8.1}$$

$$K_{IC}(\omega) = b_1 - b_2\omega + b_3\omega^2 \tag{8.2}$$

$$D_0(\omega) = c_1\omega + c_2 \tag{8.3}$$

where b_1, b_2, b_3, c_1, and c_2 are rock micromechanical parameters related to rock microstructure.

X. Li et al., *Static Creep Micro-Macro Fracture Mechanics of Brittle Solids*,
https://doi.org/10.1007/978-981-99-8203-5_8

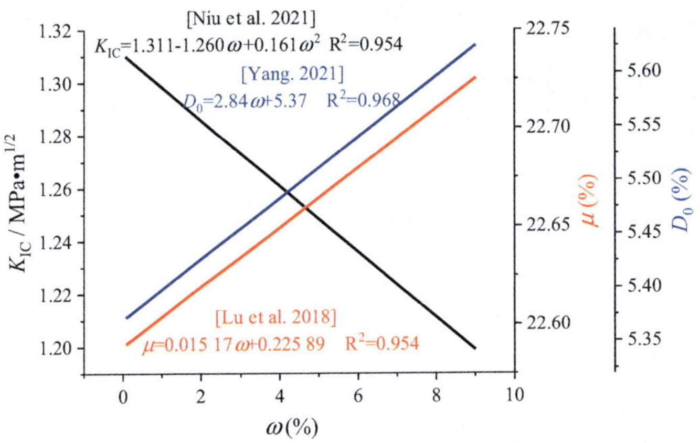

Fig. 8.1 Fracture toughness, friction coefficient, and initial damage under different water content

The influence of water content on characteristic crack rate v and stress erosion index n (Nara et al. 2010) can be expressed as (see Fig. 8.2),

$$n(\omega) = c_1\omega + c_2 \tag{8.4}$$

$$v(\omega) = \exp(d_1\omega - d_2)K_{\mathrm{IC}}^{n} \tag{8.5}$$

Equations 8.1–8.5 are put into the stress–strain relationship equation and creep evolution equation in Chapt. 5, which is explored the mechanical characteristics of progressive failure and creep failure of brittle rock under the water content.

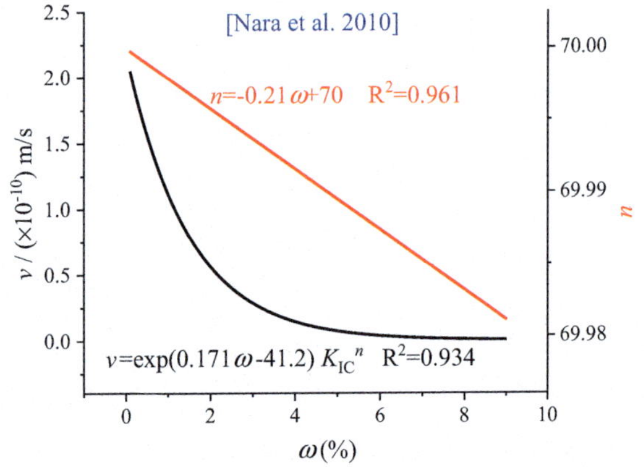

Fig. 8.2 Characteristic crack velocity and stress erosion index under different water content

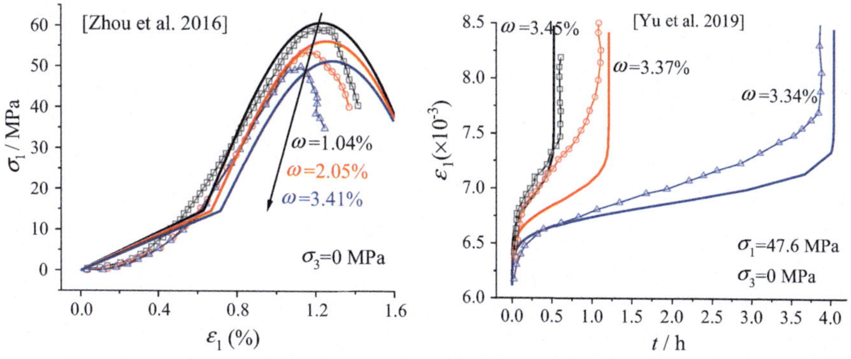

(a) Stress-strain curve of sandstone (b) Creep evolution curve of red sandstone

Fig. 8.3 Comparison of theoretical and experimental data of stress–strain and creep curves under different water content (Note: The solid lines are theoretical data, and the dotted lines are experimental data)

8.2 Results and Discussions

8.2.1 *Model Verification*

To verify the rationality of the established water content model, Fig. 8.3a compares the theoretical stress–strain of sandstone samples with the water content of 1.04, 2.05, and 3.41% and the experimental data (Zhou et al. 2016). Figure 8.3b shows the theoretical and experimental (Yu et al. 2019) comparison curve of the creep evolution process of red sandstone with water content of 3.45, 3.37, and 3.34%. The micromechanical parameters of sandstone and red sandstone is listed in Table 8.1. The comparison between theoretical and experimental data is shown in Table 8.2. The error between theoretical and practical data of curve feature points is small. There is a certain comparability between the two groups of curves under the influence of overall trend and water content. Therefore, the stress–strain relationship and creep evolution equation of water-bearing brittle rock proposed in this chapter are reasonable.

8.2.2 *The Relationship Between Water Content and Special Values in Stress–Strain*

The influence curves of water content on crack initiation stress, elastic modulus, and peak stress of brittle rock are shown in Figs. 8.4, 8.5, and 8.6. The special value of the stress–strain relationship of brittle rock decreases with the increase in water content.

For example, when the water content is 3.41%, the crack initiation stress, elastic modulus, and peak stress of sandstone specimens are 99.8%, 82.2%, and 77.9% of

Table 8.1 Selection results of micromechanical parameters or their relationship with the water content of sandstone and red sandstone

Parameter	Sandstone	Red sandstone	Note (%)
m	2	2	
ε_0	1/33	1/61	
a/mm	2.87	4.57	
$\varphi/°$	34	40	
β	0.20	0.201	
n	$0.41\omega + 19$	$-0.21\omega + 159$	
D_0 (%)	$49.84\omega + 3.71$	$49.84\omega + 2.71$	
v/m s^{-1}	$\exp(0.971\omega\text{-}21.1)K_{IC}{}^n$	$\exp(0.171\omega\text{-}41.2)K_{IC}{}^n$	
K_{IC}/MPa·m$^{1/2}$	$1.091–0.106\omega + 0.161\omega^2$	$1.231–0.106\omega + 0.161\omega^2$	
μ	$0.015\ 17\omega + 0.225\ 98$	$0.015\ 17\omega + 0.269\ 09$ $0.015\ 17\omega + 0.273\ 98$ $0.015\ 17\omega + 0.280\ 88$	$\omega = 3.45$ $\omega = 3.37$ $\omega = 3.34$

Table 8.2 Comparison of theoretical and experimental data on characteristic mechanical parameters

Curve type	Special value	Water content (%)	Theoretical data	Experimental data	Error (%)
Stress–strain curve	Peak stress/MPa	1.04	60.72	58.83	3.21
		2.05	56.36	53.66	5.03
		3.41	51.51	50.08	2.86
	Elastic modulus/GPa	1.04	2.33	2.19	6.39
		2.05	2.20	2.15	2.33
		3.41	2.05	2.10	– 2.38
Creep evolution curve	Creep failure time/s	3.34	14,539	13,981	3.99
		3.37	4349	3985	9.14
		3.45	1875	2146	– 12.6

those in the dry state (i.e., $\omega = 0\%$), and the decreases are 0.2%, 17.8%, and 22.1% respectively. The compressive strength of water-bearing rock is more sensitive than the crack initiation stress under the action of water, and the elastic modulus is also greatly affected.

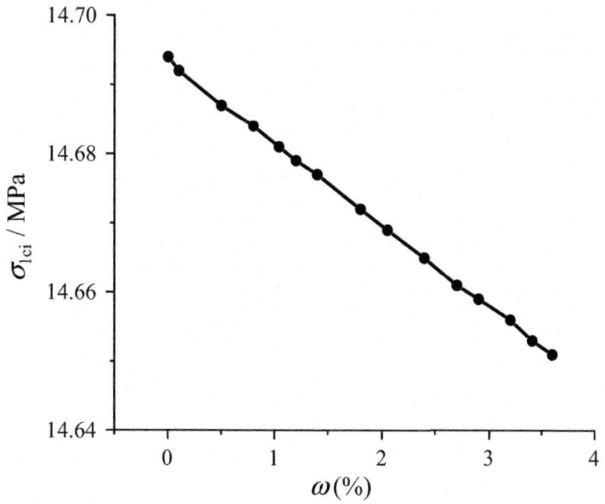

Fig. 8.4 Effect of water content on initiation stress of crack

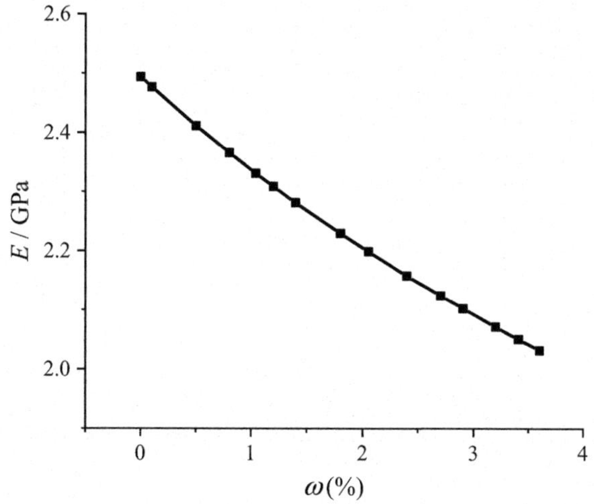

Fig. 8.5 Effect of water content on the elastic modulus

8.2.3 Creep Failure Evolution of Brittle Rock Under the Influence of Water Content

The stability of rock mass engineering under the load and water environment is often affected by the long-term mechanical properties of rock. Therefore, the creep evolution law of water-bearing rock (in Fig. 8.7) is discussed in this chapter. It can be seen from the evolution curve of axial strain over time (in Fig. 8.7a) that the higher the water content, the larger the axial strain. In Fig. 8.8, the higher the water content, the shorter the creep failure time. It shows that the creep strain of rock under the action of water is large, and creep failure occurs more quickly. The creep rate evolution curve of brittle rock with time under the influence of different water content is shown in

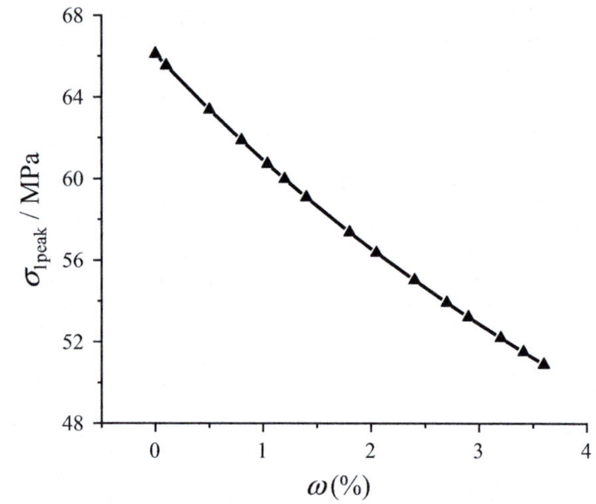

Fig. 8.6 Effect of water content on the peak stress

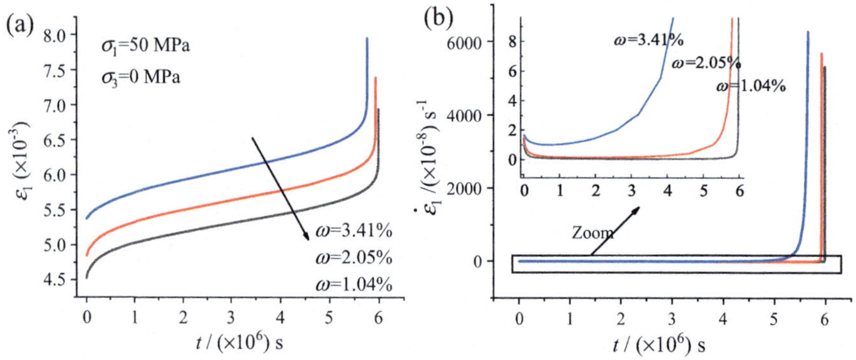

Fig. 8.7 Curve of time-dependent **a** axial strain and **b** strain rate under different water contents

Fig. 8.7b. It can be found that the steady-state creep rate of brittle rock gradually increases with the increase of water content, and the greater the creep rate, the more pronounced the effect (in Fig. 8.9).

8.3 Conclusions

Based on the wing crack extension model, the subcritical crack law, and the relationship between crack and strain, this chapter introduces the water content and the micromechanical parameters of brittle rock. It obtains a micro–macro mechanical model of creep failure of brittle rock considering water content.

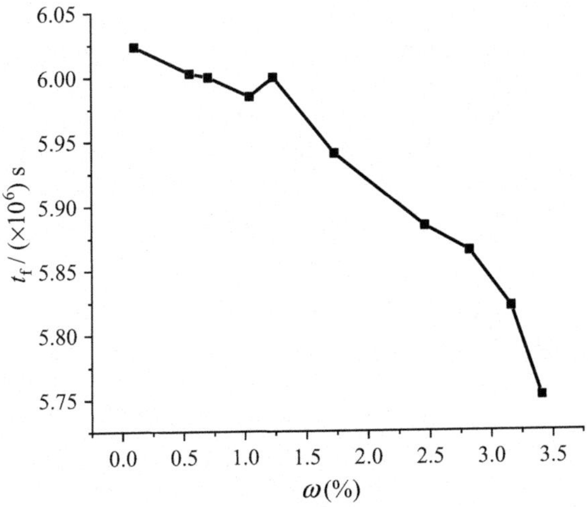

Fig. 8.8 Relationship between water content and creep failure time

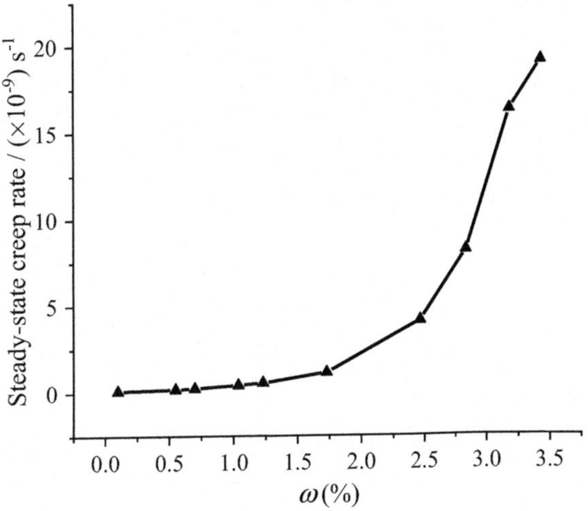

Fig. 8.9 Relationship between water content and steady-state creep rate

(1) The water content mainly greatly influences the micromechanical parameters of brittle rocks, such as fracture toughness, initial damage, friction coefficient, stress erosion index, and characteristic crack propagation velocity. The corresponding relationship equations are obtained.
(2) The water content and stress–strain constitutive relationship of brittle rock were mainly analyzed, including crack initiation stress, peak stress, and elastic modulus. The crack initiation stress, peak stress, and elastic modulus gradually decreased with the increase in water content, and the compressive strength of water-bearing rock is the most sensitive.

(3) Through the discussion of the creep evolution law of water-bearing rocks, it is found that the higher the water content, the greater the axial strain, the higher the moisture content, and the shorter the creep failure time of the rock. It shows that the creep strain of rock under the action of water is large, and creep failure is more likely to occur. The steady-state creep rate of brittle rock gradually increases with the increase of moisture content, and the greater the creep rate, the more obvious the effect.

References

Lu J, Zhang C, Zhong Y, Liu D. Research on macro-meso mechanical properties of sandy cobble surrounding rock with different moisture content. Chin J Undergr Space Eng. 2018;14(06):1564–70.

Nara Y, Takada M, Mori D, Owada H, Yoneda T, Kaneko K. Subcritical crack growth and long-term strength in rock and cementitious material. Int J Fract. 2010;164(1):57–71.

Niu C, Zhu Z, Wang F, Ying P, Deng S. Effect of water content on dynamic fracture characteristic of rock under impacts. KSCE J Civ Eng. 2021;25(1):37–50.

Yang DX. Acoustic emission behavior characteristics of rock micro-fracture evolution based on deep learning. Jiangxi University of Science and Technology, 2021. (In Chinese)

Yu C, Tang S, Tang C. Experimental investigation on the effect of water content on the short-term and creep mechanical behaviors of red sandstone. J China Coal Soc. 2019;44(02):473–81.

Zhou Z, Cai X, Cao W, Li X, Xiong C. Influence of water content on mechanical properties of rock in both saturation and drying processes. Rock Mech Rock Eng. 2016;49(8):3009–25.

Chapter 9
Constant Seepage Pressure Effect on Creep Fracture

High seepage pressure promotes the growth of microcracks in brittle rock, which is of great significance to the micro–macro mechanism of rock. Based on Chap. 5, this chapter proposes a micro–macro mechanism model of the progressive and creep failure of brittle rock, considering the crack propagation driven by seepage pressure. The model considers the influence of seepage pressure on the initial crack surface of the rock and the newly generated wing crack surface. The seepage pressure weakens the normal stress of the initial crack surface and strengthens the driving force of the newly generated wing crack propagation. The two kinds of seepage pressure on the crack surface formed the mechanism of crack growth driven by seepage pressure. In the theoretical analysis of the newly generated wing crack propagation driving force, a variable called F_{p} is introduced, which is caused by seepage pressure and plays a crucial role in inducing brittle rock crack propagation until failure due to seepage pressure.

9.1 Theory

9.1.1 Seepage Pressure Effect on Stress-Induced Crack and Strain Increment

Considering the seepage pressure P_{p}, the initial crack surface shear stress τ and normal stress σ_{n} in the compression rock are (Fig. 9.1a):

$$\tau = \frac{\sigma_3 - \sigma_1}{2} \sin 2\varphi \tag{9.1}$$

$$\sigma_{\mathrm{n}} = \frac{\sigma_1 + \sigma_3}{2} + \frac{\sigma_3 - \sigma_1}{2} \cos 2\varphi - P_{\mathrm{p}} \tag{9.2}$$

X. Li et al., *Static Creep Micro-Macro Fracture Mechanics of Brittle Solids*,
https://doi.org/10.1007/978-981-99-8203-5_9

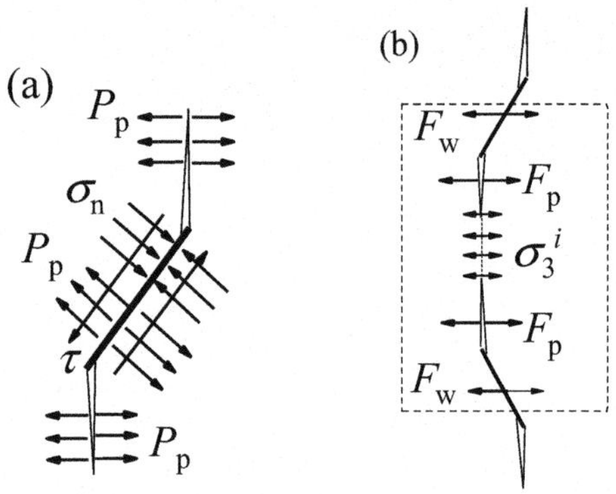

Fig. 9.1 Schematic of **a** single crack, **b** the force balance between cracks under high seepage pressure

Equation (9.2) explains that high seepage pressure leads to stress weakening in the direction of the initial crack surface of the rock, promotes the sliding of the initial crack surface, and then leads to the generation of wing crack propagation along the direction of the maximum principal stress σ_1. In this case, the existence of high seepage pressure has a strengthening effect on the crack propagation of the newly formed wing crack. The strengthening mechanism is explained by introducing the effective seepage pressure F_p parameter into the microcracks growth model. Seepage pressure P_p is denoted as negative in theoretical derivation.

By introducing the seepage pressure P_p in Fig. 9.1 into the wing crack model of Ashby and Sammis (1990), an improved stress intensity factor of the open crack at the tip of the wing crack considering the effect of seepage pressure is,

$$K_I = \frac{F_w}{[\pi(l+\beta a)]^{\frac{3}{2}}} + \frac{2}{\pi}(\sigma_3 + \sigma_3^i)\sqrt{\pi l} \tag{9.3}$$

where,

$$\sigma_3^i = \frac{F_w + F_p}{S - \pi(l+\alpha a)^2} \tag{9.4}$$

$$S = \pi^{1/3}\left[3/(4N_V)\right]^{2/3} \tag{9.5}$$

$$F_w = (\tau + \mu\sigma_n)\pi a^2 \sin\varphi \tag{9.6}$$

$$F_p = |P_p|\left[\pi(l+\alpha a)^2 - \pi(\alpha a)^2\right] \tag{9.7}$$

where the symbol of | | represents the absolute value, F_P is the tensile force acting on the newly generated wing crack surface caused by pore pressure in Fig. 9.1b and Eq. (9.7), and F_w is the wedge force at the initial crack caused by external loadings. Both forces are balanced by the internal stress $\sigma^i{}_3$ working on the surface between wing crack tips (i.e., Eq. (9.4) and Fig. 9.1b). Figure 9.2 shows the relationship

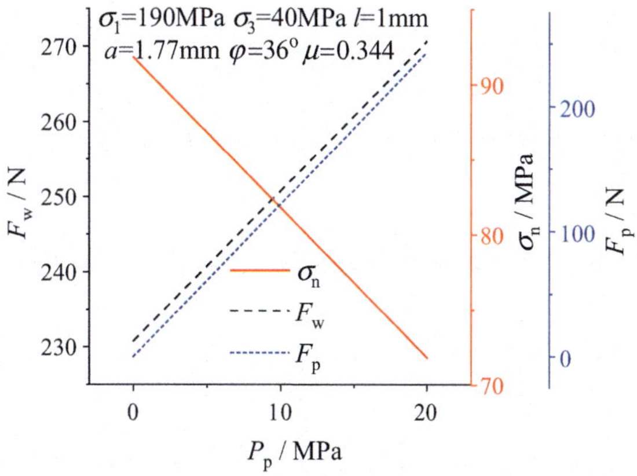

Fig. 9.2 Correlations of the normal stress and tensile force on the initial crack surface, the effective seepage force on the wing crack surface, and the seepage pressure

between seepage pressure and normal stress σ_n of the initial crack surface, tensile force F_w of the initial crack surface, and effective osmotic pressure F_p of the wing crack surface.

According to Eq. (9.3), the stress applied to brittle rock under seepage pressure and crack length is as follows,

$$\sigma_1(l) = \frac{K_{IC} - \sigma_3\left(A_1\pi a^2\chi + 2\sqrt{l/\pi}\right) + A_3 - A_4}{A_2\pi a^2\chi} \tag{9.8}$$

where,

$$A_1 = [(\sin 2\varphi + \mu + \mu\cos 2\varphi)\sin\varphi]/2 \tag{9.9}$$

$$A_2 = [(\mu - \mu\cos 2\varphi - \sin 2\varphi)\sin\varphi]/2 \tag{9.10}$$

$$A_3 = a^2\mu\pi\chi P_p \sin\varphi \tag{9.11}$$

$$A_4 = 2F_p\sqrt{l/\pi}[S - \pi(l + \alpha a)^2] \tag{9.12}$$

$$\chi = [\pi(l + \beta a)]^{-3/2} + 2\sqrt{l/\pi}[S - \pi(l + \alpha a)^2]^{-1} \tag{9.13}$$

The axial stress of rock crack initiation (i.e., $l = 0$) is,

$$\sigma_{1ci} = \left[\left(\beta^{3/2}K_{IC}\right)/\sqrt{a/\pi} - \sigma_3 A_1 + \mu P_p \sin\varphi\right]/A_2 \tag{9.14}$$

Equations (9.13) and (9.14) are judgment equations for peak stress and crack initiation stress of rocks under different confining pressures. These two critical values of stress provide essential help for the selection of constant stress in rock creep evolution. According to the theoretical method in Chaps. 2 or 5, the relationship between crack length and strain (i.e., Eq. (6.2) is combined with Eqs. (9.8) and (9.14), respectively, to obtain the stress–strain relationship under seepage pressure.

9.1.2 *Seepage Pressure Effect on the Creep Crack and Strain Evolution*

By substituting Eq. (9.3) into the subcritical crack growth velocity and crack intensity factor equation $dl/dt = v(K_I/K_{IC})^n$, the correlation of stress, crack length, crack growth velocity and time can be derived,

$$l' = \frac{dl}{dt} = v\left[\frac{\sigma_1 A_2 \pi a^2 \chi + \sigma_3\left(A_1 \pi a^2 \chi + 2\sqrt{l/\pi}\right) - A_3 + A_4}{K_{IC}}\right]^n \tag{9.15}$$

The time-dependent crack growth result $l(t)$ calculated by Eq. (9.15) is put into Eq. (6.2) of crack length and strain relation. When the applied stress is greater than crack initiation stress σ_{1ci} and less than rock strength σ_{1peak}, the axial creep strain induced by crack growth is,

$$\varepsilon_1(t) = 2E^{-1}\gamma|\sigma_3| + \varepsilon_o\left(-\ln\left[1 - D_o\left(l(t)/a + 1\right)^3\right]\right)^{\frac{1}{m}}, \left(\sigma_{1ci} < \sigma_1 < \sigma_{1peak}\right) \tag{9.16}$$

According to the rock compression test results (Jiang et al. 2014; Peng et al. 2013), seepage pressure has a significant influence on the mechanical properties of the rock in the plastic deformation stage but little influence on the mechanical properties of the rock in the elastic stage. When the axial stress applied to the rock is less than the crack initiation stress σ_{1ci}, the elastic strain of the rock can be obtained by ignoring the influence of the seepage pressure on the elastic modulus and Poisson's ratio,

$$\varepsilon_1 = |\sigma_1 - 2\gamma\sigma_3|/E, \quad (0 \le \sigma_1 \le \sigma_{1ci}) \tag{9.17}$$

Therefore, by combining the approximate linear elastic equation (i.e., Eq. (9.17)) about rock before crack propagation and the nonlinear creep equation (i.e., Eq. (9.16)) about rock after crack propagation, the complete creep evolution equation of rock can be written,

$$\varepsilon_1(t) = \begin{cases} |\sigma_1 - 2\gamma\sigma_3|/E, (0 \le \sigma_1 \le \sigma_{1c}) \\ 2E^{-1}\gamma|\sigma_3| + \varepsilon_o\left(-\ln\left[1 - D_o\left(l(t)/a + 1\right)^3\right]\right)^{\frac{1}{m}}, \\ \left(\sigma_{1c} < \sigma_1 < \sigma_{1peak}\right) \end{cases} \tag{9.18}$$

In addition, this chapter focuses on the influence of seepage pressure on the creep behavior of brittle rocks after crack initiation (i.e., $\sigma_{1c} < \sigma_1 < \sigma_{1peak}$).

9.2 Results and Discussions

9.2.1 Seepage Pressure Effect on Progressive Failure

In the past, mechanical parameters of Beishan granite (Hu et al. 2016) were selected to verify the rationality of the proposed stress–strain constitutive relationship of rock under seepage pressure. $D_0 = 0.013$, $\mu = 0.61$, $a = 3.8$ mm, $m = 2$, $\varepsilon_0 = 1/38$, $K_{IC} = 0.79$ MPa·m$^{1/2}$, $\varphi = 30.87°$, $\beta = 0.366$, $E = 30$GPa, $\gamma = 0.2$.

The stress–strain curves describing the strain-hardening and strain-softening phases under different pore pressures are illustrated in Fig. 9.3a, and the peak stress and strain of the stress–strain curve at the distinct pore pressures are drawn in Fig. 9.3b, c. The comparison between theoretical and experimental data (Hu et al. 2016) verifies the rationality of the progressive failure model under seepage pressure. Figure 9.3d shows the variation of crack length with axial stress, corresponding to the curve of axial stress and strain in Fig. 9.3a.

According to the stress–strain curve drawn based on the model proposed, Figs. 9.4 and 9.5 extract the crack initiation stress and peak strength values of rock under different parameters, respectively, for discussion. The crack initiation stress values in the rock under different confining pressures and fracture toughness are given in Fig. 9.4. The crack initiation stress decreases with the increase of seepage pressure. When the seepage pressure is constant, the crack initiation stress increases with the addition of confining pressure or fracture toughness. In Fig. 9.4a, when the confining pressure is 0, and the seepage pressure is greater than about 16 MPa, the rock initiation stress is approximately 0, which indicates that when the seepage pressure reaches a certain value, the rock internal crack initiation will also occur without applying external load.

The fracture toughness K_{IC} represents a maximum resistance value before the crack propagation of mode I crack, and the peak strength of rock represents the maximum compressive strength of rock. Both of them are important mechanical parameters of rock and have important significance for the evaluation of rock mechanical properties. Figure 9.5 shows the relationship between rock peak strength and seepage pressure under different confining pressures or fracture toughness. The peak strength of rock decreases with the increase of seepage pressure. For a certain seepage pressure, the peak strength increases with the increase of confining pressure or fracture toughness. In Fig. 9.5a, when the seepage pressure is 0, the peak strength is about 140 MPa, 270 MPa, and 430 MPa when the confining pressure is 0, 10 MPa, and 20 MPa, respectively.

By comparison with Figs. 9.4 and 9.5, a linear relationship exists between initial crack stress and seepage pressure, while a nonlinear relationship exists between peak

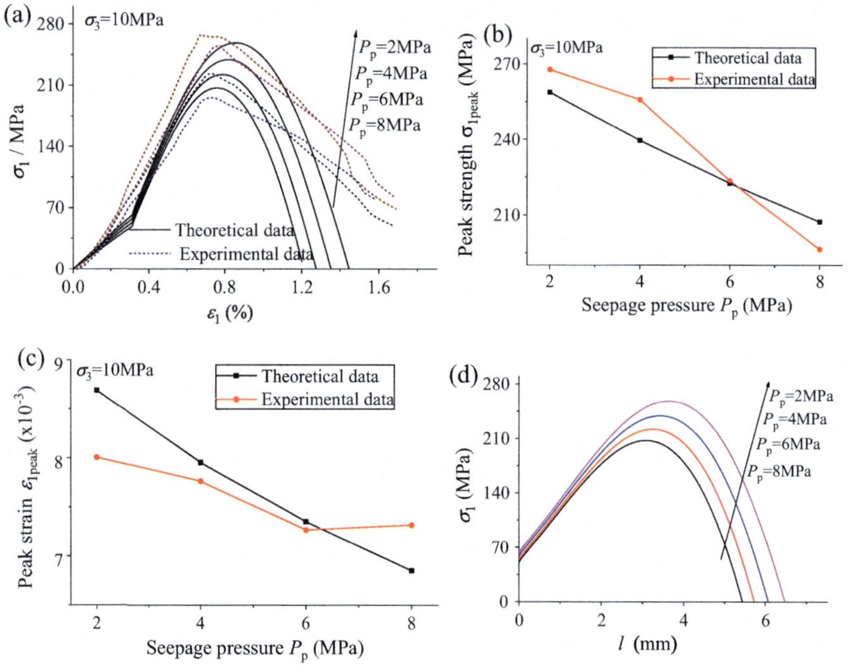

Fig. 9.3 The theoretical and experimental **a** stress–strain curves, **b** peak strength, and **c** peak strain, and **d** the theoretical stress-crack length relationship under different pore pressures

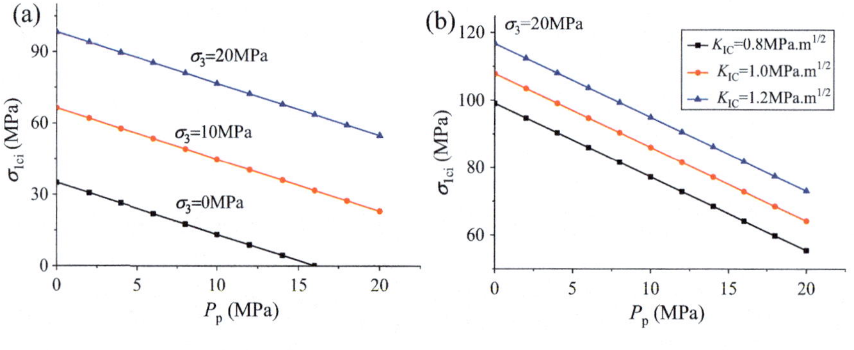

Fig. 9.4 Effects of **a** confining pressures or **b** fracture toughness on the relationship between seepage pressure and crack initiation stress

strength and seepage pressure. This phenomenon may be caused by rock exhibiting approximately elastic behavior before crack initiation stress and inelastic behavior after crack initiation.

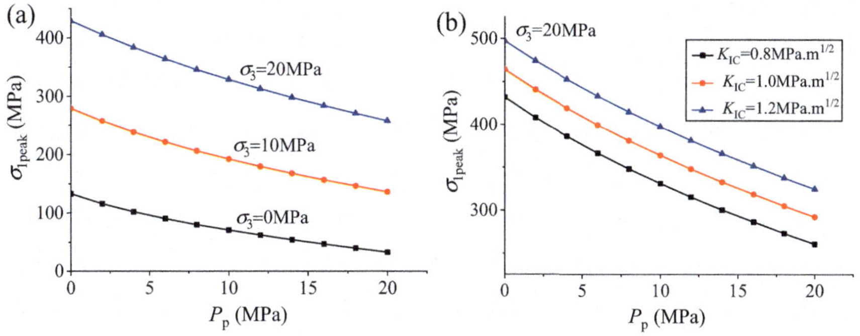

Fig. 9.5 Effects of **a** confining pressures or **b** fracture toughness on the relationship between seepage pressure and peak strength

9.2.2 Constant Seepage Pressure Effect on Creep Failure

The selection of mechanical parameters in this part refers to the parameters of deep sandstone (Jiang et al. 2014; Wang et al. 2016; Li et al. 2020; Liu et al. 2017). E = 25.03GPa, $\gamma = 0.13$, $K_{IC} = 0.69$ MPa·m$^{1/2}$, $D_o = 0.059$, $v = 1.7 \times 10^{-13}$, $\mu = 0.344$, $a = 1.77$ mm, $\varphi = 36°$, $m = 2$, $\varepsilon_o = 1/58$, $\beta = 0.29$. The stress erosion index of rock without seepage pressure is $n = 20$, the characteristic crack growth velocity is $v = 8 \times 10^{-18}$, and the rock with seepage pressure saturation is $n = 16$, $v = 1.7 \times 10^{-13}$.

Based on Eq. (9.14), it can be calculated that when confining pressure $\sigma_3 =$ 40 MPa, crack initiation stress $\sigma_{1c} = 100$ MPa in rock without seepage pressure. Therefore, when graded axial pressure is 30 and 70 MPa in Fig. 9.6, the creep strain of rock is shown as elastic strain. It corresponds to the first stage (i.e., $0 < \sigma_1 < \sigma_{1c}$) of creep time evolution Eq. (9.18). The stage of graded loading with axial pressure greater than or equal to 110 MPa is the stage of nonlinear rock creep, which corresponds to the deformation in the second stage of Eq. (9.18) (i.e., $\sigma_{1c} < \sigma_1 < \sigma_{1peak}$). Figure 9.6 shows the comparison curves between theoretical creep and experimental sandstone results under fractional axial stress loading at seepage pressure 0, 1 MPa, and 10 MPa, respectively.

The theoretical results are comparable to the experimental results (Jiang et al. 2014). The variation trend is similar, which verifies the rationality of the macro–micro mechanics model of rock creep under seepage pressure.

However, the numerical difference between the theoretical and experimental results may be due to (1) Random distribution of micro-cracks cannot be simulated in the model in this paper, but the size and angle of initial cracks inside specific kinds of rocks are assumed to be a fixed value by averaging method. (2) The influence of water-bearing rocks' chemical action on some mechanical parameters of brittle rocks is ignored in this model. (3) The value of graded axial compression in the theoretical results is not completely consistent with the value of the graded axial compression step in the test results, and there is a certain difference. (4) The selection error of model parameters.

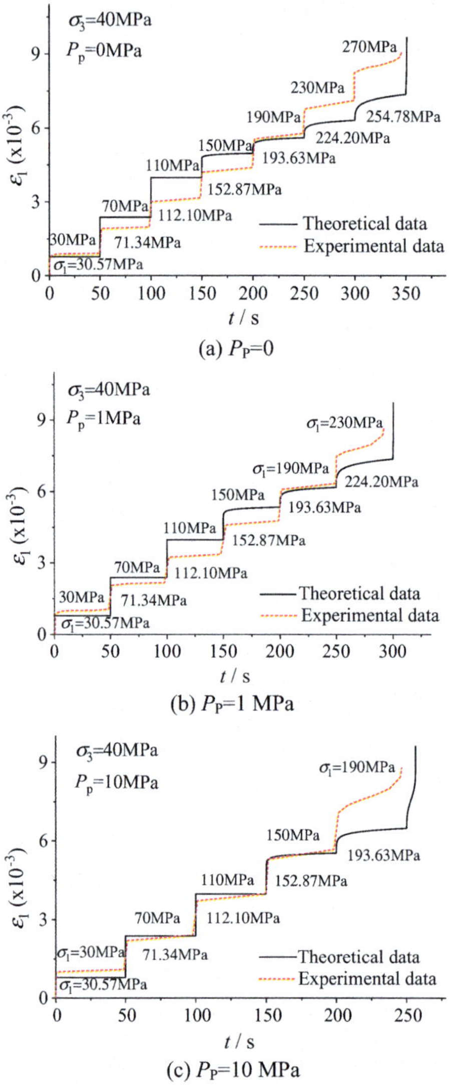

Fig. 9.6 Comparison of theoretical and experimental creep curves under step axial loading

Figure 9.7 shows the time evolution curves of rock crack length, crack growth velocity, axial strain, and axial strain rate in the process of complete creep under constant axial pressure, confining pressure, and seepage pressure. The crack length and axial strain experienced three stages deceleration, steady state, and acceleration. The crack growth and axial strain rates explain the change rate of crack length and axial strain with time, respectively. The trend of crack length and strain evolution over time is similar, which is also due to the directly proportional relationship between crack length and strain. Figure 9.8 shows the evolution trend of effective seepage pressure on the newly formed wing crack surface in the creep process of brittle rock

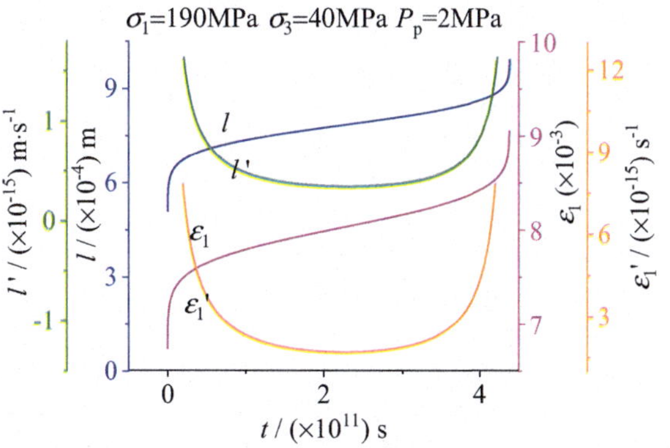

Fig. 9.7 Evolutions of crack length, crack growth velocity, axial strain, and axial strain rate under constant loadings and seepage pressure during total rock creep

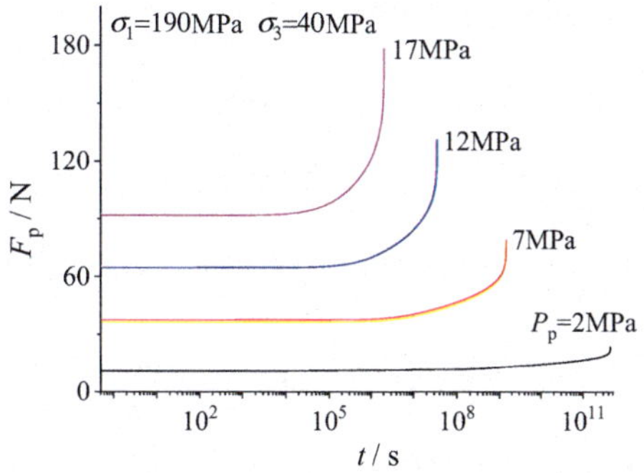

Fig. 9.8 Evolution of effective seepage pressure on wing crack surface

under different seepage pressure. The effective seepage pressure increases with time and increases at an accelerated rate in the later stage.

This phenomenon of effective seepage pressure evolution also indicates that during the stage of accelerated creep of rocks, the damage degree of internal cracks accelerates and increases, which leads to the accelerated expansion of the contact area between permeable water and the internal crack surface of rocks, and thus the accelerated failure of rocks. Since the creep failure time of rocks varies greatly with different seepage pressure, to present the evolution curve of effective seepage pressure under different seepage pressure in a figure, the logarithmic coordinate form was selected for abscissa.

In Fig. 9.9a–d, the influences of different seepage pressures on crack length, crack growth velocity, axial strain, and the evolution trend of axial strain rate in intact creep of rocks are studied, respectively. The failure time of rock creep decreases with the increase of seepage pressure. The steady creep crack propagation rate and axial strain rate increase with the rise of seepage pressure. In Fig. 9.9c, the variable between different seepage pressures is 5 MPa, which greatly influences the magnitude of rock creep failure time so that the influence of different seepage pressures on rock creep strain at the same time cannot be accurately identified. Therefore, Fig. 9.10 shows the influence of seepage pressure with variable 1 MPa on the complete creep failure curve of rocks. With the increase of seepage pressure or the decrease of effective confining pressure P_e, the axial strain at the same time keeps increasing, which is consistent with the research results of the literature (Xu et al. 2018).

Figure 9.11a–b shows the relationship curve between seepage pressure and rock creep failure time under different confining and axial pressures. The failure time of rock creep decreases with the increase of seepage pressure. Under the same seepage pressure, the creep failure time increases with the increase of confining pressure or the decrease of axial pressure. In addition, when confining pressure or axial pressure produces the same stress variable of 5 MPa, the influence of confining pressure on the magnitude of rock creep failure time is greater than that of axial pressure on rock creep failure time. That is, the impact of the sensitivity of confining pressure on rock creep failure is greater than that of axial pressure. In Fig. 9.11a, it can be seen that when the confining pressure is 45 MPa, and the seepage pressure is 0 MPa, there is no data point in the creep failure time of the rock. The reason is that under the stress state at this time, the creep failure time of the rock is infinite, which means that the rock will not fail at this time.

9.2.3 Step Seepage Pressure Effect on Creep Failure

The influence of constant seepage pressure on rock creep behavior was given in the previous section. However, in a deep underground environment, with the interference of external factors such as construction, rocks' internal seepage pressure may show a constant change trend. Therefore, this section will study the influence of the path of seepage pressure increasing with time on the evolution trend of rock creep.

Figure 9.12 shows the evolution trend of rock crack length, crack growth velocity, axial strain, and axial strain rate over time under constant axial pressure, confining pressure load, and graded seepage pressure. The initial seepage pressure P_{p1} of the graded seepage pressure path is 2 MPa, the seepage pressure increment ΔP_p of each stage is 2 MPa, and the duration of each stage is 50 h. The crack length and axial strain experienced deceleration and increased many times. The accelerated increase occurred at the last stage of seepage pressure, at which time the rock was destroyed. With time, the evolution curve of crack growth velocity and axial strain rate can explain the evolution trend of crack length and axial strain classification change rate more clearly. The minimum crack growth velocity or axial strain rate increases with the stage seepage pressure steps increase. The variation trend of the research results is consistent with the experimental results in the literature (Yan et al. 2010; Yang et al. 2015).

Figure 9.13a–d shows the creep crack length, crack growth velocity, axial strain, and axial strain rate evolution curves of rock under graded seepage pressure with different seepage pressure increments ΔP_p, respectively. With the increase of seepage pressure increment, the number of creep failure steps decreases, the creep failure time decreases continuously, and the crack length, axial strain, crack growth velocity, and axial strain rate of rock at a given moment increase continuously.

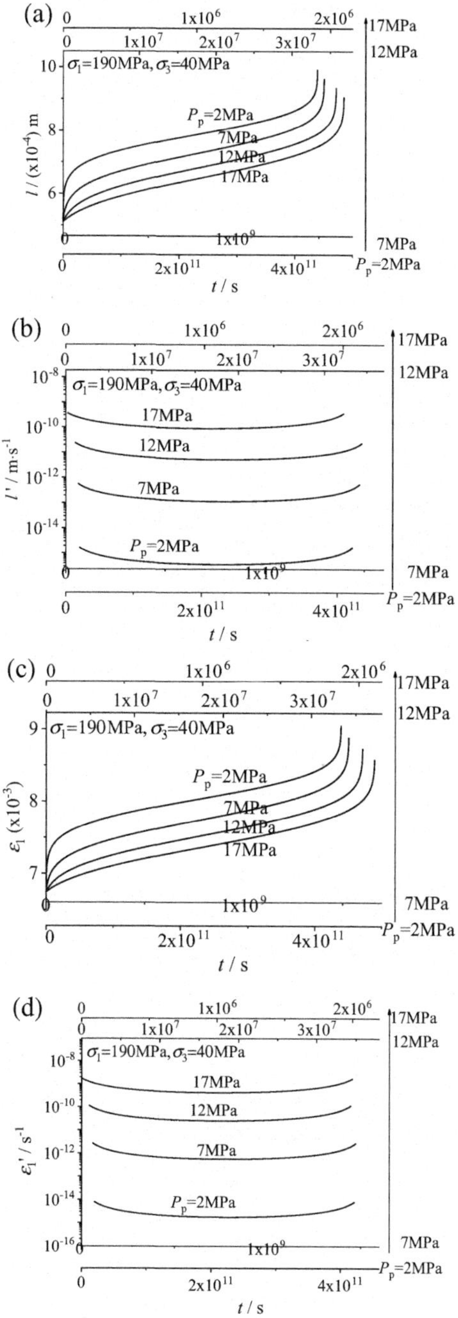

Fig. 9.9 Effects of seepage pressure on evolutions of **a** crack length, **b** crack growth velocity, **c** axial strain, and **d** axial strain rate during total creep

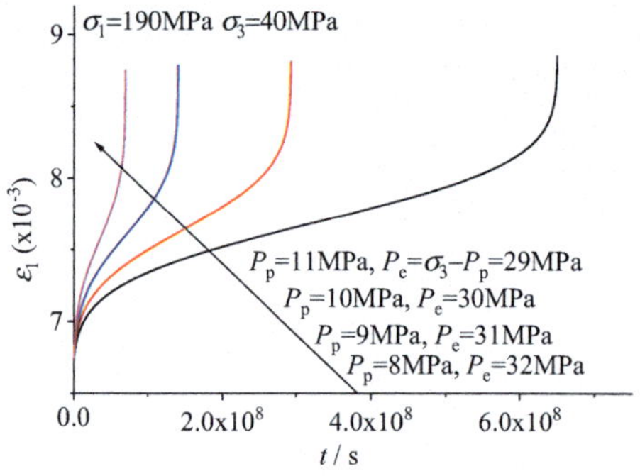

Fig. 9.10. Effects of seepage pressure P_p or effective confining pressure P_e on creep behaviors

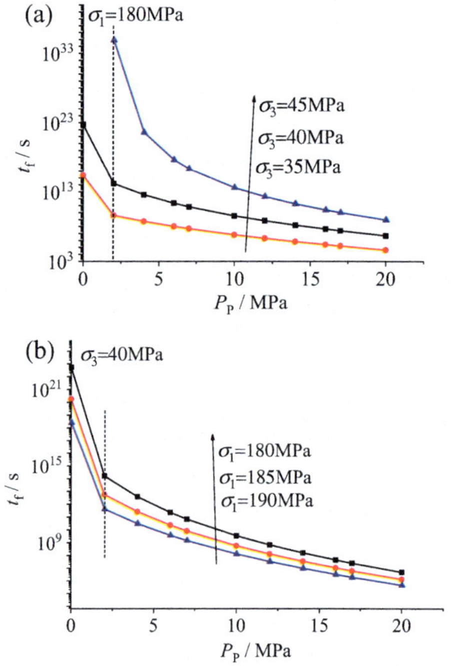

Fig. 9.11 Relationships between seepage pressure and creep failure time under different **a** confining pressures or **b** axial stresses

9.3 Conclusions

Based on the fracture damage mechanics theory, a macro and micro–macro mechanism of progressive creep failure of brittle rock was established, considering the influence of seepage pressure. The result of seepage pressure on the stress–strain relationship curve was studied, and the influence of seepage pressure on the creep microcracks' length, crack growth velocity, effective seepage pressure, strain, and strain rate of brittle rock was explained. The specific results are as follows,

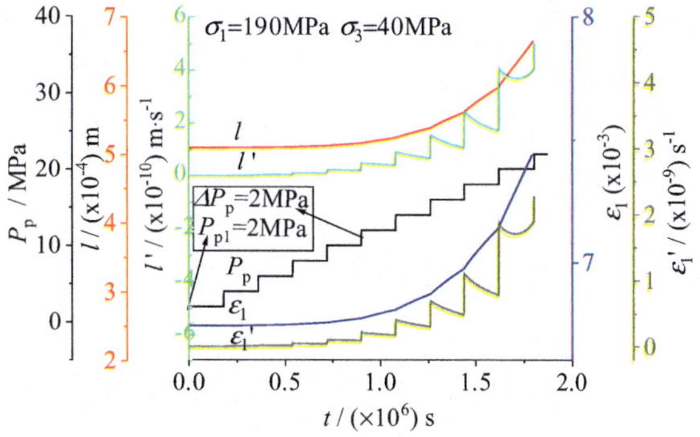

Fig. 9.12 Evolutions of crack length, crack growth velocity, axial strain, and axial strain rate under constant loadings and step seepage pressure during rock creep

(1) The seepage pressure weakens the normal stress of the initial crack surface inside the rock and increases the tension driving force of the newly formed wing crack surface, which leads to the weakening effect of the seepage pressure on the rock strength. With the increase of seepage pressure, the normal stress of the initial crack surface in rock decreases linearly, the tensile tension of the initial crack surface increases linearly, and the effective seepage pressure of the newly formed wing crack surface increases linearly.

(2) The initiation stress and peak strength of rock cracks decrease with the increase of seepage pressure. The crack initiation stress and peak strength increase with the increase of confining pressure or fracture toughness when the seepage pressure is constant.

(3) Under constant seepage pressure, with the increase of seepage pressure, the creep failure time of the rock slows down and decreases continuously. The steady-state creep crack growth velocity and axial strain rate of rock increase continuously, and the axial strain increases continuously at the same time. Under the same seepage pressure, the creep failure time increases with the increase of confining pressure or the decrease of axial pressure. The influence of confining pressure on the magnitude of rock creep failure time is greater than that of axial pressure on rock creep failure time. That is to say, the influence sensitivity of confining pressure on rock creep failure is greater than axial pressure. The effective seepage pressure increases with the creep time evolution, and in the creep acceleration stage, the effective seepage pressure also accelerates.

(4) Under graded seepage pressure, the crack length and axial strain experienced deceleration and increased several times. The accelerated increase occurred at the last stage of seepage pressure until the rock failure. The minimum crack growth velocity or axial strain rate increases with the stage seepage pressure steps increase. With the increment of graded seepage pressure increase, the number of creep failure steps decreases, the creep failure time decreases continuously, and the crack length, axial strain, crack growth velocity, and axial strain rate of rock at a given moment increase continuously.

Fig. 9.13 Effects of step seepage pressure on evolutions of **a** crack length, **b** crack growth velocity, **c** axial strain, and **d** axial strain rate during creep

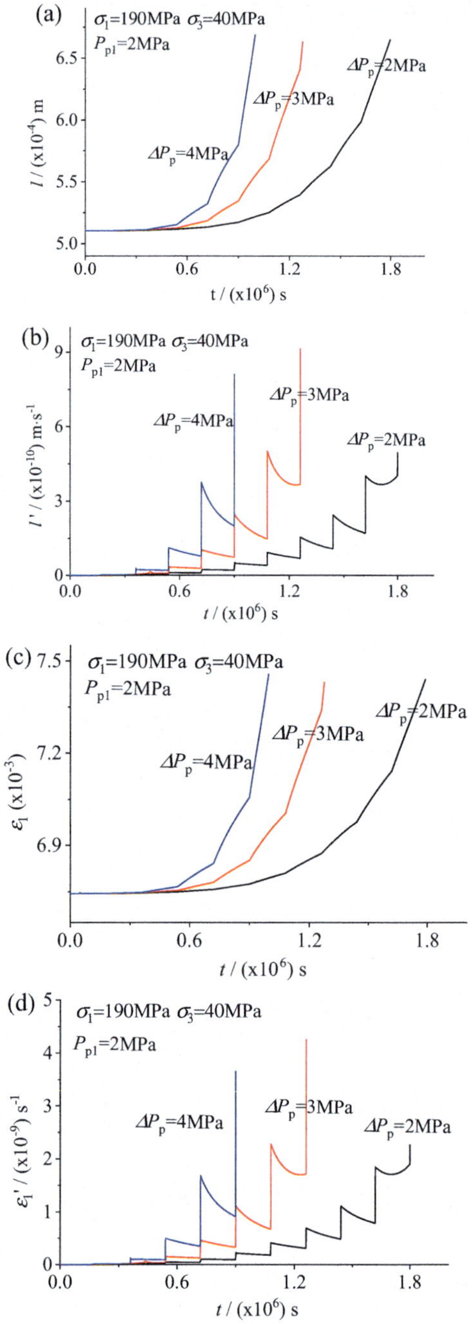

References

Ashby MF, Sammis CG. The damage mechanics of brittle solids in compression. Pure Appl Geophys. 1990;133(3):489–521.

Hu SH, Zhang G, Zhao SL, Jiang XL, Chen YF, Zhou CB. Laboratory tests on deformation characteristics of thermally treated Beishan granite under high pore water pressure. Chin J Geotech Eng. 2016;38(s2):330–5.

Jiang HF, Liu DY, Huang W, Liu FY. Influence of high pore water pressure on creep properties of rock under high confining pressure. J China Coal Soc. 2014;39(7):1248–56.

Li B, Huang D, Ma WZ. Study on the fracturing behavior of sandstone influenced by bedding plane. Chin J Rock Soil Mech. 2020;41(3):1–12.

Liu X, Liu J, Li D, Wang J, Wang Z, Zhong Z. Unloading mechanical properties and constitutive model of sandstone under different pore pressures and initial unloading levels. J China Coal Soc. 2017;42(10):2592–600.

Peng J, Rong G, Zhou C, Wang X, Hou D. Experimental study of effect of water pressure on progressive failure process of rocks under compression. Chin J Rock Soil Mech. 2013;34(04):941–6.

Wang JB, Liu XR, Liu J, Song ZP. Mechanical properties of sandstone and an improved Duncan-Chang constitutive model. Chin J Rock Mech Eng. 2016;35(12):2388–97.

Xu T, Zhou G, Heap MJ, Yang SQ, Konietzky H, Baud P. The modeling of time-dependent deformation and fracturing of brittle rocks under varying confining and pore pressures. Rock Mech Rock Eng. 2018;51:3241–63.

Yan Y, Wang EZ, Wang SJ, Hu Y. Study of seepage-rheology coupling experiment of rocks. Chin J Rock Soil Mech. 2010;31(7):2095–103.

Yang HW, Xu J, Nie W, Peng SJ. Experimental study on creep of rocks under step loading of seepage pressure. Chin J Geotech Eng. 2015;37(9):1613–9.

Chapter 10
Cyclic Seepage Pressure Effect on Creep Fracture

Deep underground projects often face complex geological environments such as high stress and pore pressure, making the disasters of underground engineering rocks induced by rock instability widespread (Tu et al. 2021). Pore pressure, as an inseparable influencing factor of the stability of slope engineering, has also received more and more attention. The slopes and foundations around the reservoir often undergo displacements or even landslide damage, due to the change of pore pressure and reservoir level in Fig. 10.1. Therefore, it is cyclic pore pressure that plays an important role in rock deformation (Vulliet et al. 1988; Zangerl et al. 2010; Preisig et al. 2016). Based on the Chap. 9, this chapter introduces the cyclic pore pressure path into the Hooke-Kelvin and wing crack propagation models. Time-dependent elastic modulus evolution of rock under cyclic pore pressure is obtained, which affects viscoelastic strain under cyclic pore pressure. A micro–macro mechanical model of brittle rock creep is established under cyclic pore water pressure. The influence of cyclic pore pressure, axial stress, and confining pressure on mechanical rock properties such as creep strain, strain rate, and failure time of brittle rock are studied.

10.1 Theory

Based on Chap. 5, the cycle loading and unloading path of pore pressure P_P shown in Fig. 10.2a, b are inserted into the Hooke-Kelvin and wing crack propagation models. The cyclic pore pressure effect on the creep failure is studied.

X. Li et al., *Static Creep Micro-Macro Fracture Mechanics of Brittle Solids*,
https://doi.org/10.1007/978-981-99-8203-5_10

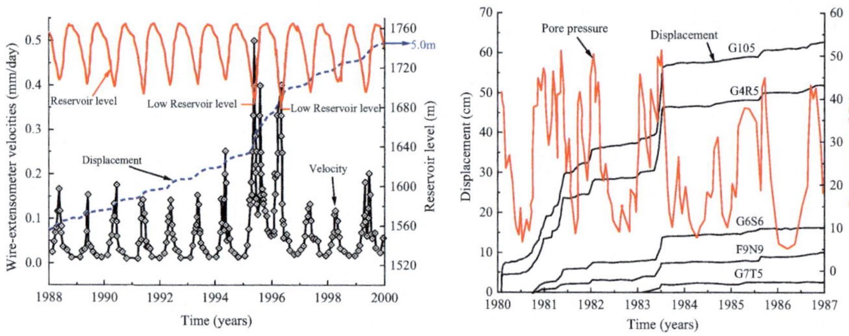

(a) Time-dependent variation of slope velocity and reservoir level.

(b) Time-dependent displacement and pore pressure

Fig. 10.1 Time-dependent movement of two landslides: **a** Hochmais-Atemkopf landslide (Zangerl et al. 2010) and **b** Salledes landslide (Vulliet et al. 1988)

Fig. 10.2 Schematic of the micro–macro model in the creep fracture under cyclic pore pressure

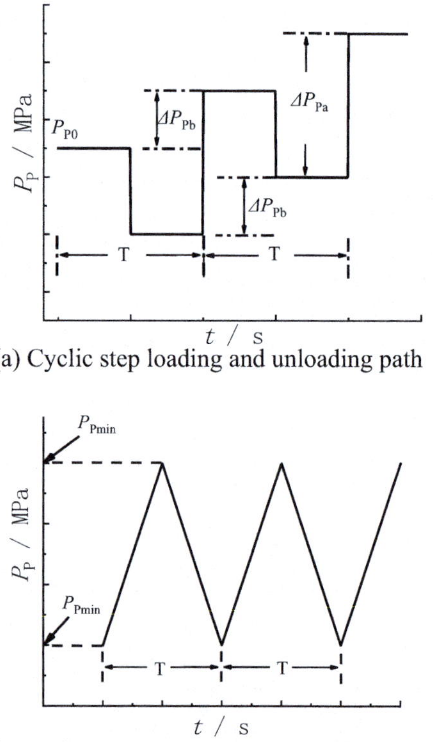

(a) Cyclic step loading and unloading path

(b) Cyclic triangle loading and unloading path

10.1.1 Elastic and Viscoelastic Creep Strain Under Cyclic Pore Pressure

The Hooke-Kelvin model, made up of Hooke springs and Kelvin styles, is a good description of the deformation of the viscoelastic phase can be expressed as:

$$|\sigma_1| + \frac{\eta}{E_{\text{vet}} + E(P_{\text{P}})}|\dot{\sigma}_1| = \frac{E_{\text{Vet}}E(P_{\text{P}})}{E_{\text{Vet}} + E(P_{\text{P}})}\varepsilon_1 + \frac{\eta E(P_{\text{P}})}{E_{\text{Vet}} + E(P_{\text{P}})}\dot{\varepsilon}_1 \tag{10.1}$$

where σ_1 is axial stress, ε_1 is axial strain, E_{vet} is the viscoelastic modulus, η is viscosity coefficient, || is an absolute value, E is the elastic modulus of brittle rocks.

Using the model of wing microcrack before wing crack extension with pore pressure in Chap. 9 and the time-dependent pore pressure in Fig. 10.2a, b, the modulus of elasticity E relating to pore pressure is defined as:

$$E(P_{\text{P}}(t)) = \frac{\sigma_{1\text{ci}}}{\varepsilon_{1\text{ci}}} = \frac{\sigma_{1\text{ci}} = \left[K_{\text{IC}}\beta^{3/2}\sqrt{\pi/a} - \sigma_3 B_2 + \mu P_{\text{P}}(t)\sin\varphi\right]/\,B_1}{\varepsilon_{1\text{ci}} = \varepsilon_0[-\ln(1 - D_0)]^{\frac{1}{m}}} \tag{10.2}$$

where

$$B_1 = (\mu - \mu\cos 2\varphi - \sin 2\varphi)\sin\varphi/2 \tag{10.3}$$

$$B_2 = (\sin 2\varphi + \mu + \mu\cos 2\varphi)\sin\varphi/2 \tag{10.4}$$

Among them, $\sigma_{1\text{ci}}$ is the critical stress of the viscoelastic and elastic deformation stage at wing crack initiation ($l = 0$), and $\varepsilon_{1\text{ci}}$ is the elastic strain corresponding to that critical stress, which is related to cyclic pore pressure (i.e., $P_{\text{P}}(t)$). The Elastic modulus E is an essential parameter in the creep process of brittle rock under the cycle pore pressure. The elastic modulus changes with the cycle pore pressure and is inversely proportional to the pore pressure. That is, the elastic modulus during pore pressure loading is less than the elastic modulus at unloading.

The existing conditions are $|\dot{\sigma}|$=0 and $\varepsilon_1(t = 0) = |\sigma_1|/E_0$, where E_0 is the initial modulus of elasticity obtained under the action of the initial pore pressure P_{P0}. Thus the viscoelastic strain under constant pore pressure as:

$$\varepsilon_{1\text{veL}}(t) = \frac{|\sigma_1|}{E_{\text{vet}}E_0}\left\{E_{\text{vet}}(e^{-tE_{\text{vet}}/\eta} - 1) - \left[e^{-tE_{\text{vet}}/\eta}(E_{\text{vet}} - E_0) - E_0\right]\right\} \tag{10.5}$$

By known conditions $|\dot{\sigma}|$=0 and $\varepsilon_{11} = \varepsilon_1(t = t_{\text{veL}}) = \varepsilon_{1\text{veL}}(t_{\text{veL}}) - (|\sigma_1|/E_{\text{L}}(t) - |\sigma_1|/E_{\text{UL}}(t))$, as well as where E_{L} and E_{UL} are the moduli of elasticity obtained by loading pore pressure $P_{\text{PL}}(t)$ and unloading pore pressure $P_{\text{PUL}}(t)$, respectively. Solution Eq. (10.1), where the pore pressure can be unloaded to a specific value P_{PUL}, the axial viscoelastic strain can be solved as follows:

$$\varepsilon_{1\mathrm{veUL}}(t) = \frac{\varepsilon_{1\mathrm{ci}}(1 - e^{(t_{\mathrm{veL}}-t)E_{\mathrm{vet}}/\eta})|\sigma_1|}{\sigma_{1\mathrm{ci}}(t)} + \frac{|\sigma_1| + (\varepsilon_{11}E_{\mathrm{vet}} - |\sigma_1|)e^{(t_{\mathrm{veL}}-t)E_{\mathrm{vet}}/\eta}}{E_{\mathrm{vet}}} \tag{10.6}$$

where t_{veL} is the loading time in Eq. (10.5), it should be noted that the action time of each cycle loading and unloading pore pressure is independent of each other.

The initial strain conditions for the next pore pressure loading after the first pore pressure loading and unloading cycle are obtained:

$$\varepsilon_{12} = \varepsilon_{1\mathrm{veUL}}(t_{\mathrm{veL}} + t_{\mathrm{veUL}}) + (|\sigma_1|/E_{\mathrm{L}} - |\sigma_1|/E_{\mathrm{UL}}) \tag{10.7}$$

The next viscoelastic strain over time under the pore pressure loading can be expressed as:

$$\begin{aligned} \varepsilon_{1\mathrm{veLN}}(t) = & \frac{|\sigma_1|[\varepsilon_{1\mathrm{ci}}E_{\mathrm{vet}} + \sigma_{1\mathrm{ci}}(t)]}{E_{\mathrm{vet}}\sigma_{1\mathrm{ci}}(t)} \\ & + \frac{e^{(t_{\mathrm{veL}}+t_{\mathrm{veUL}}-t)E_{\mathrm{vet}}/\eta}}{E_{\mathrm{vet}}\sigma_{1\mathrm{ci}}(t)}[\varepsilon_{1\mathrm{ci}}|\sigma_1|E_{\mathrm{vet}} + \sigma_{1\mathrm{ci}}(t)(\varepsilon_{12}E_{\mathrm{vet}} + |\sigma_1|)] \end{aligned} \tag{10.8}$$

10.1.2 Plastic Creep Strain Induced by Microcrack Growth Under Cyclic Pore Pressure

The time-dependent and plastic axial strain under cyclic pore pressure can be written as improved axial strain related to wing crack minus the crack initiation strain:

$$\varepsilon_{1\mathrm{p}}(t) = \varepsilon_{1\mathrm{wc}}(t) - \varepsilon_{1\mathrm{ci}} = \varepsilon_0\left\{-\ln\left[1 - D_0\left(l(t)/a + 1\right)^3\right]\right\}^{\frac{1}{m}} - \varepsilon_{1\mathrm{ci}} \tag{10.9}$$

where $\varepsilon_{1\mathrm{wc}}$ is the axial strain related to the wing crack. Inserting the function of time-dependent pore pressure $P_{\mathrm{P}}(t)$ of Fig. 10.2a and b into Eq. (9.15), time-dependent wing cracks $l(t)$ by the cyclic pore pressure can be solved.

10.1.3 Complete Elastic–Viscoelastic-Plastic Creep Fracture Under Cyclic Pore Pressure

Combining Eqs. (10.5–10.9), the time-dependent creep fracture of complete elastic–viscoelastic-plastic deformations under cyclic pore pressure can be acquired by:

$$\varepsilon_1(t) = \varepsilon_{1\mathrm{veL}}(t) + \varepsilon_{1\mathrm{veUL}}(t) + \varepsilon_{1\mathrm{veLN}}(t) + \varepsilon_{1\mathrm{p}}(t) \tag{10.10}$$

10.2 Results and Discussions

10.2.1 Model Validation

According to the above theoretical model, the following study is conducted on the creep evolution results of brittle rocks under cyclic pore pressure.

Based on Eq. (10.10), Fig. 10.3 shows the time-dependent strain of fine-grained sandstone considering cyclic triangle pore pressure in Fig. 10.2b. The same model parameters of sandstone are: $n = 19$, $D_0 = 0.048$, $\varepsilon_0 = 1/46$, $\mu = 0.248$, $a = 3.1$ mm, $\varphi = 42°$, $v = 1.3 \times 10^{-9}$ m/s, the rest of the different parameters are shown in Fig. 10.3. Theoretical and experimental results (Yang 2011; Yang et al. 2011) are comparable in trend and numerical value, and the rationality of the theoretical model is proved. However, the numerical difference between the theoretical and experimental results may be due to the following reasons: (1) the randomly distributed cracks cannot be simulated realistically, but the mechanical parameters such as the angle, crack friction coefficient, and initial crack in a specific type of rock are assumed to be a fixed value by an averaging method, (2) the model only studies the cyclic pore pressure of water and ignores the influence of chemical factors of water, and (3) there is a certain error in the selection of model parameters.

10.2.2 Sensitivity Analysis of Cyclic Pore Pressure Parameters

(1) **Effect of cyclic step pore pressure in P_{Pd} , P_{Pu} and P_{Pi} on creep fracture behaviors**

This section will discuss the temporal evolution of visco-elastic–plastic creep-related variables of brittle rocks under cyclic step pore pressure. The corresponding model parameters are: $K_{IC} = 1.99$ MPa·m$^{1/2}$, $D_0 = 0.048$, $m = 2$, $\varepsilon_0 = 1/78$, $\mu = 0.51$, $a = 3.1$ mm, $n = 57$, $v = 1.3 \times 10^{-9}$ m/s, $\varphi = 32.03°$, $\beta = 0.276$, $E_{vet} = 90$ GPa, $\eta = 2 \times 10^5$ GPa·s.

Firstly, according to the common pore pressure changes in the project, the cyclic value of pore pressure is divided into three kinds of situations: gradually decreasing (i.e., P_{Pd}), remaining unchanged (i.e., P_{Pu}), and progressively increasing (i.e., P_{Pi}).

Figure 10.4a is three specific descriptions of the cyclic pore stress function in Fig. 10.2a. Under the action of three stresses, Fig. 10.4c gives the pure plastic strain over the time curve caused by wing crack growth. It can be found in the figure that the pure plastic strain under cyclic loading and unloading values increasing progressively of pore pressure (i.e., ε_{1pi}) is greater than the pure plastic strain under the unchanged (i.e., ε_{1pu}), and the pure plastic strain under the gradual decrease (i.e., ε_{1pd}) is minimum.

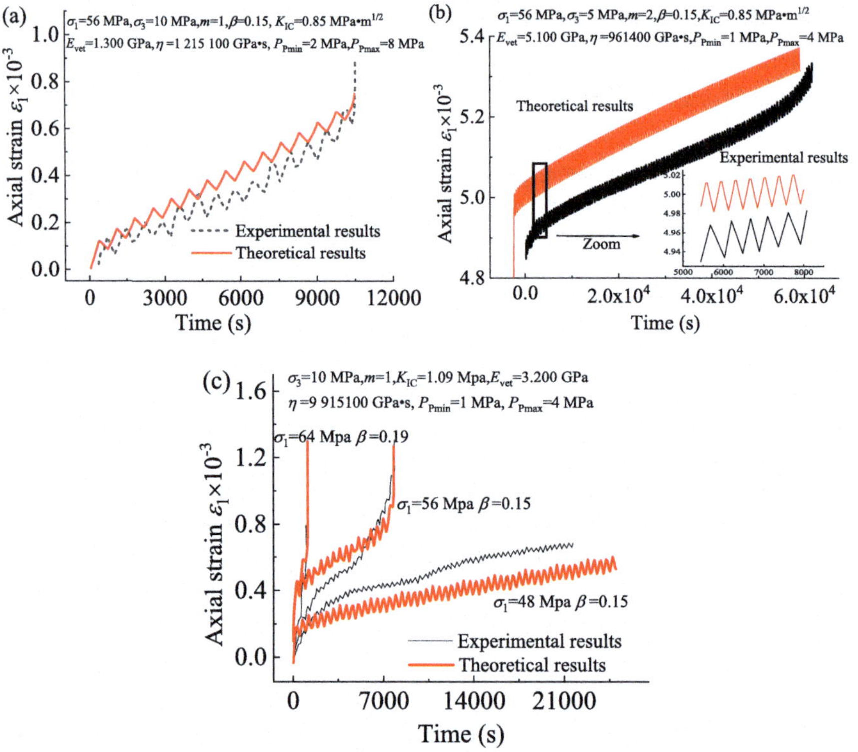

Fig. 10.3 Theoretical versus experimental (Yang 2011; Yang et al. 2011) data of fatigue creep during cyclic pore pressure

In Fig. 10.4d, the strain is decelerating increases under cyclic pore pressure loading for the viscoelastic strain evolution curve over time. It should be noted that under cycle pore pressure unloading, the strain is also decelerating and increases with time. The overall trend of the strain is the same as the changing trend of cyclic pore pressure, which corresponds to the changing direction of gradually increasing, unchanged, and decreasing.

Through the Fig. 10.4e, it can be found that under the three different cyclic pore pressure stresses, the total strain is not much additional in the early stage of the cycle (i,e., first 13 cycles), and the total strain change rule in the later period is $\varepsilon_{1\mathrm{i}} > \varepsilon_{1\mathrm{u}} > \varepsilon_{1\mathrm{d}}$ ($\varepsilon_{1\mathrm{i}}$, $\varepsilon_{1\mathrm{u}}$, $\varepsilon_{1\mathrm{d}}$ are the strain value under the cyclic pore stresses increasing, remaining unchanged, and gradually decreasing respectively). The failure time change rule is $t_{\mathrm{fi}} < t_{\mathrm{fu}} < t_{\mathrm{fd}}$ (t_{fi}, t_{fu}, t_{fd} are the failure time under the gradual increase of the cyclic pore stresses, the remaining unchanged, and the gradual decrease of the action, respectively). The creep strain increases when the cyclic value is more significant and the failure time is shorter, the same as the creep evolution law of brittle rock under constant pore pressure.

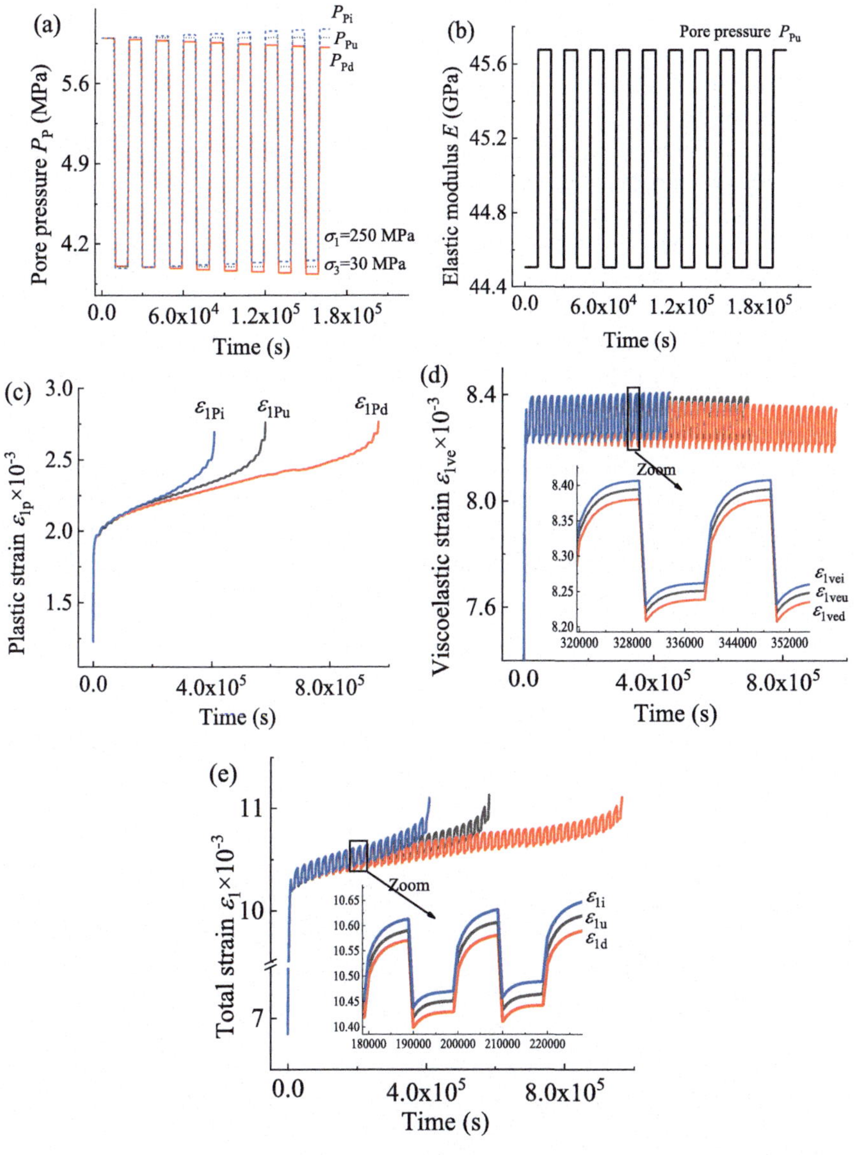

Fig. 10.4 Time-dependent **b** elastic modulus E **c** plastic axial strain ε_{1p}, **d** viscoelastic axial strain ε_{1ve}, **e** total axial strain ε_1 under **a** cyclic pore pressure

The strain rebound phenomenon generated under cyclic pore pressure is mainly caused by viscoelastic strain rebound. The crack evolution under cyclic pore pressure cannot meet the crack recovery conditions and the plastic strain rebound in Chap. 7. At the same time, according to many calculations, it is impossible to obtain an obvious plastic rebound phenomenon. So it can be concluded that the extended crack cannot be recovered, the pure plastic strain cannot be rebounded under cyclic pore pressure, and the rebound phenomenon that occurs in the total strain is mainly due to the viscoelastic strain rebound of the rock.

The evolution curves of rock plastic strain, viscoelastic strain, total strain, and corresponding strain rate over time are shown under pore pressure P_{Pi} in Fig. 10.5. Observing the overall change trend of the plastic strain rate in Fig. 10.5a, it is found that the rate is gradually reduced from a large creep rate, then becomes stable, and finally increases rapidly, mainly reflected in the cyclic pore pressure loading stage. The strain rate in the unloading stage is approximately zero, confirming that pure plastic strain cannot rebound.

Figure 10.5b shows the viscoelastic strain and strain rate change curves over time. It can be found that regardless of the pore pressure cycle loading or unloading, the strain rate has a gradually decreasing process. The initial strain rate value the in the loading stage is significantly larger than that in the unloading stage. The evolution curve of total strain and total strain rate over time and its effect can be seen as the superposition of plasticity and viscoelasticity in Fig. 10.5c. Figure 10.5d is a detailed description of region I in Fig. 10.5c, which shows a detailed comparison of the total strain rate and the total strain during the 5th and 6th cycles.

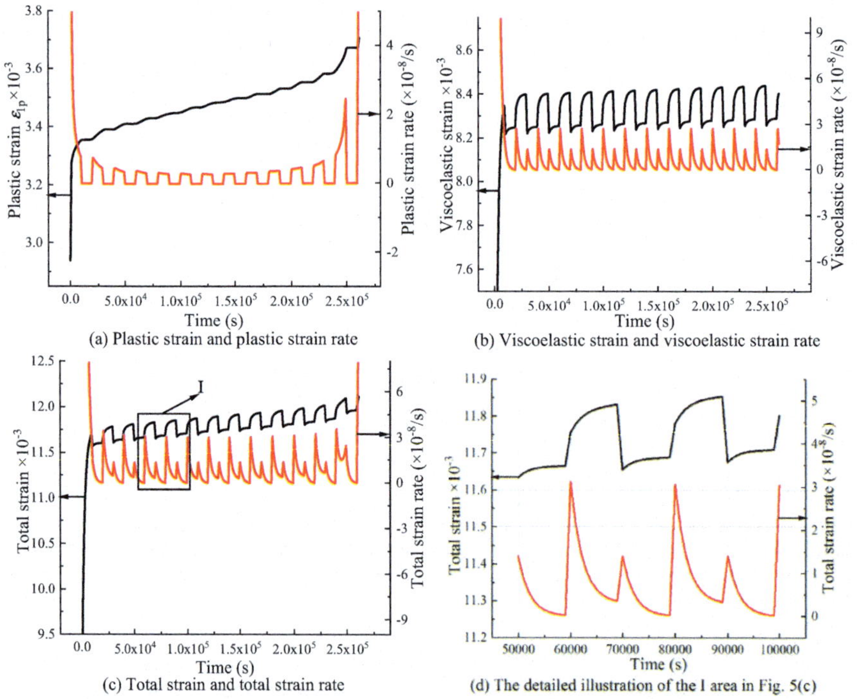

Fig. 10.5 Time-dependent three strain and strain rates under the cyclic pore pressure P_{Pi}

(2) Cyclic step pore pressure parameters ΔP_{Pa} and ΔP_{Pb} effect on creep failure behaviors

In the circulating water pressure environment, such as tidal tides, reservoir storage, and release, it is more common for the circulating pore pressure to increase gradually (Watson et al. 2006; Alzo'ubi et al. 2010; Gischig et al. 2016). Thus, the creep failure characteristics under gradually increasing value of cyclic pore pressure P_{Pi} will be further discussed in details below.

As shown in Fig. 10.6, creep strain discussion, and analysis results are obtained under increasing pore pressure (see Fig. 10.6a). It is found that when the parameters ΔP_{Pa} and ΔP_{Pb} are larger, the plastic strain related to crack propagation in Fig. 10.6b, the viscoelastic strain in Fig. 10.6c, and also the viscoelastic rebound value in Fig. 10.7 get greater. In addition, it analyzes the changes rule in the viscoelastic rebound as the cycle pore progresses in Fig. 10.7. The viscoelastic rebound increases while the increment of the number of cycles (see Fig. 10.7a), and the larger the cyclic value of pore pressure, the larger the rebound value in Fig. 10.7b. The total strain also increases, and the creep failure time is shorter.

The effect of the mechanical parameters ΔP_{Pa} and ΔP_{Pb} on the failure time in Fig. 10.8 shows that the creep failure time is significantly shortened with increased parameters. After summarizing, it can be concluded that the more considerable cyclic value of pore pressure, the viscoelasticity, plastic strain, total strain, and viscoelastic rebound increased. In contrast, the failure time is shortened.

Figure 10.9 shows the pure plastic strain rate, viscoelastic strain rate, and total strain rate change under the cyclic pore pressure in Fig. 10.6a. The analysis shows that the larger the cyclic value of pore pressure (i.e., the larger ΔP_{Pa} and ΔP_{Pb}), the higher the strain rate to a certain extent, and the impact on the plastic strain rate is more prominent. The result of pore pressure loading on the total strain rate is greater than the effect of unloading, which is more evident in the later stages of the cycle. In addition, the minimum total strain rate under cyclic pore pressure is obtained in Fig. 10.10. The minimum total strain rate decreases first and then increases under loading and unloading. The total strain rate of minimum in the loading stage is much larger than in the unloading stage.

(3) Cyclic pore pressure period T effect on creep failure behaviors

Figure 10.11 shows the effect of the cycle period T of 68 sets of cyclic pore pressure on creep failure time, and it can be observed that the creep failure time also increases with the increase of the cycle period, but with the rise of the cycle period, the failure time increase rate is slowed down. It shows that as the cyclic pore pressure rate increases, brittle rock is more prone to creep failure, and the slower the pore pressure rate, the longer the creep failure time.

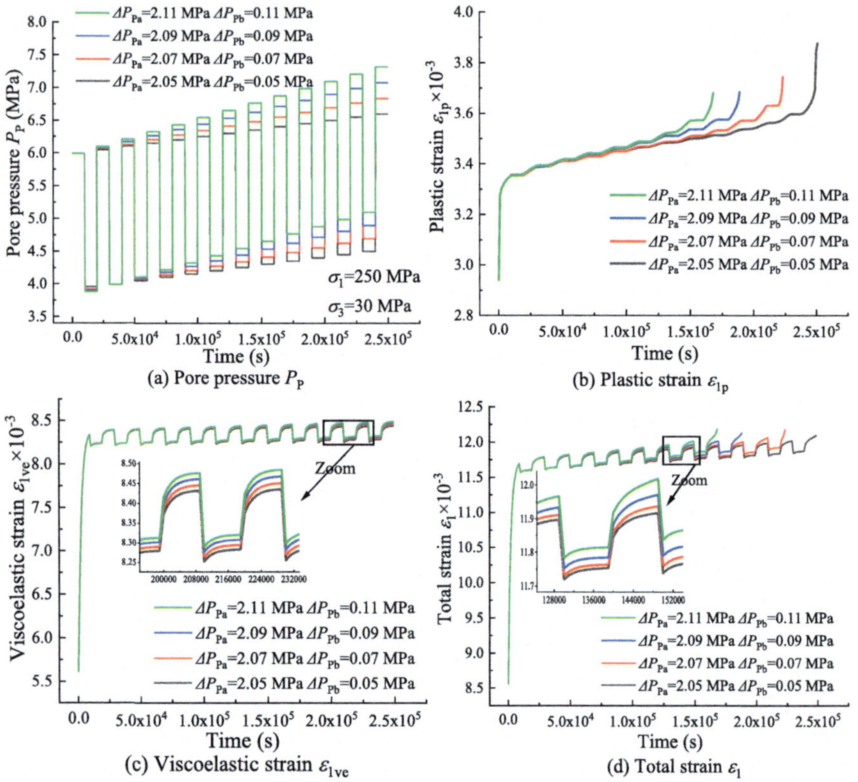

Fig. 10.6 Evolution curves of time-dependent, **b** plastic strain, **c** viscoelastic strain, and **d** total strain during **a** gradually increasing cyclic pore pressure under different parameters ΔP_{Pa} and ΔP_{Pb} (Note: $\Delta P_{\mathrm{Pa}} - \Delta P_{\mathrm{Pb}} = 2$)

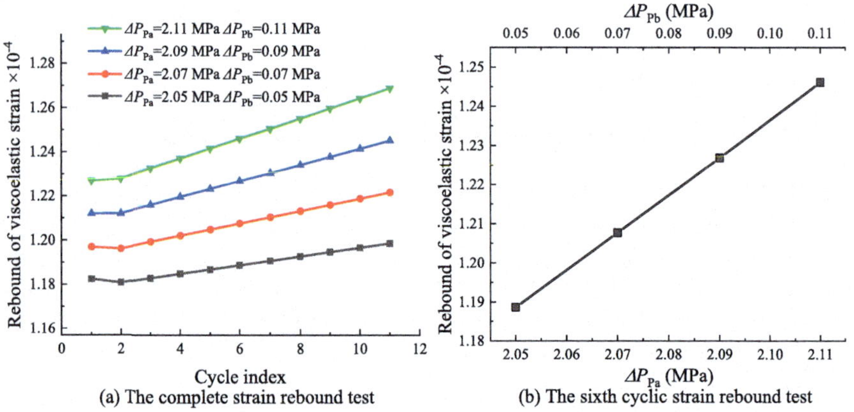

Fig. 10.7 Rebound value of computational results of viscoelastic strain

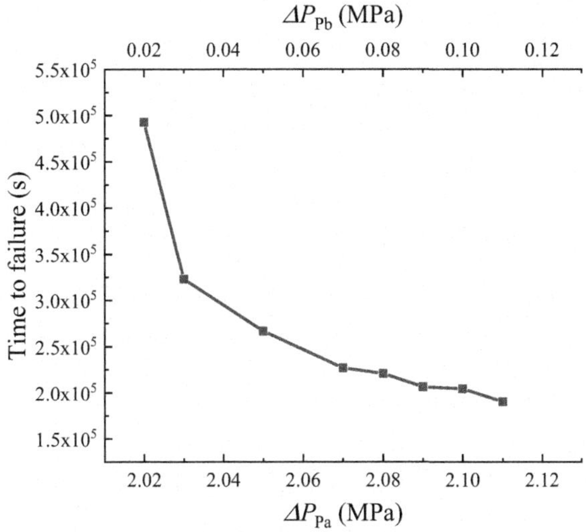

Fig. 10.8 Effect of parameters ΔP_{Pa} and ΔP_{Pb} on failure time

10.3 Conclusions

Introducing the improvement of the Hooke-Kelvin model, the wing crack extension model, the subcritical crack law, and the crack-strain model under cyclic pore pressure, a micro–macro mechanical model of brittle rock creep failure considering the complete elasticity-viscoelasticity-plasticity under the cyclic pore pressure is established. The specific conclusions are:

(1) The creep strain is the largest, and the creep failure time is the shortest under the cyclic pore pressure P_{Pi} by comparing the influence of three different cyclic pore pressure. But the creep strain is the smallest, and the time of creep failure is the longest under the cyclic pore pressure P_{Pd}. Under cyclic pore pressure, the extended crack cannot be recovered, the pure plastic strain cannot be rebounded, and the rebound phenomenon that occurs in the total strain is mainly due to the viscoelastic strain rebound of the rock. It is concluded that the total strain decelerates during the loading and unloading, and the initial value of the strain rate in the loading stage is significantly larger than that in the unloading stage by comparing the strain and strain rate under the cyclic pore pressure.
(2) Under the action of gradually increasing cyclic pore pressure, the larger the loading and unloading value of cyclic pore pressure, i.e., ΔP_{Pa} and ΔP_{Pb}, the viscoelasticity, plastic strain, and total strain are increased, and the viscoelastic rebound value also increased certain extent, and the failure time is shortened. The strain rate increases to a certain extent with the cyclic pore pressure rate, of which the plastic strain rate is more obviously affected. The minimum total strain rate decreases first and then increases under loading and unloading, and

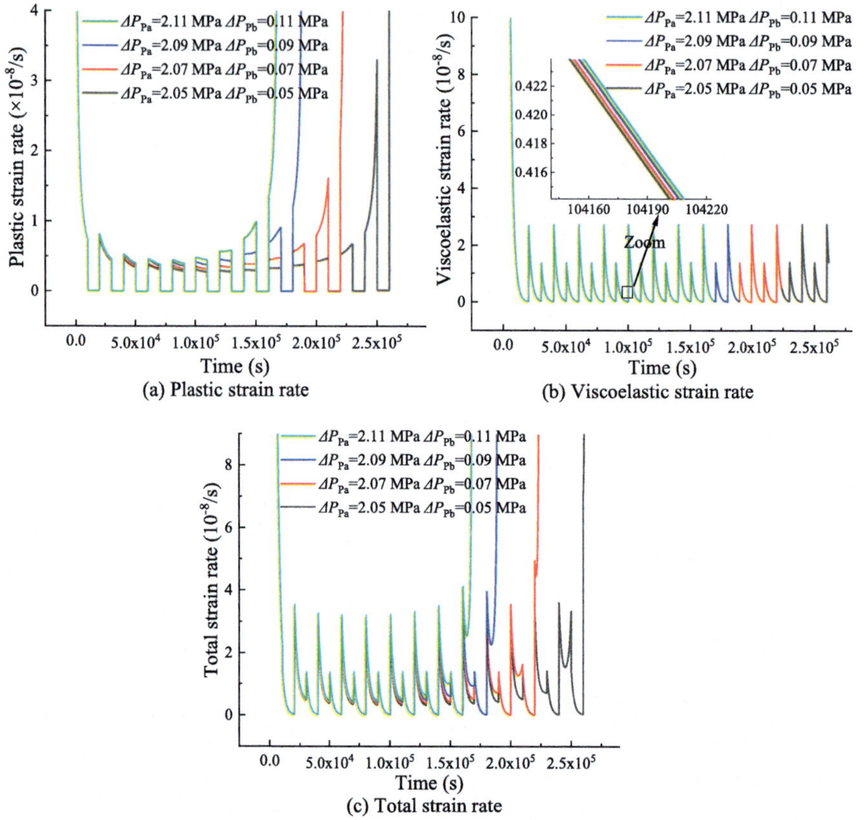

Fig. 10.9 Time-dependent three axial strain rate under cyclic pore pressure of Fig. 10.6a

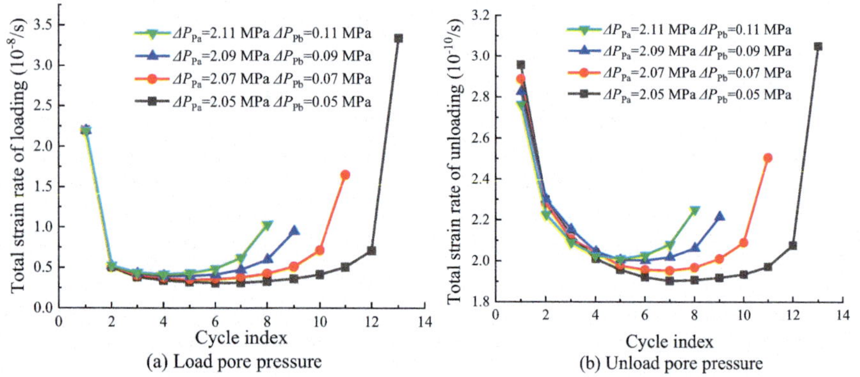

Fig. 10.10 Evolution of the minimum total strain rate with the number of cycles under **a** load pore pressure and **b** unload pore pressure

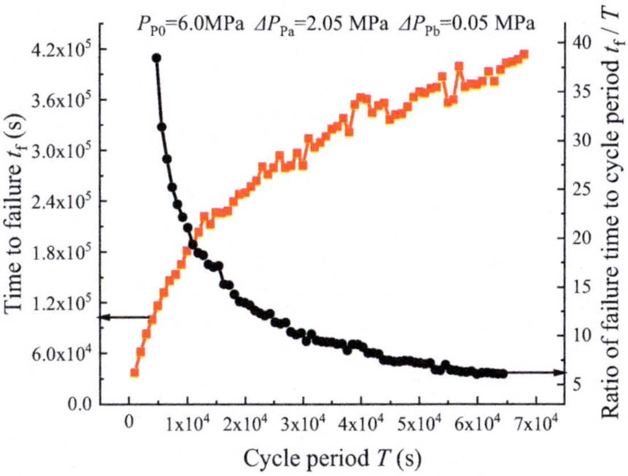

Fig. 10.11 Effect of cycle period T on creep failure time t_f

the total strain rate of the minimum in the loading stage is much larger than in the unloading stage.

(3) The increase in the cycle period T causes a certain degree of growth in the creep failure time of brittle rocks. Still, with the rise of the cycle, the failure time increase rate is slowed down, indicating that the greater the cycle pore pressure loading and unloading rate, the more brittle rock is prone to creep failure. On the contrary, reducing the loading and unloading rate can prolong the time for the creep failure of rocks.

References

Alzo'ubi AK, Martin CD, Cruden DM. Influence of tensile strength on toppling failure in centrifuge tests. Int J Rock Mech Mining Sci. 2010;47(6):974–82.

Gischig V, Preisig G, Eberhardt E. Numerical investigation of seismically induced rock mass fatigue as a mechanism contributing to the progressive failure of Deep-Seated landslides. Rock Mech Rock Eng. 2016;49(6):2457–78.

Preisig G, Eberhardt E, Smithyman M, Preh A, Bonzanigo L. Hydromechanical rock mass fatigue in Deep-Seated landslides accompanying seasonal variations in pore pressures. Rock Mech Rock Eng. 2016;49(6):2333–51.

Tu WF, Li LP, Cheng S, Chen DY, Yuan YC, Chen YH. Evolution mechanism, monitoring, and early warning method of water inrush in Deep-Buried long tunnel. Geofluids. 2021:1–16.

Vulliet L, Hutter K. Viscous-type sliding laws for landslides. Can Geotech J. 1988;25(3):467–77.

Watson AW, Martin CD, Moore DP, Stewart TWG, Lorig L. Integration of geology, monitoring and modelling to assess rockslide risk. Felsbau. 2006;24:50–8.

Yang HW, Xu J, Wu X, Peng SJ, Zhang Y. Experimental analysis of the deformation of sandstone under cyclic pore water pressure and wavelet transformation. J Chongqing Univ. 2011;34:6–12.

Yang HW. Study on coupling mechanism of rock and pore water under cyclic loading. Doctoral dissertation. Chongqing: Chongqing University. 2011:182–98.

Zangerl C, Eberhardt E, Perzlmaier S. Kinematic behaviour and velocity characteristics of a complex deep-seated crystalline rockslide system in relation to its interaction with a dam reservoir. Eng Geol. 2010;112(1–4):53–67.

Chapter 11
Dynamic Damage Effect on Creep Fracture

After the excavation or operation stage of deep underground engineering caverns, the surrounding rock masses bear constant initial ground stress loads, and rock deformation at this time is creep deformation. During the process of rock creep, if subjected to dynamic disturbances (such as explosions, earthquakes, vibration waves, etc.), the rock will remain in a stable state for a certain period of time, but the accumulated microcracks and damages caused by the dynamic disturbances will change the local stress environment. The larger the energy of the dynamic disturbances and the more frequent they occur, the greater the cumulative crack damage in the rock. When the damage reaches a certain degree, the rock will undergo accelerated creep failure. Therefore, the static creep characteristics of brittle rocks under stress wave dynamic disturbances are of great practical significance for the evaluation of the deformation of surrounding rock in deep underground engineering. Based on Chap. 5, this paper introduces a macro–micro mechanical model of the creep fracture characteristics of brittle rocks under stress wave dynamic disturbances, which includes a dynamic disturbance damage evolution function, a static-dynamic load evolution path function, and a viscoelastic constitutive model. The dynamic damage is achieved by controlling the number of microcracks inside the rock. The influence of stress wave amplitude and period on rock dynamic damage is determined.

11.1 Theory

This chapter proposed a macro–micro mechanical model for the static creep and fracture behavior of brittle rocks under dynamic loading and damage. The authors describe the viscoelastic deformation of rocks using the Hooke-Kelvin model, and the plastic deformation using a wing-shaped crack propagation model. They then

X. Li et al., *Static Creep Micro-Macro Fracture Mechanics of Brittle Solids*,
https://doi.org/10.1007/978-981-99-8203-5_11

introduce a function describing the time-dependent evolution of axial stress and a dynamic damage function based on the number of micro-cracks to establish the macro–micro mechanical model.

11.1.1 Evolution of Static Axial Viscoelastic Strain

The Hooke-Kelvin model can well reflect the stress–strain relationship of viscoelastic materials.

$$\sigma_1 + \frac{\eta}{E_v + E_w}\dot{\sigma}_1 = \frac{E_v E_w}{E_v + E_w}\varepsilon_1 + \frac{\eta E_w}{E_v + E_w}\dot{\varepsilon}_1 \tag{11.1}$$

where σ_1 represents the axial stress. In order to facilitate the reading of the calculation results, the compressive stress in the theoretical derivation is set as negative, but the absolute value is taken in the graphs. Ev is the viscoelastic modulus, E_w is the elastic modulus of the rock, η is the viscosity coefficient, and $\dot{\sigma}_1$ represents the rate of change of the axial stress. ε_1 is the axial strain of the rock (Note: in this chapter, since compressive stress is defined as negative, the calculated compressive strain is also negative, but it is indicated as a positive value in the charts and graphs). During preloading, the load is constant, so the rate of change of the axial stress is zero. Based on Eq. (11.1), the Hook-Kelvin model can be expressed as:

$$\sigma_1 = \frac{E_v E_w}{E_v + E_w}\varepsilon_1 + \frac{\eta E_w}{E_v + E_w}\dot{\varepsilon}_1 \tag{11.2}$$

Based on the initial condition that the viscoelastic strain of the rock is 0 at time t = 0, and the axial stress is σ_{1L}, the relationship between the axial viscoelastic strain ε1ve and time under static loading can be solved based on Eq. (11.2):

$$\varepsilon_{1ve}(t) = \left(\frac{1}{E_w} + \frac{1 - e^{-E_v t/\eta}}{E_v}\right)\sigma_{1L} \tag{11.3}$$

11.1.2 Axial Viscoelastic Strain Evolution of Rock Under Dynamic Loading

The dynamic loading stress σ_1 as a function of time can be expressed as:

$$\sigma_1(t) = \frac{\Delta\sigma}{2}\cos\left(\frac{2\pi}{T}t\right) - b \tag{11.4}$$

In the equation, T represents the period of the stress wave, $\Delta\sigma$ represents the stress wave amplitude, and the constant b can be determined based on $\Delta\sigma$ and the axial stress σ_{1L}.

The slope of the stress wave function in Eq. (11.4) is:

$$\dot{\sigma}_1(t) = -\frac{\pi\Delta\sigma}{T}\sin\left(\frac{2\pi}{T}t\right) \tag{11.5}$$

The axial viscoelastic strain of rock at the time of dynamic loading, after a duration of constant load precompression of t_1, can be expressed as:

$$\varepsilon_{1\text{ve}}(t_1) = \left(\frac{1}{E_\text{w}} + \frac{1 - e^{-E_\text{v}t_1/\eta}}{E_\text{v}}\right)\sigma_{1\text{L}} \tag{11.6}$$

The relationship between the viscoelastic strain of the rock and time during the stress wave loading stage can be solved based on Eqs. (11.1), (11.4), (11.5), and (11.6):

$$\varepsilon_{1\text{ve}}(t) = \frac{e^{-E_\text{v}t/\eta}}{E_\text{w}E_\text{v}A_1}\left\{A_1\left(A_3 - \frac{\Delta\sigma}{2}E_\text{v}(A_4 + e^{E_\text{v}t_1/\eta}A_5)\right)\right\} \tag{11.7}$$

where

$$A_1 = E_\text{v}^2T^2 + 4\pi^2\eta^2 \tag{11.8}$$

$$A_2 = e^{E_\text{v}t_1/\eta}(E_\text{w} + E_\text{v}) \tag{11.9}$$

$$A_3 = b[e^{E_\text{v}t/\eta}(E_\text{w} + E_\text{v}) - A_2] + (A_2 - E_\text{w})\sigma_{1\text{L}} \tag{11.10}$$

$$A_4 = e^{E_\text{v}t/\eta}(A_1 + E_\text{w}E_\text{v}T^2)\cos(2\pi t/T) + 2e^{E_\text{v}t/\eta}E_\text{w}\pi T\eta\sin(2\pi t/T) \tag{11.11}$$

$$A_5 = -A_{4(t=t_1)} \tag{11.12}$$

After the stress wave loading is completed, the viscoelastic strain at the end of the stress wave loading stage, $\varepsilon_{1\text{ve(t2)}}$, is taken as the initial value for the viscoelastic strain during the second static axial compression loading stage, which can be obtained by solving Eq. (11.2) with this initial value.

11.1.3 Plastic Strain Evolution Under Microcrack Damage Caused by Dynamic Disturbance

Substituting the dynamic damage evolution function $D_{dy}(t)$ into Eq. (5.17) in Chap. 5 and replacing the initial damage variable D_0 in the equation, the time evolution equation of plastic strain under dynamic perturbation damage is obtained as follows:

$$\varepsilon_{1wcg}(t) = \varepsilon_0\{-\ln[1 - D_{dy}(t)(l(t)/a + 1)^3]\}^{1/m} \tag{11.13}$$

The definition of D_{dy}(t) is as follows:

When subjected to dynamic impact disturbances, the dynamic damage inside the rock increases sharply, resulting from the activation of microscopic cracks inside the rock, and the number of cracks increases sharply. In the following study, it was found that the dynamic damage mutation occurred at the initial moment of stress wave loading, and the form of dynamic damage increase had no effect on the creep failure time of the rock. Therefore, dynamic damage is defined to occur at the initial moment of stress wave loading, and the form of dynamic damage increase is a step mutation. The specific time evolution path of dynamic damage is shown in Fig. 11.2a, and the expression for dynamic damage D_{dy} evolution is:

$$D_{dy}(t) = \begin{cases} D_0 & 0 \le t \le t_1 \\ D_0 + \mathrm{x}\Delta D_0 & t_\mathrm{x} \le t \le t_{\mathrm{x}+1} \end{cases} \tag{11.14}$$

In the equation, x represents the number of consecutive dynamic perturbations of stress waves, and ΔD_0 is the sudden change in dynamic damage caused by each loading of dynamic damage. Since the initial damage D_0 is equal to $N_V a^3$, where N_V is the number of cracks per unit volume of rock, it can be seen that dynamic damage is related to the number and initial size of cracks. Essentially, this paper assumes that the crack size remains unchanged and controls the sudden change in the number of cracks to achieve dynamic damage mutation (see Fig. 11.2b).

Note: (1) Fig. 11.2 only shows the dynamic damage change characteristics of rocks after being subjected to dynamic disturbances on the basis of initial damage, and does not include the damage caused by the extension of wing cracks in rocks under static conditions. Formula (11.13) comprehensively considers the microcrack damage characteristics caused by the joint action of dynamic and static forces. (2) The dynamic damage caused by a single continuous stress wave dynamic perturbation corresponds to a sudden damage mutation value ΔD_0. That is, whether it is a single-cycle stress wave or a continuous cyclic stress wave dynamic perturbation, the model only focuses on the total dynamic damage after this perturbation ends. (3) Assuming that a rock is subjected to dynamic disturbances of the same amplitude, period, and duration, the damage caused should be different as the rock deforms over time.

11.1.4 Complete Viscoelastic-Plastic Creep Under Dynamic Disturbance

In conclusion, the total static creep strain of rock under stress wave dynamic disturbance varies with time according to:

$$\varepsilon_1(t) = \varepsilon_{1p}(t) + \varepsilon_{1ve}(t) \tag{11.15}$$

where ε_{1p} is the pure plastic strain, which can be determined by the difference between the rock strain caused by wing crack propagation and the strain ε_{1ci} before the expansion of the wing crack, as expressed in Eq. (11.13):

$$\varepsilon_{1p}(t) = \varepsilon_{1wcg}(t) - \varepsilon_{1ci} \tag{11.16}$$

$$\varepsilon_{1ci} = \varepsilon_0(-\ln[1 - D_0])^{1/m} \tag{11.17}$$

Assuming that there is no plastic strain in the rock before the crack initiation, the elastic modulus can be determined by the ratio of the crack initiation stress and crack initiation strain at the point where the wing crack is zero, using formulas (5.8) and (5.11), respectively:

$$E_w = \frac{\sigma_{1\,ci}}{\varepsilon_{1\,ci}} = \frac{K_{IC}\beta^{3/2}\sqrt{\pi/a} - \sigma_3 B_2}{\varepsilon_o B_1(-\ln[1-D_o])^{1/m}} \tag{11.18}$$

11.2 Results and Discussions

11.2.1 Model Validation

Based on the micromechanical stress wave perturbation creep model mentioned above, the creep failure experimental results of sandstone (Zhu et al. 2019) and gneiss (Wang et al. 2019) under different dynamic parameters will be selected to verify the rationality of the model. The specific parameter values are shown in Table 11.1.

11.2.1.1 Single-Cycle Stress Wave Dynamic Impact Perturbation

Figure 11.3 presents the theoretical and experimental creep curves of sandstone under single-cycle stress wave perturbations with different static and dynamic loads (Zhu et al. 2019). Under initial constant stress, the rock undergoes deceleration until it reaches a steady-state creep deformation. When subjected to sudden single-cycle

Table 11.1 Model parameters

Parameters	Sandstone	Gneiss
K_{IC} (MPa·m$^{1/2}$)	1.29	1.89
M	1	1
φ (°)	27.39	32.03
μ	0.41	0.51
a (m)	0.0119	0.0036
β	0.126	0.26
v (m·s^{-1})	0.29	0.16
n	40	57
Ev (GPa)	290	290
η (GPa)	4×10^5	3×10^5
ε_0	0.0126	0.0017
σ_3	0	0

impact loading, the axial strain of the rock experiences a step-jump. After the impact load disappears, the axial strain gradually undergoes deceleration until it reaches a steady-state creep deformation again. Subsequently, under the current deformation, the rock undergoes the above deformation process repeatedly until it ultimately fails. In the theoretical calculation results, there is a certain strain rebound phenomenon of the rock's axial strain under each impact load, which is consistent with the numerical simulation results reported in (Yang et al. 2022). The specific rebound process will be analyzed in detail in the following discussion chapter. However, the experimental results cannot capture the transient strain rebound phenomenon well.

11.2.1.2 Continuous Cyclic Stress Wave Dynamic Disturbance

Figure 11.4a–c present the comparison curves of theoretical and experimental strain evolution under cyclic loading with different amplitudes of stress waves (Wang et al. 2019), along with the evolution curve of dynamic damage D_{dy}. The strain increases with the increase of stress wave amplitude, while the creep failure time decreases. Figure 11.6a shows the relationship between stress wave amplitude and dynamic damage mutation value ΔD_0. Assuming that the dynamic damage mutation value ΔD_0 is the same during the continuous disturbance stage of stress waves with the same amplitude in multiple creep processes, ΔD_0 increases rapidly with the increase of stress wave amplitude.

Figure 11.5a–c present the comparison of theoretical and experimental results of rock strain as a function of time under stress waves of the same amplitude but different periods, along with the evolution of dynamic damage D_{dy} as a function of time. Rock strain decreases with an increase in stress wave period, while the time to failure due to creep increases with an increase in stress wave period. Figure 11.6b shows the relationship between stress wave period and dynamic damage threshold ΔD_0.

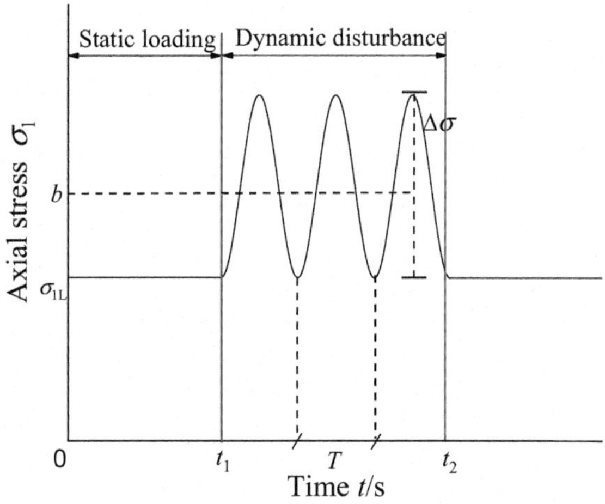

Fig. 11.1 Schematic diagram of stress wave

Assuming that during the process of creep under multiple stress wave perturbations, the dynamic damage threshold remains the same during consecutive perturbation stages of stress waves of the same period, ΔD_0 accelerates its decrease with an increase in stress wave period. In addition, the stress-time evolution path in Figs. 11.4 and 11.5 is consistent with the standard path defined in Fig. 11.1, except for a larger number of cycles, which resulted in slightly irregularities in the graph.

By comparing the dynamic creep curves of sandstone and gneiss obtained from theoretical and experimental studies, it can be seen that although the two types of experiments have different numbers of loading cycles, stress wave amplitudes, and periods, both produce dynamic damage. The difference lies in the definition of the stress-time evolution path function and dynamic damage evolution path function under different experimental conditions. The theoretical and experimental results show good similarity in terms of trend and rock creep failure time, and have certain comparability in terms of rock strain values, but there are still some differences overall.

After analysis, the reasons for these differences can be attributed to the following points: (1) The theoretical model proposed in this paper is based on an isotropic homogeneous elastic body. (2) The proposed model cannot realize the non-uniform and anisotropic mechanical behavior caused by randomly distributed cracks in brittle rocks, but assumes that all crack sizes and angles are average values, and approximates the influence of microcrack geometric behavior on the overall mechanical properties of rocks. (3) It is difficult to accurately describe the specific form of dynamic damage change when the rock is subjected to an impact load. The theoretical model defines the moment and value of dynamic damage mutation as an approximate equivalent presentation of the real situation. (4) Errors in experimental testing and model parameters.

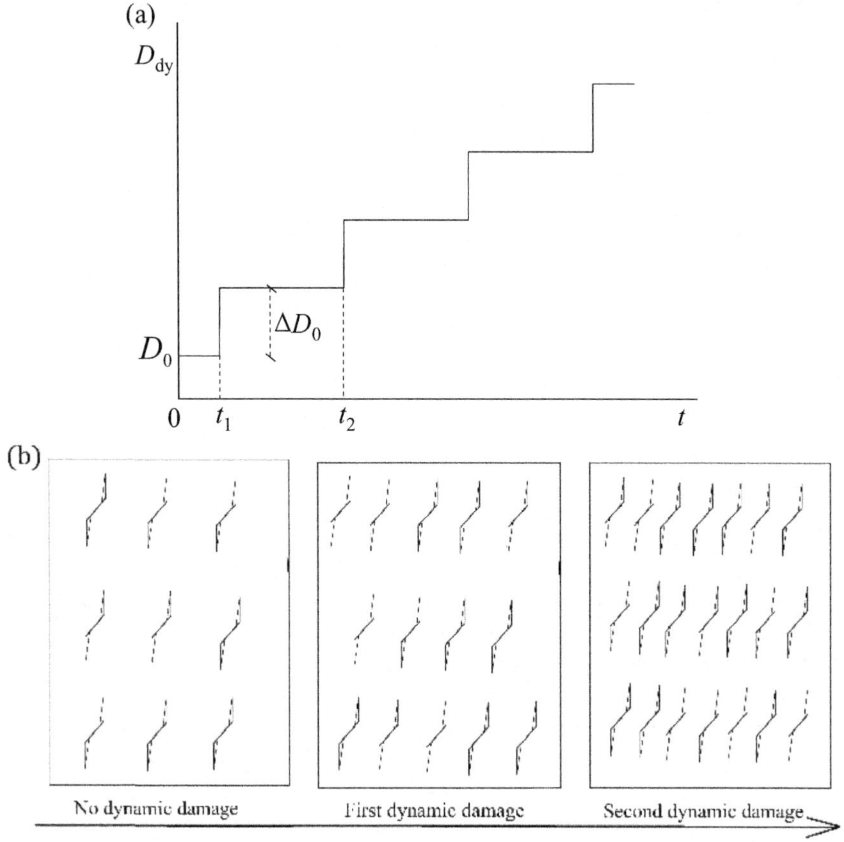

Fig. 11.2 Diagrams of **a** the evolution curve and **b** the corresponding basic theory of dynamic damage in rocks

11.2.2 Sensitivity Analysis of Parameters Related to Single Cycle Dynamic Impact Damage

11.2.2.1 The Impact of Dynamic Damage Size on Creep Failure

When brittle rocks are subjected to stress wave impact loads, the internal microcrack damage increases, leading to changes in their creep failure characteristics. This phenomenon can be equivalent to an increase in the number of internal microcracks in rocks after being subjected to stress wave disturbances, which in turn affects their creep properties. The timing, magnitude, and form of dynamic damage (i.e. damage mutation or gradual change) will be discussed to investigate the impact of dynamic damage on the creep properties of rocks.

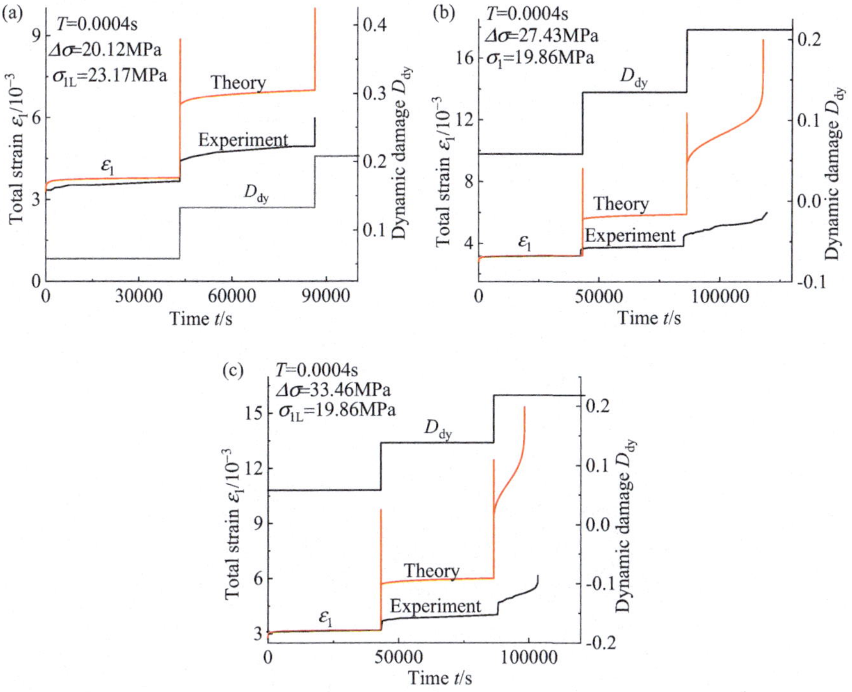

Fig. 11.3 Theoretical and experimental (Zhu et al. 2019) creep curves of sandstone under the single period impact loading of **a** $\sigma 1L = 23.17$ MPa, $\Delta\sigma = 20.12$ MPa, **b** $\sigma 1L = 19.86$ MPa, $\Delta\sigma = 27.43$ MPa, **c** $\sigma 1L = 19.86$ MPa, $\Delta\sigma = 33.46$ MPa

After being loaded with a confining pressure of 60 MPa for two hours along the axial direction, a single cycle of stress wave dynamic loading was performed with a defined stress wave period of 0.001 s and an amplitude of 40 MPa, as shown in Fig. 11.7a. Dynamic damage was defined to undergo an instantaneous mutation at the moment of stress wave loading, with dynamic damage changing from an initial damage value of 0.048 to 0.067, 0.068, and 0.069, as shown in Fig. 11.7b. Figure 10.7c, d show the evolution curves of viscoelastic and plastic strain under one cycle of stress wave dynamic disturbance, respectively. The total strain evolution curve shown in Fig. 11.7e was obtained based on Fig. 11.7c and d. The larger the dynamic damage mutation value, the shorter the rock creep failure time.

If the dynamic damage value is not defined, that is, $\Delta D_0 = 0$, the rock only undergoes viscoelastic strain and does not undergo plastic strain mutation, and brittle rock does not undergo creep failure under dynamic perturbation, as shown in Fig. 11.8. However, when the dynamic damage undergoes a sudden mutation at the moment when the rock is subjected to the stress wave impact, the rock ultimately undergoes creep failure, as shown in Fig. 11.7e, indicating that dynamic impact perturbation accelerates the creep failure of the rock.

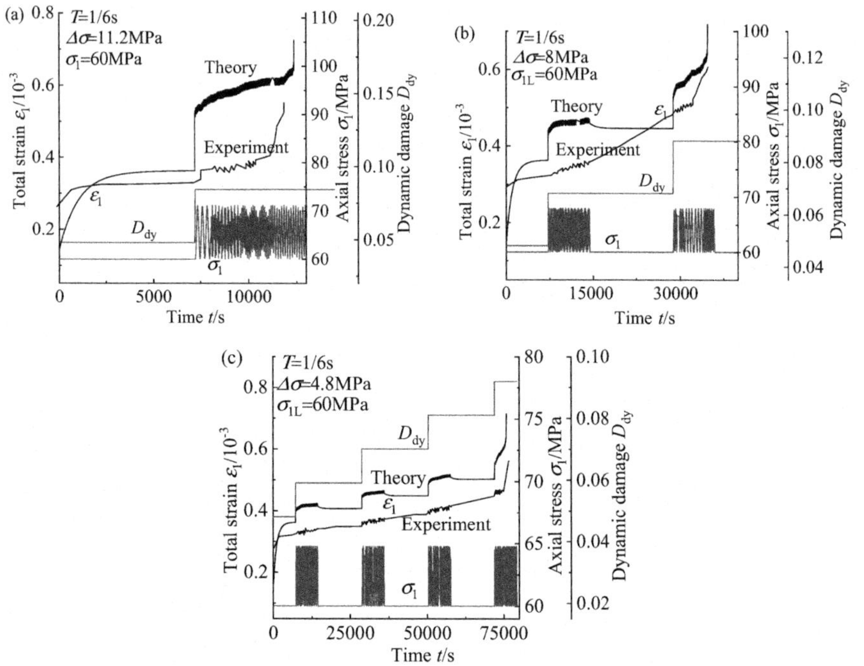

Fig. 11.4 Theoretical and experimental creep curves of gneiss under continuous stress wave loading of $\Delta\sigma$ **a** 11.2 MPa, **b** 8 MPa and **c** 4.8 MPa

Figure 11.9 shows that with an initial dynamic damage value (D_0) of 0.048 and a mutation value (ΔD_0) of 0.009, the instantaneous mutation of dynamic damage in the rock during stress wave loading at different times ($t_z = 0$ s, 0.0003 s, 0.0005 s, 0.0008 s) has no effect on the subsequent strain values. The analysis indicates that the time and form of dynamic damage mutation have no effect on the failure time of rock creep. Additionally, Fig. 11.9b illustrates the relationship between axial strain and time for rock subjected to stress wave loading with a dynamically damaged evolution path of linear increase or mutation. The evolution of dynamic damage has no effect on the subsequent strain values, and the analysis shows that it has no effect on the failure time of rock creep.

In summary, it can be analyzed that the timing and form of dynamic damage mutation have no effect on creep failure. The dynamic damage mutation value is the main influencing factor of rock creep failure. The larger the dynamic damage mutation value, the larger the axial strain of the rock, and the shorter the time for the rock to undergo creep failure.

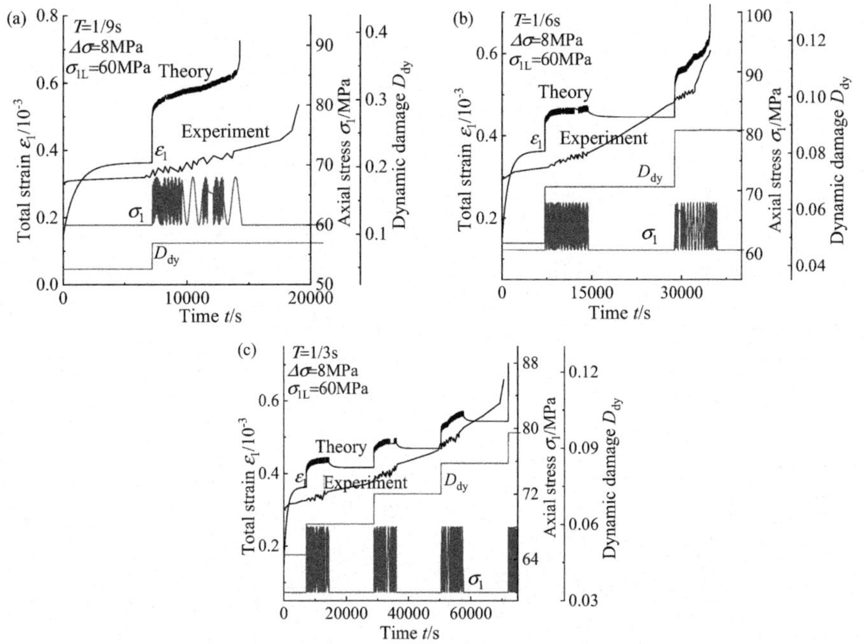

Fig. 11.5 Theoretical and experimental creep curves of gneiss under continuous stress waves with periods of **a** 1/9 s, **b** 1/6 s and **c** 1/3 s

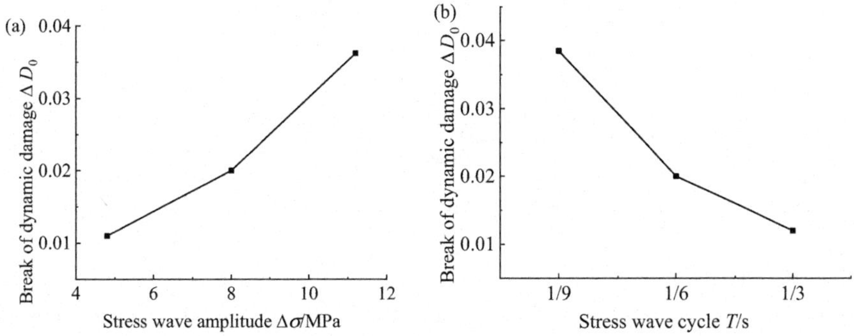

Fig. 11.6 Relationship between the stress wave **a** amplitude and **b** period and the dynamic damage break value

11.2.2.2 The Effect of Stress Wave Amplitude on Rock Creep Failure

Under the condition of axial compression of 60 MPa, after constant loading for two hours, rock specimens were subjected to stress waves with frequencies of 0.001 s and amplitudes of 40 MPa, 30 MPa, and 20 MPa, respectively. The stress path is shown in Fig. 11.10a. Figure 11.10b shows the relationship between axial viscoelastic

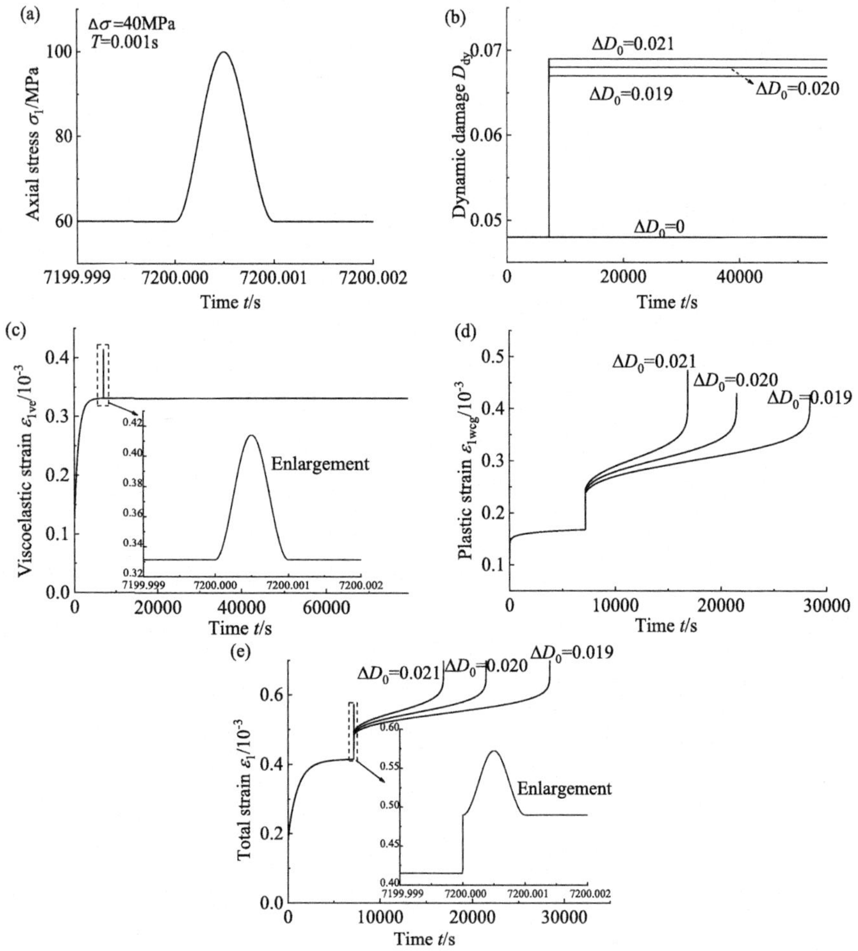

Fig. 11.7 Influence of **b** dynamic damage caused by **a** stress wave disturbance on creep evolution curves of the **c** viscoelastic, **d** plastic and **e** total strain

strain and time under different amplitudes. It can be seen that the rock undergoes strain increase and viscous rebound during stress wave loading and unloading, and the amount of viscous rebound is greatly influenced by the amplitude of the stress wave. The maximum viscous rebound occurs at an amplitude of 40 MPa. When the dynamic damage mutation is the same, the time to creep failure of the rock remains constant, which contradicts the experimental rule presented earlier. Therefore, when the amplitude of the stress wave increases, the dynamic damage mutation value should be appropriately increased to obtain consistent experimental results. The initial damage D_0 is 0.048, and the dynamic damage mutation values ΔD_0 are 0.016, 0.018, and 0.020, respectively, as shown in Fig. 11.10c. Figure 11.10d shows the total creep strain evolution curves of brittle rock under different disturbance amplitudes.

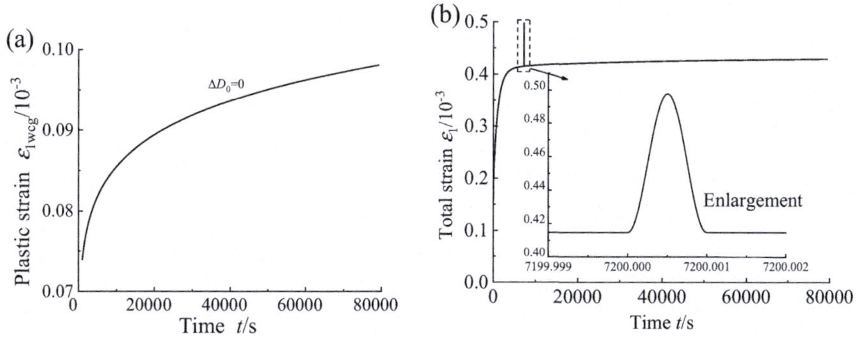

Fig. 11.8 **a** Axial plastic strain and **b** total strain evolution curve of rock without dynamic damage

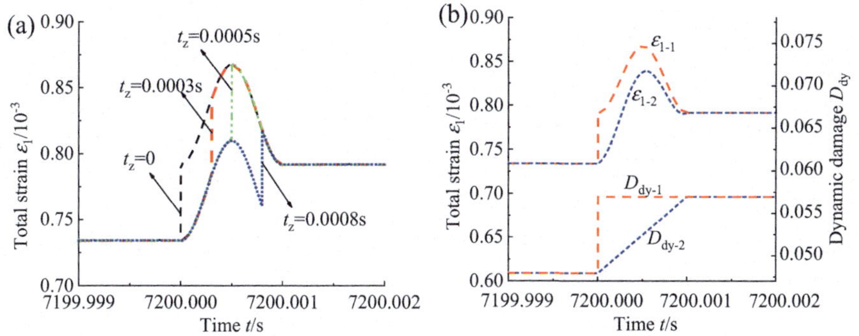

Fig. 11.9 Evolution curves of the axial strain at different **a** times and **b** evolutionary paths of dynamic damage

The rock undergoes attenuated creep and steady-state creep stages before being subjected to stress wave disturbance. After stress wave disturbance, the axial strain of the rock increases sharply, and then the creep strain gradually increases, exhibiting an accelerated creep stage, and finally completely fails. The creep failure times of the rock under amplitudes of 40 MPa, 30 MPa, and 20 MPa are 21,505 s, 38,901 s, and 78,852 s, respectively.

Figure 11.11 shows the axial strain–time curves of rocks under stress wave loading with different amplitudes, without considering the dynamic damage effect. The rocks did not experience creep failure, and only the axial strain varied in magnitude during the stress wave loading stage. The dynamic damage size affects the creep characteristics by changing the stress wave amplitude. The stress wave impact disturbance did not immediately cause rock failure, but caused a sudden increase in local micro-cracks inside the rock (i.e., the dynamic damage mutation in this paper), leading to an increase in damage and a decrease in creep failure time.

In addition, the peak intensity of the stress–strain curve under dynamic impact damage of rocks and the stress wave amplitude under dynamic disturbance, where

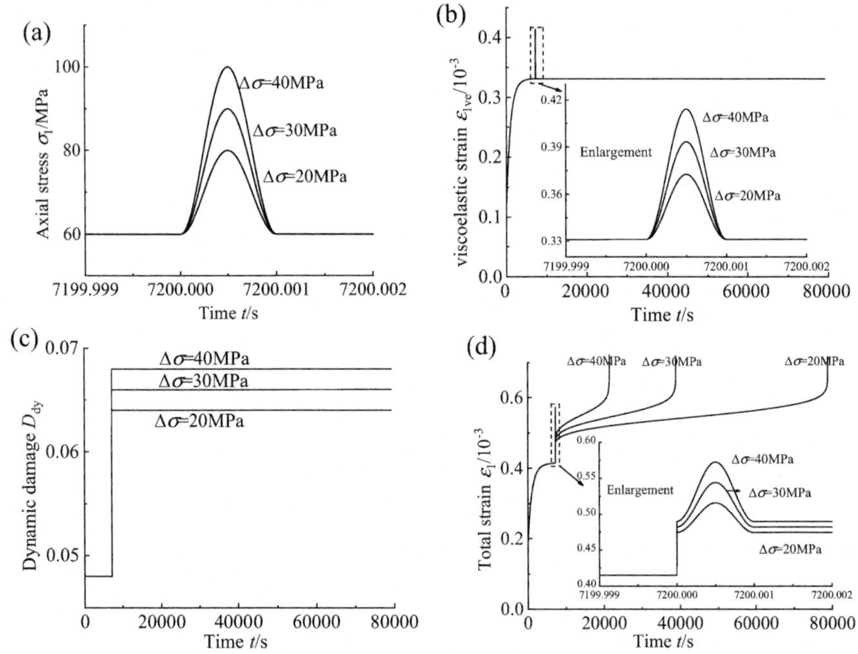

Fig. 11.10 The influence of **a** stress wave amplitude on **c** the dynamic damage and the creep evolution curve of **b** the viscoelastic and **d** total strain

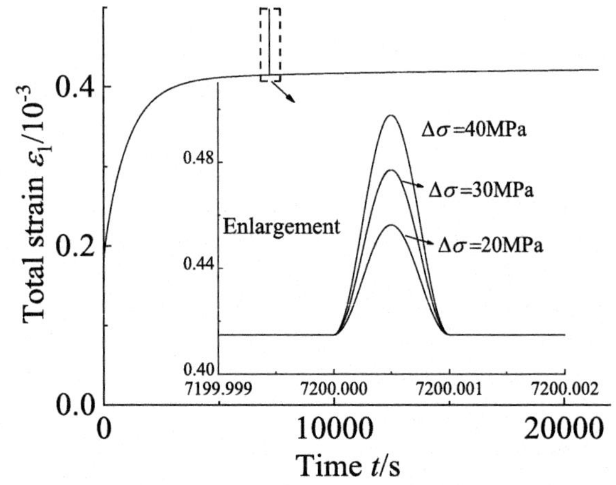

Fig. 11.11 Evolution of axial strain under stress wave loading with different amplitudes without considering dynamic damage effect

the rock did not immediately fail, can be compared, but they are not exactly the same. They can be approximately understood as causing the same level of rock damage. That is, the larger the peak value of the stress–strain curve under dynamic impact to the rock, the greater the degree of rock fragmentation. Similarly, under stress wave

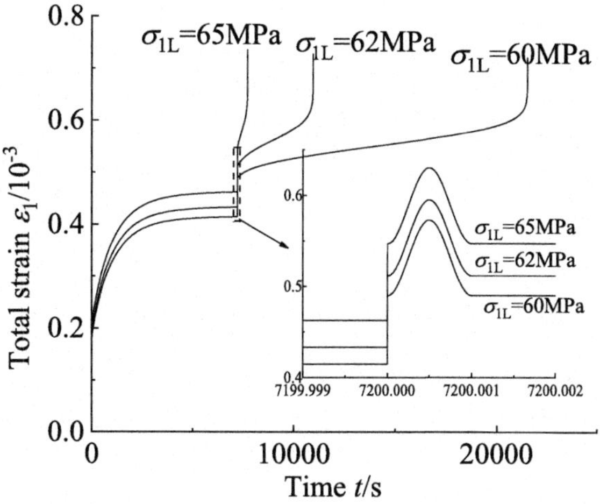

Fig. 11.12 Evolution of axial strain of rock under different axial compression stresses considering dynamic damage effect

disturbance studied in this paper, the larger the amplitude, the more microcracks activated inside the rock and the greater the local damage to the rock.

11.2.2.3 The Influence of Axial Compression on Rock Creep Damage Under Dynamic Disturbance

At the moments when the axial stresses were loaded for two hours at 65, 62, and 60 MPa, a dynamic stress wave disturbance was applied once. In the evolution path of dynamic damage, the initial damage D0 was 0.048, and the dynamic damage mutation value ΔD_0 was 0.020. Figure 11.12 shows the relationship between the axial strain and time under different axial stresses. It can be seen that as the axial stress increases, the creep deformation of the rock increases, and the influence of the stress wave disturbance on the rock becomes more pronounced, resulting in a significant reduction in the creep failure time.

11.2.2.4 The Effect of Stress Wave Period on Rock Creep Damage

After applying a axial stress of 60 MPa for two hours, the rock was subjected to cyclic stress waves with amplitudes of 5.6 MPa and periods of 1/6, 1/9, and 1/12 s for two hours. The creep failure time of the rock did not decrease with decreasing period, contrary to the experimental results reported by Wang et al. (2019). The trend of the dynamic damage mutation value with the evolution of the stress wave period, as obtained from Fig. 11.6b, indicated that the initial damage D_0 was 0.048 and the dynamic damage mutation values ΔD_0 were 0.020, 0.0202, and 0.0205, respectively. The dynamic damage increased slightly with decreasing stress wave period. Figure 11.13a shows the axial strain evolution of the rock subjected to continuous

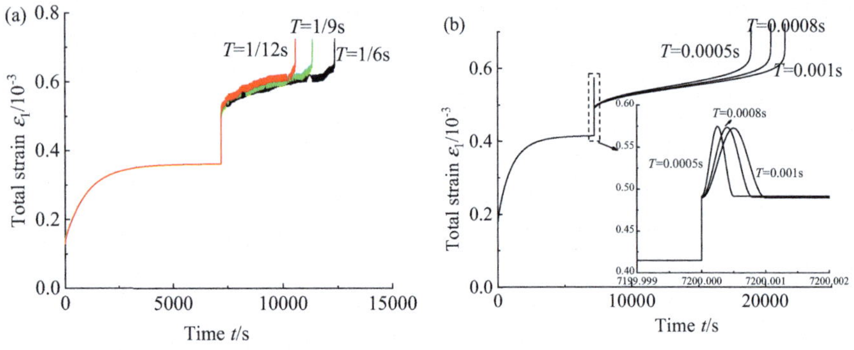

Fig. 11.13 Creep curves during **a** continuous and **b** single dynamic disturbance under different periodic stress waves considering dynamic damage effect

cyclic stress wave perturbations with different periods. After undergoing attenuation creep and steady-state creep stages, the axial creep deformation of the rock gradually increased and exhibited an accelerated creep stage after stress wave perturbation, ultimately resulting in complete failure. The initial damage D_0 was 0.048 and the dynamic damage mutation values ΔD_0 were 0.0362, 0.0368, and 0.0376, respectively, for periods of 1/6 s, 1/9 s, and 1/12 s. The creep failure times of the rock were 12,369 s, 11,353 s, and 10,576 s, respectively. Figure 11.13b shows the relationship between axial strain and time during single stress wave impact loading with different stress wave periods of 0.001 s, 0.0008 s, and 0.0005 s, respectively. The creep failure times were 21,505 s, 20,424 s, and 18,958 s, respectively. The stress wave period had a significant effect on the creep characteristics of the rock, with a smaller stress wave period resulting in a more pronounced dynamic damage effect and faster creep failure.

11.2.3 Optimization of Parameters Related to Cyclic Dynamic Damage

11.2.3.1 Study on Dynamic Damage of Rock Under One-Cycle and Multi-Cycle Stress Wave Disturbance

After stable axial loading preloading, the rock is subjected to a stress wave load consisting of multiple complete cycles, followed by continued axial loading. According to the research of this chapter above, a complete cycle of stress wave loading can cause a sudden increase in dynamic damage transition. Therefore, it can be assumed that during a multi-cycle stress wave loading process, dynamic damage will undergo a stepwise transition increase at the beginning of each cycle of stress wave loading. However, in practical studies, a single stress wave load may contain

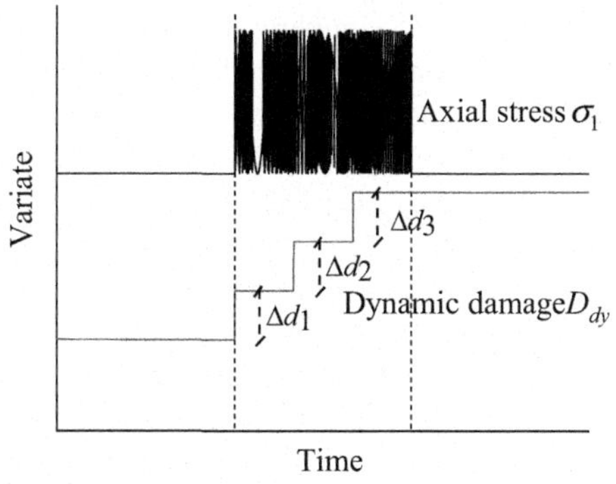

Fig. 11.14 Dynamic damage evolution curve corresponding to a multi-cycle stress wave path

a large number of complete cycles, and it is impossible to present all the dynamic damage transitions that occur in each cycle. Thus, a multi-cycle stress wave load can be equivalent to the continuous loading of n identical complete cycles of stress waves, with an equivalent cycle time of n divided by the stress wave loading time. The dynamic damage is defined as having a stepwise transition increase at each equivalent stress wave loading, so dynamic damage will undergo n transitions during the loading process, with the nth transition resulting in a sudden change in the dynamic damage of Δdn. Figure 11.14 shows the dynamic damage evolution curve corresponding to a single multi-cycle stress wave load path.

(1) **The effect of equivalent stress cycle number**

Under the condition of axial pressure of 60 MPa, the rock was subjected to constant loading for 2 h. Then, a stress wave with a magnitude of 11.2 MPa and a frequency of 6 Hz was applied to the rock for 2 h of continuous disturbance. According to the experimental curve, the axial strain of the rock initially decreased and then gradually increased during the constant loading stage, and then entered a stable development stage. After being subjected to the continuous disturbance of stress waves, the axial strain suddenly increased and had viscous rebound with the change of axial stress. At around 11,000 s, it started to accelerate creep until failure occurred at around 12,000 s. Dynamic damage is defined as an equal step transition increase at each equivalent stress wave loading. Table 11.2 shows the dynamic damage value for each step transition and the total dynamic damage value for a multi-cycle loading under different n values.

Table 11.2 Dynamic damage values corresponding to different damage break numbers

Break number of dynamic damage n	Each break value of dynamic damage Δd_0	Total break value of dynamic damage D_{dy}
1	0.0362	0.0362
3	0.015	0.045
5	0.010	0.050

Figure 11.15 compares the axial total strain–time curves with experimental results under different n values. The initial damage of the rock is 0.048. When $n = 1$, the dynamic damage undergoes a stepwise transition mutation at 7200 s, with a mutation value of $\Delta d_0 = 0.0362$. The axial total strain suddenly increases at 7200 s and the rock undergoes creep failure at 12,369 s. When $n = 3$, the dynamic damage undergoes stepwise transition mutations at 7200, 9600, and 12,000 s, with a mutation value of $\Delta d_0 = 0.015$ each time. The axial total strain increases sharply at the time when the dynamic damage increases and the rock undergoes creep failure at 12,021 s. When $n = 5$, the dynamic damage undergoes stepwise transition mutations at 7200, 8640, 10,800, 11,520, and 12,960 s, with a mutation value of $\Delta d_0 = 0.010$ each time. The axial total strain increases sharply at the time when the dynamic damage increases, and after 11,520 s, the axial strain accelerates to failure at 12,042 s. As the value of n increases, the mutation value of dynamic damage Δd_0 decreases, the total change value of dynamic damage Δd increases, the value of axial strain at the mutation point decreases, the axial strain–time curve becomes more coherent, and the axial strain value, rock creep failure time, and strain–time curve approach the experimental results more closely.

(2) **Effect of trend of increase in dynamic damage value**

When a rock is subjected to multi-cycle stress wave loading, it can be equivalent to n identical and continuous stress wave loadings, and dynamic damage occurs at the time of each equivalent stress wave loading. However, as the rock crack damage intensifies, the size of dynamic damage caused by each identical continuous equivalent stress wave loading is not necessarily the same. Taking $n = 3$ and three dynamic damage mutation values are Δd_1, Δd_2, and Δd_3, the effects of four different situations on the creep failure characteristics of rocks are discussed, including the constant dynamic damage mutation value, the increasing dynamic damage mutation value in turn, the increasing dynamic damage mutation value first and then decreasing, and the

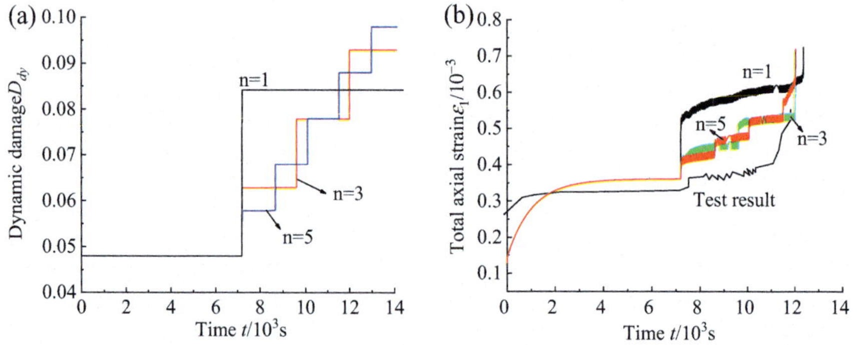

Fig. 11.15 The **b** total axial strain–time curve of rock under the influence of **a** different equivalent damage break numbers

decreasing dynamic damage mutation value in turn. Table 11.3 gives the change value of dynamic damage under each working condition.

Figure 11.16 shows the comparison between the rock strain–time curves and experimental results under different magnitudes of dynamic damage threshold values.

As can be seen, when the four conditions have a similar impact on time, the total damage value can also remain unchanged. When $\Delta d_1 < \Delta d_2 < \Delta d_3$, the axial total strain–time curve most closely matches the experimental results. When $\Delta d_1 > \Delta d_2 > \Delta d_3$, the axial total strain is larger than the other conditions, indicating that the previous dynamic damage mutation is greater and will cause the cumulative increase of axial strain under the next stress wave disturbance. To ensure that the creep failure time remains unchanged, the next damage mutation should be reduced. This provides a reference for selecting the value of each damage mutation. Under the condition of ensuring that the creep failure time of the four working conditions of the rock is similar to the experimental results, when $\Delta d_1 < \Delta d_2 < \Delta d_3$, the axial strain is closest to the experimental results. It is concluded that under the loading of cyclic stress waves, the increase of dynamic damage mutation is closest to the actual situation. It can be known that as the stress wave loads, the damage will accelerate due to the extension of microscopic cracks.

Based on the research results, a theoretical result that is closest to the trend of the experimental strain–time curve was obtained by calculating $n = 5$, $\Delta d_1 = 0.002$, $\Delta d_2 = 0.003$, $\Delta d_3 = 0.007$, $\Delta d_4 = 0.016$, $\Delta d_5 = 0.021$. Figure 11.17 shows the theoretical and experimental comparison results at this time. The axial strain of the rock first

Table 11.3 Dynamic damage values correspond to different working conditions

	Δd_1	Δd_2	Δd_3	D_{dy}
$\Delta d_1 = \Delta d_2 = \Delta d_3$	0.015	0.015	0.015	0.045
$\Delta d_1 < \Delta d_2 < \Delta d_3$	0.010	0.015	0.020	0.045
$\Delta d_1 < \Delta d_2 < \Delta d_3$	0.010	0.020	0.015	0.045
$\Delta d_1 < \Delta d_2 < \Delta d_3$	0.020	0.015	0.010	0.045

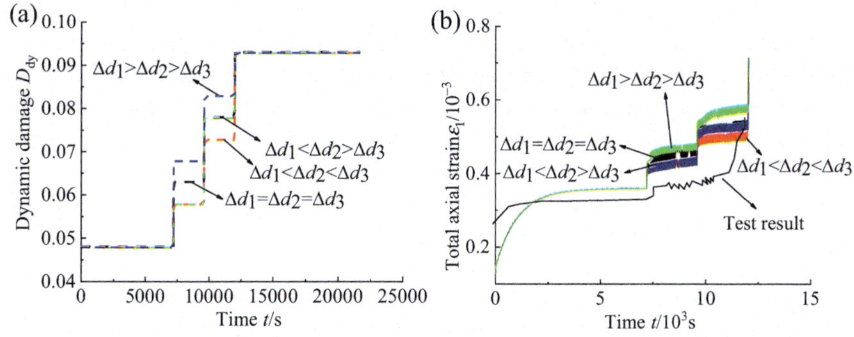

Fig. 11.16 **b** Strain–time curves of rock under **a** different dynamic damage mutation values

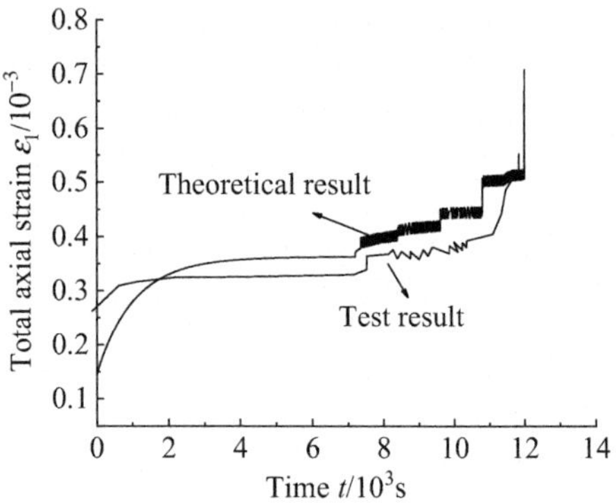

Fig. 11.17 Comparison between theory and experiment

increases and then tends to be stable during the preloading stage, and undergoes a mutation at the moment of stress wave loading, and then increases steadily during the stress wave loading. The theoretical curve suddenly accelerates around 10,800 s, producing a mutation of approximately 0.0005, which corresponds to the mutation in the experimental curve and occurs slightly earlier than the time of the mutation in the experiment. It is inferred that by adjusting the time of each equivalent mutation occurrence, without ensuring that the interval between each mutation occurrence is consistent, the theoretical curve can better match the experimental results.

11.2.3.2 Study on Dynamic Damage Due to Multiple Multi-cycle Stress Wave Perturbations

(1) **Effect of the abrupt increase trend of dynamic damage.**

Under the condition of axial compression of 60 MPa, the rock was subjected to a continuous stress wave with a magnitude of 11.2 MPa and a frequency of 6 Hz for 2 h after a constant load for two hours. After four hours, the same stress wave was applied for another 2 h, and this cycle was repeated four times. The stress path is shown in Fig. 11.18.

Dynamic damage is defined as the occurrence of a sudden jump in damage at each stress wave loading, where the magnitude of dynamic damage remains constant between two consecutive jumps. The magnitudes of the jumps are denoted as ΔD_1, ΔD_2, ΔD_3, and ΔD_4, respectively. The evolution path of dynamic damage is shown in Fig. 11.19.

The initial damage of the rock is 0.048, and four working conditions are studied: $\Delta D_1 = \Delta D_2 = \Delta D_3 = \Delta D_4$, $\Delta D_1 < \Delta D_2 < \Delta D_3 < \Delta D_4$, $\Delta D_1 < \Delta D_2 = \Delta D_3 > \Delta D_4$, and $\Delta D_1 > \Delta D_2 > \Delta D_3 > \Delta D_4$ in Table 11.4.

Figure 11.20 shows the comparison between the rock axial total strain–time curve and experimental results under different sizes of dynamic damage mutation values. As shown in the figure, the rock axial total strain experiences a sudden increase at the moment when dynamic damage undergoes mutation, and accelerates rapidly in the

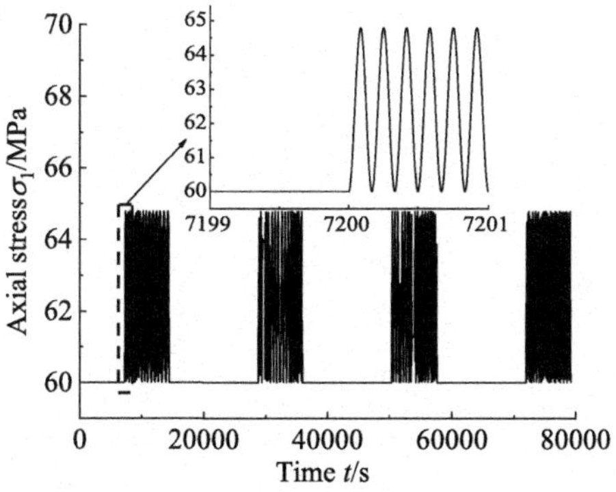

Fig. 11.18 Axial stress-time relationship curve

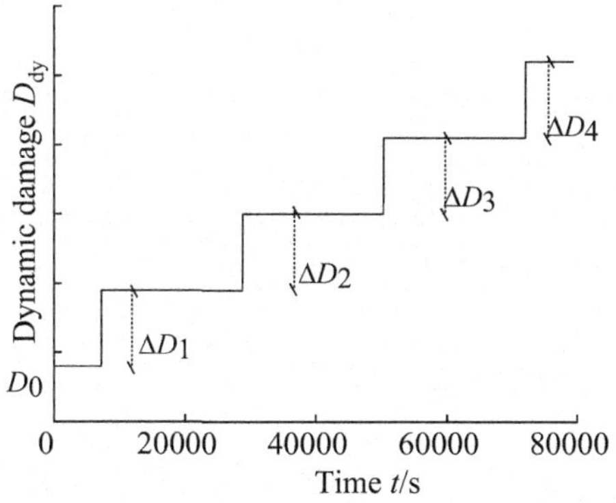

Fig. 11.19 Evolution curve of dynamic damage with abrupt change under multiple and multi-cycle stress wave disturbance

Table 11.4 Dynamic damage values correspond to different working conditions

	ΔD_1	ΔD_2	ΔD_3	ΔD_4	D_{dy}
$\Delta D_1 = \Delta D_2 = \Delta D_3 = \Delta D_4$	0.011	0.011	0.011	0.011	0.044
$\Delta D_1 < \Delta D_2 < \Delta D_3 < \Delta D_4$	0.007	0.009	0.011	0.018	0.045
$\Delta D_1 < \Delta D_2 = \Delta D_3 > \Delta D_4$	0.009	0.013	0.013	0.010	0.045
$\Delta D_1 > \Delta D_2 > \Delta D_3 > \Delta D_4$	0.012	0.011	0.010	0.0095	0.0425

last loading process of multiple stress waves until creep failure occurs. When $\Delta D_1 < \Delta D_2 < \Delta D_3 < \Delta D_4$, the dynamic damage mutation values increase in turn, which are 0.007, 0.009, 0.011, and 0.018 respectively. The rock experiences creep failure at 74,088 s, which is closest to the experimental axial total strain–time curve. This result is consistent with the conclusion obtained from the loading of stress waves for multiple cycles. It indicates that with the increase of stress wave loading cycles, the microscopic crack propagation of rock will lead to the acceleration of dynamic damage.

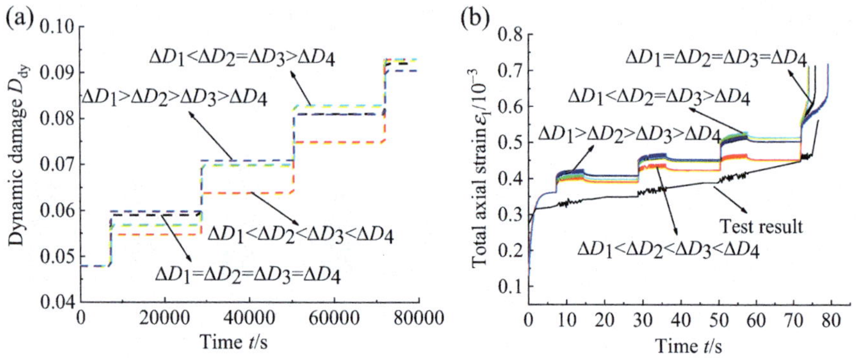

Fig. 11.20 **b** Strain–time curves of rock under different **a** dynamic damage mutation values

(2) **Influence of dynamic damage linear increment trend**

Based on the conclusions drawn from the previous research, dynamic damage will continue to change under stress wave impact. Therefore, the increase in dynamic damage can be defined as a linear increase during the stress wave loading process, with mutation values of ΔD_1, ΔD_2, ΔD_3, and ΔD_4. The evolution path of dynamic damage is shown in Fig. 11.21.

The initial damage of the rock is 0.048, and four working conditions are studied: $\Delta D_1 = \Delta D_2 = \Delta D_3 = \Delta D_4$,, $\Delta D_1 < \Delta D_2 < \Delta D_3 < \Delta D_4$, $\Delta D_1 < \Delta D_2 = \Delta D_3 > \Delta D_4$, and $\Delta D_1 > \Delta D_2 > \Delta D_3 > \Delta D_4$, as described in Table 11.5.

Based on the research in the previous two sections, it can be concluded that when rocks are subjected to multiple, multi-cycle stress waves, the change in dynamic damage increases sequentially, i.e., $\Delta D_1 < \Delta D_2 < \Delta D_3 < \Delta D_4$. Under this condition, the creep failure curve of rocks under the influence of dynamic damage effects is closest to the experimental curve in terms of failure time, strain magnitude, and trend of change. Therefore, by comparing the curves under the influence of linearly

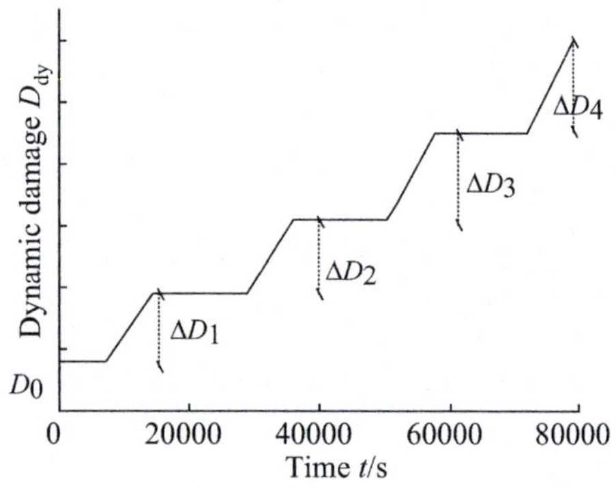

Fig. 11.21 Evolution curve of dynamic damage with linear increase under multiple and multi-cycle stress wave disturbance

Table 11.5 Dynamic damage values correspond to different working conditions

	ΔD_1	ΔD_2	ΔD_3	ΔD_4	D_{dy}
$\Delta D_1 = \Delta D_2 = \Delta D_3 = \Delta D_4$	0.013	0.013	0.013	0.013	0.052
$\Delta D_1 < \Delta D_2 < \Delta D_3 < \Delta D_4$	0.008	0.009	0.014	0.023	0.054
$\Delta D_1 < \Delta D_2 = \Delta D_3 > \Delta D_4$	0.011	0.014	0.014	0.013	0.052
$\Delta D_1 > \Delta D_2 > \Delta D_3 > \Delta D_4$	0.015	0.014	0.012	0.0011	0.052

increasing and abruptly increasing dynamic damage with the experimental results, the evolution law of dynamic damage under the influence of multiple stress wave loading can be obtained, which is most closely related to engineering practice. Figure 11.22 presents a comparison between the theoretical curve of rock axial strain–time relationship under different dynamic damage increase modes and the experimental results.

(3) **Comparison of linear increase and abrupt increase in dynamic damage**

Based on the research in the previous two sections, it can be concluded that when rocks are subjected to multiple, multi-cycle stress waves, the change in dynamic damage increases sequentially, i.e., $\Delta D_1 < \Delta D_2 < \Delta D_3 < \Delta D_4$. Under this condition, the creep failure curve of rocks under the influence of dynamic damage effects is closest to the experimental curve in terms of failure time, strain magnitude, and trend of change. Therefore, by comparing the curves under the influence of linearly increasing and abruptly increasing dynamic damage with the experimental results, the evolution law of dynamic damage under the influence of multiple stress wave loading can be obtained, which is most closely related to engineering practice. Figure 11.23 presents a comparison between the theoretical curve of rock axial strain–time relationship under different dynamic damage increase modes and the experimental results.

In Fig. 11.23, it can be seen that both curves can describe well the increase in axial strain and viscous rebound caused by stress wave loading, and the rock undergoes

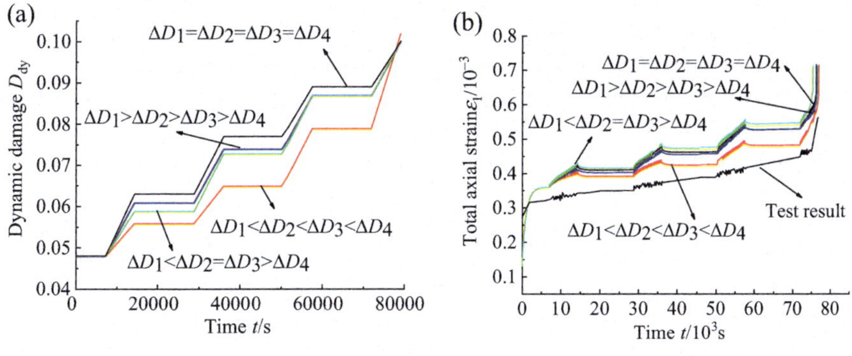

Fig. 11.22 **b** Strain–time curve of rock under **a** linear increment of different dynamic damage

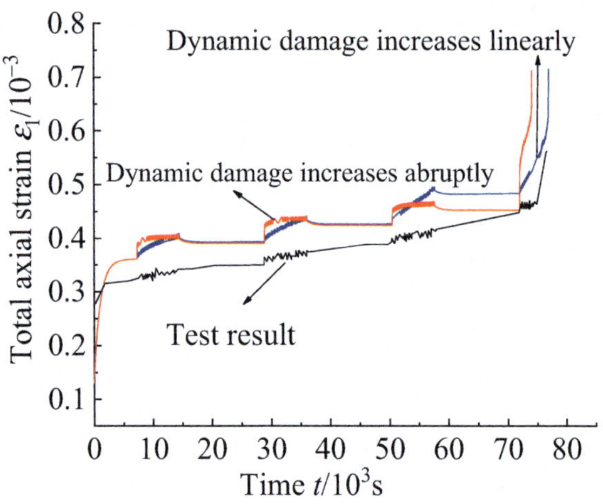

Fig. 11.23 Comparison of experimental results and theoretical curves of axial strain–time relationship of rock affected by different dynamic damage increasing modes

failure during the last stage of multiple stress wave loading. Before failure, the axial strain accelerates to increase, and the failure time is comparable to the experimental results. However, by comparison, it can be found that when the dynamic damage linearly increases during the stress wave loading stage, the total axial strain of the rock gradually increases during the stress wave loading, and the strain curve is more continuous, which is more in line with the actual results compared to the sudden increase in strain at the onset of stress wave loading. Additionally, the magnitude of the axial strain and the rock's creep failure time are closer to the experimental results.

(4) **Comparison between linear-abrupt coupling increase and linear increase in dynamic damage**

Based on the research results of this chapter, it is concluded that in the dynamic damage path, a larger value of n, and an increase in the magnitude of each dynamic damage change results in a linear increase in dynamic damage (i.e., $\Delta D_1 < \Delta D_2 < \Delta D_3 < \Delta D_4$), which is closer to the final experimental results. Therefore, in the case of multiple cyclic stress wave loadings, it is considered that the dynamic damage undergoes a sudden increase coupled with a linear increase in each loading, and there are two sudden increases and two linear increases in dynamic damage for each loading. The magnitude of the sudden increase in dynamic damage is equal to the magnitude of the linear increase. The dynamic damage path under these conditions is shown in Fig. 11.24.

Figure 11.25 compares the strain–time curves of rock under the influences of linear and coupled increases of dynamic damage, as well as experimental results. The coupled curve shows the time of rock's creep failure at 77,208 s, which is consistent with the creep failure time of the rock under the two linear increases of dynamic damage. Compared with linear increase, the coupled increase curve is closer to the experimental results in terms of the axial stress value, and it can better reflect

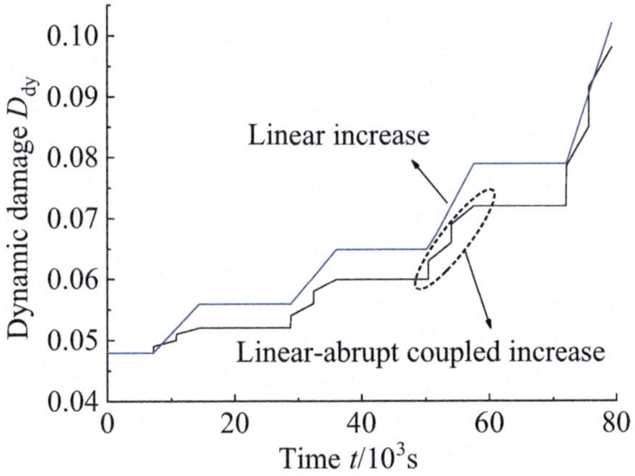

Fig. 11.24 Damage evolution paths under the linear increase and linear-abrupt coupled increase in dynamic damage

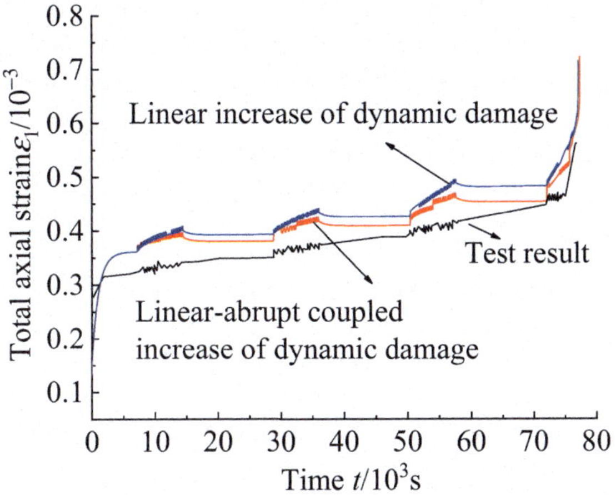

Fig. 11.25 Comparison of strain–time curve between the experimental results and the theoretical results of linear increase and linear-abrupt coupled increase in dynamic damage

the slow increase trend of axial strain during one loading process. Therefore, the coupled increase can better describe the creep failure characteristics of rock under dynamic disturbance than linear increase of dynamic damage. It can be inferred that as the number of dynamic damage mutations and linear increases set in each stress wave loading process increases, the theoretical results of the axial strain–time curve of the rock will become more accurate. When the number of changes is equal to the number of stress wave cycles, the result is consistent with the actual situation. Due to the complexity of the setting, it can be simplified in theoretical research by increasing the number of changes as much as possible to improve the accuracy of theoretical results.

11.3 Conclusion

Based on the rock viscoelastic model, microcrack propagation mechanics model, and dynamic damage effect, a macro and micro mechanical model for creep of brittle rock under stress wave disturbance was established. The creep failure characteristics of rock under dynamic disturbance were mainly studied, and the rationality of the model was verified by experimental results. The conclusions are as follows:

(1) With the increase of dynamic damage, the greater the axial strain of rock, the shorter the time of creep failure. Under single-pulse stress wave loading, the timing and form of dynamic damage change have almost no effect on the strain and creep failure time of rock after stress wave loading. Dynamic damage can be equivalent to instantaneous mutation occurring at the moment of stress wave loading. Moreover, the amount of dynamic damage change accelerates with the increase of stress wave amplitude and the decrease of period.
(2) With the increase of stress wave amplitude, the axial strain of rock increases, and the creep failure time of dynamic disturbance decreases. With the decrease of stress wave period, the axial strain of rock increases, and the creep failure time of dynamic disturbance decreases. The greater the static pre-compression stress on the rock, the greater the strain of the rock, and the faster the strain changes with time, resulting in a shorter creep failure time of dynamic disturbance.
(3) By defining different dynamic damage evolution curves under the same stress path, the creep evolution curve of rock under multi-cycle stress wave disturbance affected by dynamic damage effect was established. By comparing with experimental results, the evolution form of dynamic damage was optimized. The more frequent the dynamic damage mutation is, the greater the mutation value increases with time, and the damage changes from mutation form to linear increase form, all of which promote the theoretical results to be closer to the experimental results.

References

Wang JG, Liang B, Yang PJ. Creep experiment and nonlinear disturbance creep model of gneiss under dynamic and static loads. J China Coal so. 2019;44(1):192–8.
Yang Z, Zhu WC, Guan K. Influence of dynamic disturbance on rock creep from time, space and energy aspects. Geomat Nat Hazards and Risk. 2022;13(1):1065–86.
Zhu W, Li S, Li S, Niu L. Influence of dynamic disturbance on the creep of sandstone: An experimental study. Rock Mech Rock Eng. 2019;52:1023–39.

Chapter 12
High Temperature Heat-Treated Effect on Creep Fracture

The creep failure behavior of brittle rocks subsequent to high-temperature heat treatment holds significant implications for the advancement of deep subsurface resource utilization. The immediate and prolonged mechanical responses of rocks post-heat treatment may manifest diverse tendencies amid varying confining pressures accompanying temperature escalation. However, the underlying causalities remain obscured. Research focusing on elucidating macroscopic and microscopic mechanical models that expound the correlation between microcrack propagation and macroscopic deformation during both short-term deformation and long-term creep processes in brittle rocks subject to high-temperature heat treatment is scant. Thus, this chapter proposed a micro–macro model to solve the above problems on basis of Chap. 5 by introducing five temperature-dependent microcrack model parameters, i.e., initial crack damage (D_0), fracture toughness (K_{IC}), crack extension stress corrosion index (n), characteristic crack propagation rate (v), and initial crack friction coefficient (μ). These parameters are deduced via dedicated experimentation. The model elucidates the evolution of the strain–time curve for heat-treated rocks, while discerning the coupled influences of distinct temperatures and confining pressures on parameters such as rock initiation stress, peak strength, long-term strength, and expiration time.

12.1 Theory

Based on Chap. 5, we introduce two experimentally derived variations in friction coefficients (μ) associated with different heat treatment temperatures (Tang and Zhang 2020; Tang 2020). Furthermore, we integrate experimental data (Fan et al. 2018; Huang et al. 2017; Nasseri et al. 2007; Xu and Zhang 2018; Li et al. 2021; Sha et al. 2020; Srinivasan et al. 2020) that elucidate the temperature-dependent fluctuations in initial damage (D_0), fracture toughness (K_{IC}), stress corrosion index

X. Li et al., *Static Creep Micro-Macro Fracture Mechanics of Brittle Solids*,
https://doi.org/10.1007/978-981-99-8203-5_12

(n), and characteristic crack propagation rate (v). These data serve as the foundation for constructing both macroscopic and microscopic mechanical models for brittle rocks following heat treatment-induced damage. This model account for the diverse trends observed in the variations of friction coefficients. The specific theoretical framework is articulated as follows:

12.1.1 Heat Treated Temperature-Related Parameters

Presently, a multitude of scholars have undertaken experimental investigations into rocks following exposure to high-temperature heat treatments (Fan et al. 2018; Huang et al. 2017; Feng et al. 2019; Nasseri et al. 2007; Xu and Zhang 2018; Li et al. 2021; Sha et al. 2020; Srinivasan et al. 2020). These investigations aim to shed light on mechanical properties such as the elastic modulus (E), initial damage (D_0), fracture toughness (K_{IC}), longitudinal wave velocity (V_P), and certain long-term mechanical parameters. Among these enduring mechanical parameters are the characteristic crack propagation rate (v) and the stress corrosion index (n) within the subcritical crack growth equation $dl/dt = v(K_I/K_{IC})^n$. Notably, these parameters exhibit variation in response to heat treatment temperature. The preponderance of research findings indicates that, up to a certain critical temperature threshold, internal rock damage remains minimal. However, beyond this threshold temperature, the internal rock structure experiences a rapid deterioration. Consequently, the trends observed in these relevant parameters with respect to temperature typically follow a discernible pattern characterized by an initial phase of relative stability, followed by a steep increase or decrease as the temperature surpasses the threshold, culminating in a return to a state of stability. This pattern is illustrated in Fig. 12.1 and can be described collectively using a fitting function:

$$Y = A_2 + \frac{(A_1 - A_2)}{1 + e^{(T - A_3)/A_4}} \tag{12.1}$$

Nonetheless, an expanding body of experimental evidence (Shao et al. 2014; Liu and Xu 2015; Chen et al. 2017) has demonstrated that while heat treatment leads to the deterioration of the internal structure of rocks, certain mechanical properties of specific rocks manifest distinct trends as temperature rises. Notably, the frictional attributes of rocks (Tang et al. 2019; Tang and Zhang 2020; Tang 2020; Meng et al. 2021) (as depicted in Fig. 12.2) and even certain mechanical parameters, such as peak strength (Wu et al. 2022) and long-term strength (Liu et al. 2023), can exhibit an increase in response to higher heat treatment temperatures, particularly under specific confining pressures.

In Fig. 12.2, the basic friction angle, determined through experiments conducted by the references (Tang and Zhang 2020; Tang 2020), displays a linear correlation with heat treatment temperature, aligning with the linear equation $\varphi_b = A_1 T + A_2$. By amalgamating this basic friction angle with the connection between the friction

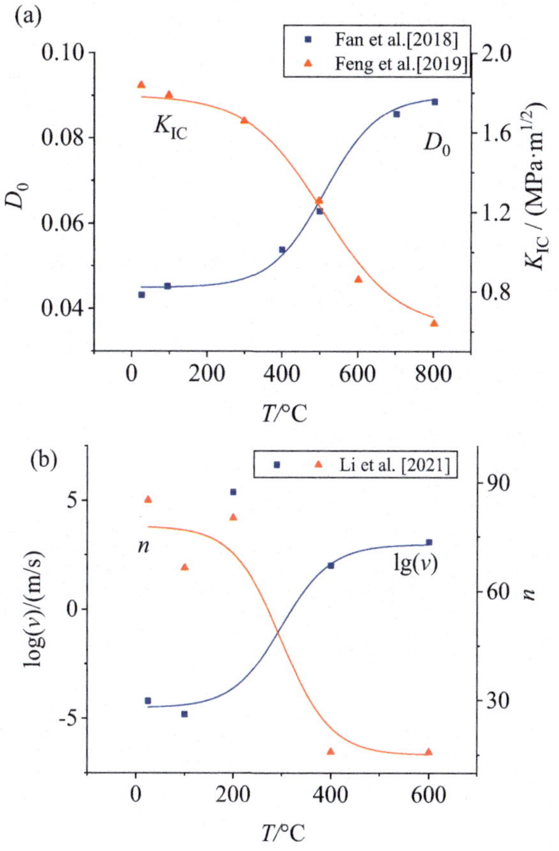

Fig. 12.1 Curves of initial damage (D_0), fracture toughness (K_{IC}), stress corrosion index (n), and characteristic crack propagation rate (v) with changing temperatures (Fan et al. 2018; Feng et al. 2019; Li et al. 2021)

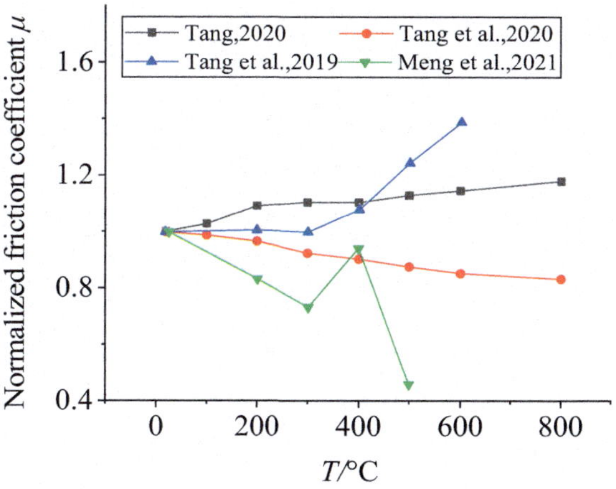

Fig. 12.2 Normalized curves of friction coefficient with heat treatment temperature (Tang and Zhang 2020; Tang 2020; Tang et al. 2019; Meng et al. 2021)

angle and the friction coefficient, we can express the friction coefficient as follows:

$$\mu = \tan(A_1 T + A_2) \tag{12.2}$$

where A_1, A_2, A_3, and A_4 are material constants for the fitting equation.

12.1.2 *Stress–Strain Relation for Brittle Rocks After Heat Treatment*

To establish the relationship between the stress state and crack propagation length in heat-treated rocks, Eq. (5.8) can be integrated with the temperature-dependent fracture toughness $K_{\text{IC}}(T)$, initial damage $D_0(T)$, and friction coefficient $\mu(T)$ from formulas (12.1) and (12.2) as follows:

$$\sigma_l(l, T) = \frac{K_{IC}(T) - \sigma_3\left(B_1\pi\left(\frac{D_0(T)}{N_v}\right)^{\frac{2}{3}}\eta + 2\sqrt{l/\pi}\right)}{B_2\pi\left(\frac{D_0(T)}{N_v}\right)^{\frac{2}{3}}\eta} \tag{12.3}$$

where

$$B_1 = \frac{(\sin 2\varphi + \mu(T) + \mu(T)\cos 2\varphi)\sin 2\varphi}{2} \tag{12.4}$$

$$B_2 = \frac{(\mu(T) - \mu(T)\cos 2\varphi - \sin 2\varphi)\sin 2\varphi}{2} \tag{12.5}$$

$$\eta = [\pi(l + \beta a)]^{-3/2} + 2\sqrt{l/\pi}\left[S - \pi(l + \alpha a)^2\right]^{-1} \tag{12.6}$$

When $l = 0$, based on Eq. (12.3), the temperature-dependent crack initiation stress σ_{1ci} for heat treatment is given by:

$$\sigma_{\text{1ci}}(T) = \frac{K_{\text{IC}}(T) - \sigma_3 B_1 \pi^{-\frac{1}{2}}\beta^{-\frac{3}{2}}\left(\frac{D_0(T)}{N_v}\right)^{\frac{1}{6}}}{B_2\pi^{-\frac{1}{2}}\beta^{-\frac{3}{2}}\left(\frac{D_0(T)}{N_v}\right)^{\frac{1}{6}}} \tag{12.7}$$

Introducing formula (12.1) into Eq. (5.12), the relationship between axial strain and wing crack length after heat treatment can be expressed as:

$$l = a\left[D_0^{-1/3}(T)\left(1 - e^{-(\varepsilon_1/\varepsilon_0)^m}\right)^{1/3} - 1\right] \tag{12.8}$$

Substituting (12.8) into (12.3), the relationship between axial stress and strain with respect to heat treatment temperature is as follows:

$$\sigma_1(\varepsilon_1, T) = \frac{K_{IC}(T) - \sigma_3\left(B_1\pi\left(\frac{D_0(T)}{N_v}\right)^{\frac{2}{3}} J_1 + 2\sqrt{J_2/\pi}\right)}{B_2\pi\left(\frac{D_0(T)}{N_v}\right)^{\frac{2}{3}} J_1} \tag{12.9}$$

where

$$J_1 = [\pi(J_2 + \beta a)]^{-3/2} + \frac{2\sqrt{J_2/\pi}}{S - \pi(J_2 + \alpha a)^2} \tag{12.10}$$

$$J_2 = a\left[D_0^{-1/3}(T)\left(1 - e^{-(\varepsilon_1/\varepsilon_0)^m}\right)^{1/3} - 1\right] \tag{12.11}$$

12.1.3 *Creep Model for Heat-Treated Brittle Rocks*

Substituting the temperature-dependent fracture toughness $K_{IC}(T)$, initial damage $D_0(T)$, characteristic crack velocity $v(T)$, stress corrosion index $n(T)$, and friction coefficient $\mu(T)$ from formulas (12.1) and (12.2) into Eq. (5.16), we deduce the following relationships as a function of heat treatment temperature:

$$l\prime(T) = v(T)\left(\left(\pi\eta D_0^{\frac{2}{3}}(T)\right)\left(\frac{\sigma_1 B_2 + \sigma_3\left(B_1 + 2\sqrt{l/\pi}\right)}{N_v^{\frac{2}{3}} K_{IC}(T)}\right)\right)^{n(T)} \tag{12.12}$$

By incorporating the obtained time evolution results of crack length into the equation derived from transformation in Eq. (12.8) representing strain in terms of crack length, we can subsequently obtain the equation governing the time evolution of rock strain induced by stress corrosion at different heat treatment temperatures:

$$\varepsilon_1(t, T) = \varepsilon_0\left(-\ln\left[1 - D_0(T)(l(t, T)/a)^3\right]\right)^{\frac{1}{m}} \tag{12.13}$$

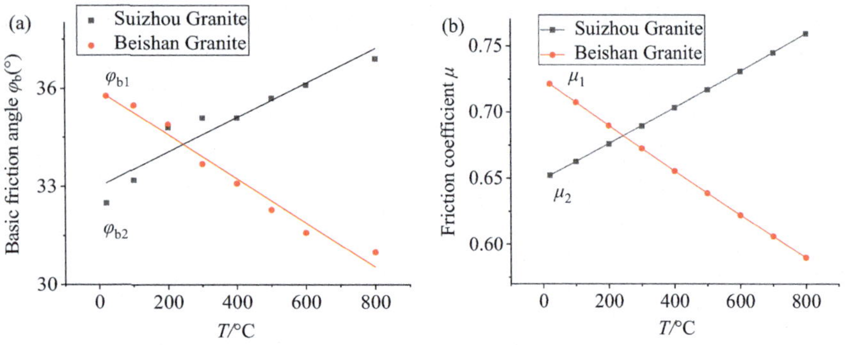

Fig. 12.3 Curves of **a** basic friction angle and **b** friction coefficient with temperature (Tang and Zhang 2020; Tang 2020)

12.2 Results and Discussion

12.2.1 Parameter Selection

This study focuses on granite-specific parameters. Experimental data (Fan et al. 2018; Huang et al. 2017; Feng et al. 2019; Nasseri et al. 2007; Xu and Zhang 2018; Li et al. 2021; Sha et al. 2020; Srinivasan et al. 2020) reveal that formula (1) aptly describe the temperature-dependent changes in various parameters such as initial damage (D_0), fracture toughness (K_{IC}), stress corrosion index (n), and logarithmic function of characteristic crack propagation rate log(v). The behavior of the basic friction angle with respect to temperature is intricate, demonstrating both increasing and decreasing trends, as illustrated in Fig. 12.3a. Considering the relationship between the basic friction angle and the friction coefficient, $\mu = \tan(\varphi_b)$, Fig. 12.3b presents the function describing the friction coefficient's variation with temperature, in accordance with formula (12.2). For clarity in subsequent discussions, we denote the increasing and decreasing trends of the friction coefficient with temperature as μ_1 and μ_2, respectively. The fitting parameters for the temperature-dependent model characteristic parameters can be found in Table 12.1. Furthermore, the initial crack size and angle within the rock were determined using a computational approach: $a = 2.1$ mm and $\varphi = 33°$. The material constants for the rock are as follows: $m = 2$, $\varepsilon_0 = 1/48$, and $\beta = 0.19$.

12.2.2 Model Validation

Figure 12.4 provides a comparative analysis, aligning theory with experimentation, of short-term peak strength $\sigma_{1\text{peak}}$ in relation to temperature across various friction coefficient scenarios. Notably, discernible differences emerge in the trends of

Table 12.1 Fitting curve parameters

Parameter	A_1	A_2	A_3	A_4
D_0	0.06	0.07	470.39	40.70
K_{IC}	1.72	1.3	472.88	49.19
n	75	40	470	50
$\log(v)$	−2.7	0.6	470	50
φ_{b1}	−0.00676	35.96	–	–
φ_{b2}	0.0106	27.00	–	–

σ_{1peak} concerning temperature, contingent on the specific friction coefficient values. Specifically, when the friction coefficient μ equals μ_1, σ_{1peak} exhibits a consistent decline as temperature rises. However, when μ equals μ_2, the trend varies: initially, σ_{1peak} rises with temperature before subsequently declining (it's worth noting that the threshold temperature may vary among rock samples, and these adjusted parameters are clearly labeled in Fig. 12.4). Drawing from this comparison, the ensuing sections will delve into an in-depth analysis of the creep fracture characteristics of the rock, taking into account these two distinct friction profiles.

After meticulous numerical analysis, the selection of μ_2 emerges as a better fit to the experimental outcomes. Consequently, this choice yields graded loading creep curves at distinct heat treatment temperatures, as visually depicted in Fig. 12.5a–c. These curves, whether theoretical or experimental, exhibit a consistent pattern: prior to reaching the threshold temperature of 300 °C, corresponding to minor thermal damage to the rock's internal structure, negligible disparities manifest in the graded loading curves. However, when the heat treatment temperature surpasses 600 °C, exceeding the threshold, more pronounced structural deterioration within the rock occurs. This manifests initially as a relatively stable failure time (Fig. 12.5d) followed by a precipitous decline.

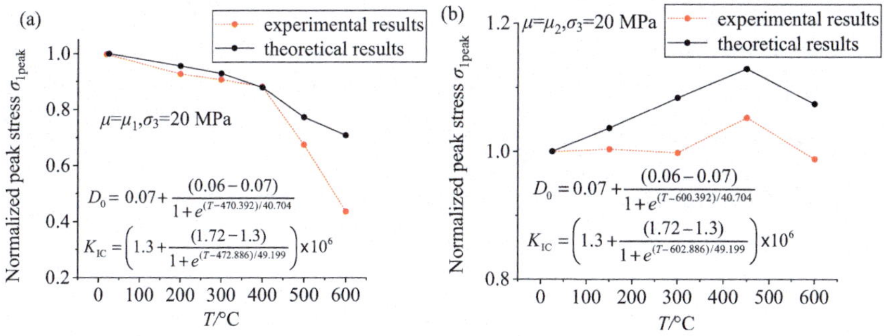

Fig. 12.4 Curves of peak stress comparison for **a** $\mu = \mu_1$, **b** $\mu = \mu_2$, after heat treatment (Wu et al. 2022; Zhu et al. 2021)

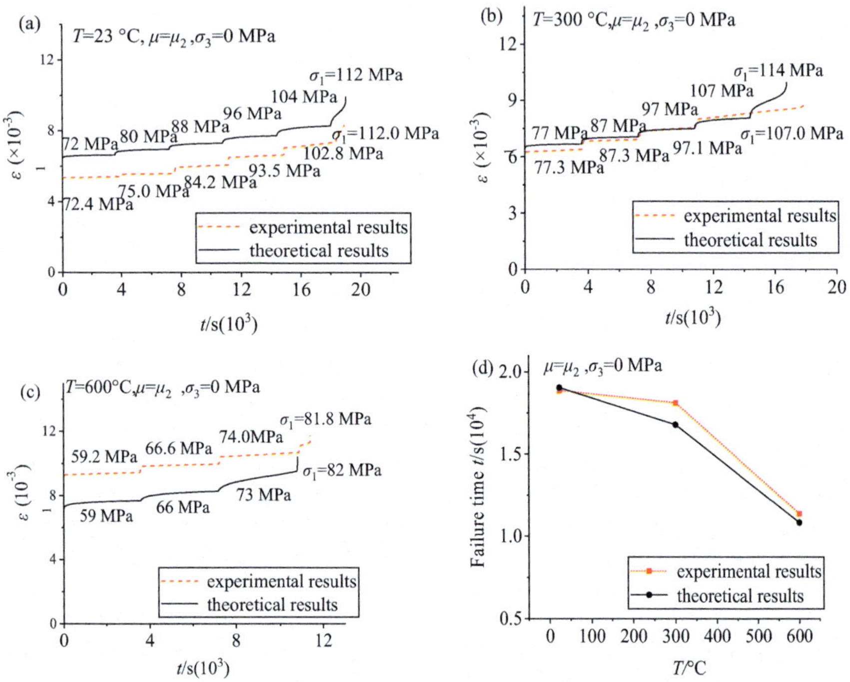

Fig. 12.5 Comparison curves of theoretical and experimental (Liu et al. 2023) for **a**–**c** staged loading and **d** failure time at different heat treatment temperatures

Figure 12.6 further illustrates normalized theoretical–experimental comparison curves, delineating the long-term strength trends under varying confining pressures (all data are normalized by dividing by the initial data point). As heat treatment temperature rises, distinctive trends emerge in the long-term strength response to different confining pressures. When $\sigma_3 = 0$, long-term strength initially stabilizes and subsequently experiences rapid deterioration. Conversely, under a confining pressure of 25 MPa, long-term strength generally increases with rising temperature, suggesting that the rock's lateral confinement is subject to the influence of the friction coefficient.

12.2.3 Comparative Analysis of Short-Term Mechanical Properties of Heat-Treated Rocks

Figure 12.7 presents the constitutive diagrams for heat-treated rocks under different friction coefficient trends. Comparing $\mu = \mu_1$ with $\mu = \mu_2$, it can be observed that under $\mu = \mu_2$, the range of variation in peak strength $\sigma_{1\text{peak}}$ and crack initiation stress $\sigma_{1\text{ci}}$ at different heat treatment temperatures is reduced, indicating that the sensitivity of the rock's short-term mechanical properties to temperature decreases.

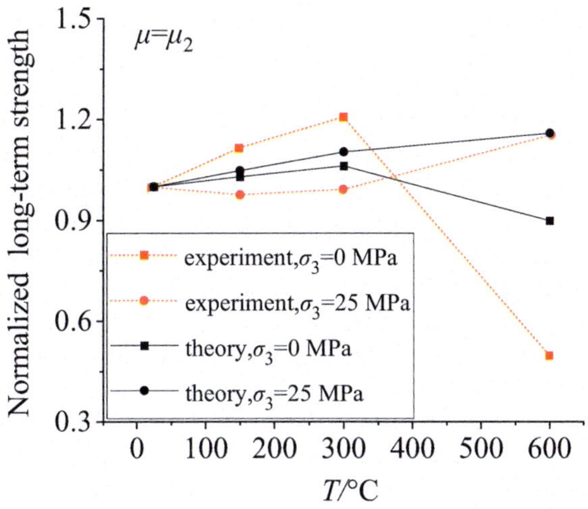

Fig. 12.6 Curves of theoretical experimental comparison for long-term strength after heat treatment (Liu et al. 2023)

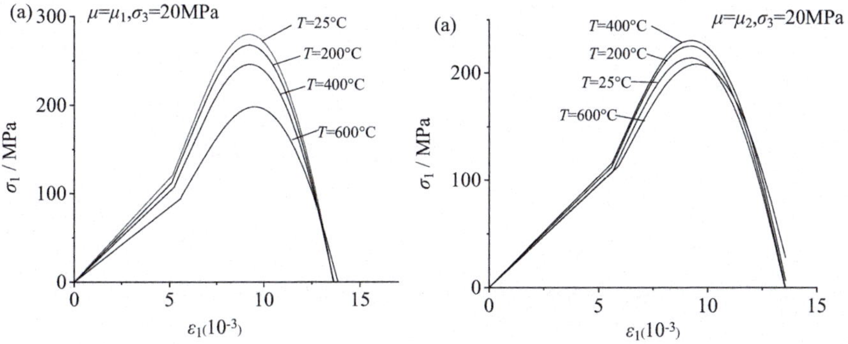

Fig. 12.7 Constitutive curves for **a** $\mu = \mu_1$, **b** $\mu = \mu_2$, after heat treatment

Peak strength ($\sigma_{1\text{peak}}$) and crack initiation stress ($\sigma_{1\text{ci}}$) concerning heat treatment temperature at various confining pressures (σ_3) are shown in Figs. 12.8 and 12.9. Notably, it is observed that confining pressure (σ_3) can influence the scope and tendencies of the rock's short-term mechanical characteristics with temperature under different friction coefficient trends. Specifically, for $\mu = \mu_1$, escalating confining pressure (σ_3) results in a wider spectrum of changes in peak strength ($\sigma_{1\text{peak}}$) and initiation stress ($\sigma_{1\text{ci}}$) with temperature, both exhibiting a consistent decline. However, under $\mu = \mu_2$, these properties display distinct behavior under varying confining pressures: $\sigma_{1\text{peak}}$ initially ascends with temperature before declining. Notably, the trajectory of initiation strength ($\sigma_{1\text{ci}}$) with temperature alters with increasing confining pressure: at lower confining pressures (σ_3), $\sigma_{1\text{ci}}$ initially rises and then falls with temperature, but when $\sigma_3 = 20$ MPa, σ_{1ci} consistently increases with temperature, mirroring trends seen in the preceding section concerning long-term strength shifts. This hints at a potential link between σ_{1ci} and long-term strength, a connection we will delve into further in the subsequent analysis of long-term mechanical properties.

Creep failure in rocks necessitates that axial stress (σ_1) falls within the range: $\sigma_{1ci} \le \sigma_1 < \sigma_{1peak}$. Figure 12.10 illustrates the variation in the span of axial stress values when subjected to triaxial loading under different friction coefficient trends and temperatures. It depicts σ_{1peak}, σ_{1ci} (Fig. 12.10a, b), and $\sigma_{1peak} - \sigma_{1ci}$ (Fig. 12.10c), representing the range of axial stress values associated with creep failure. When $\mu = \mu_1$, this range gradually contracts, while for $\mu = \mu_2$, it initially expands before contracting. Beyond the threshold temperature of heat treatment, these two curves tend to converge. This suggests that, under various friction coefficient trends, disparities in short-term mechanical performance gradually diminish, with the degradation of the internal rock structure emerging as the primary factor influencing short-term mechanical properties.

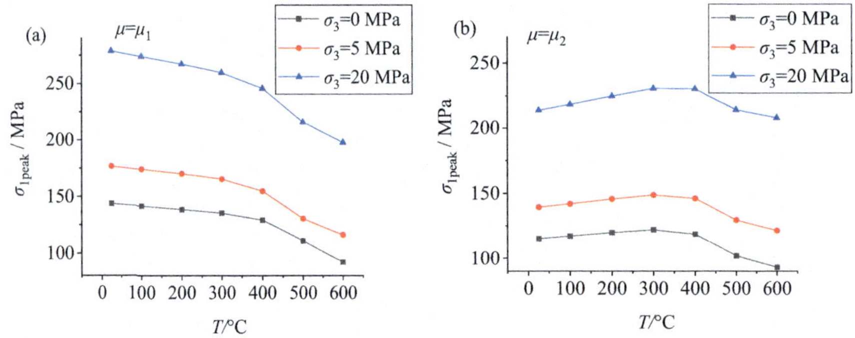

Fig. 12.8 Curves of peak strength with temperature under different confining pressures for **a** $\mu = \mu_1$, **b** $\mu = \mu_2$ after heat treatment

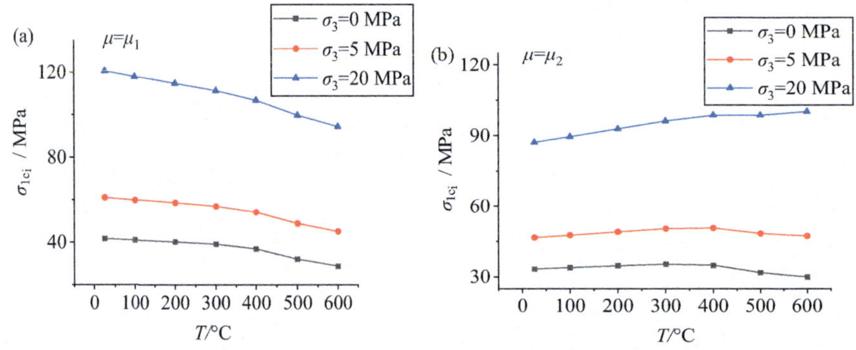

Fig. 12.9 Curves of crack initiation stress with temperature under different confining pressures for **a** $\mu = \mu_1$, **b** $\mu = \mu_2$, after treatment

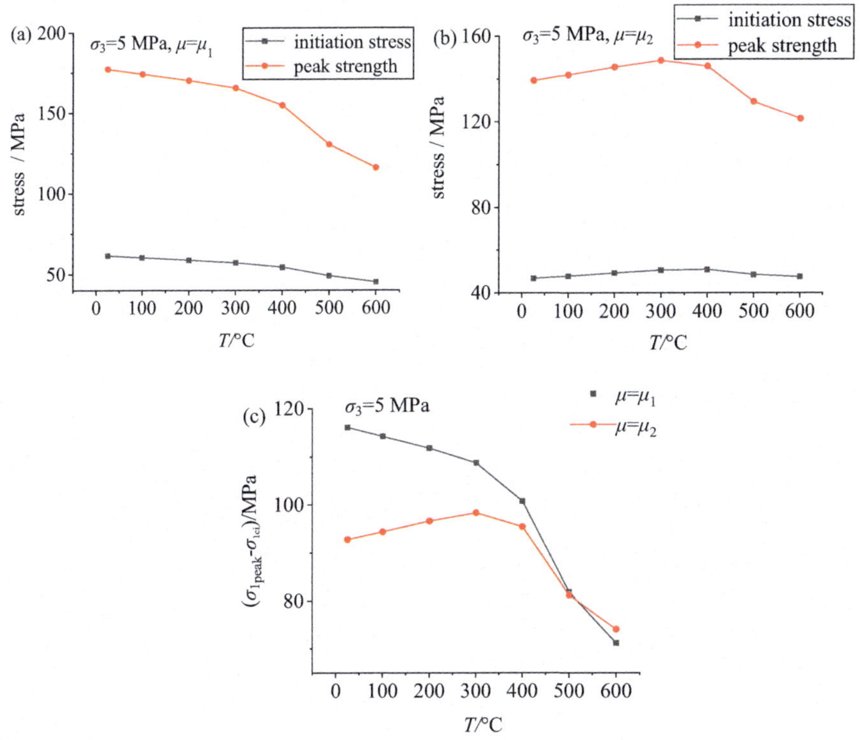

Fig. 12.10 Curves of peak strength ($\sigma_{1\text{peak}}$) and crack initiation stress ($\sigma_{1\text{ci}}$) with temperature for **a** $\mu = \mu_1$, **b** $\mu = \mu_2$, and **c** comparison curves of $\sigma_{1\text{peak}} - \sigma_{1\text{ci}}$ for $\mu = \mu_1$ and μ_2, after heat treatment

12.2.4 Comparative Analysis of Long-Term Creep Mechanical Properties of Heat-Treated Rocks

The determination of compressive peak strength and crack initiation stress ranges from Fig. 12.10 furnishes crucial criteria for examining long-term mechanical properties under constant stress conditions in this section. To delve into the long-term mechanical characteristics of brittle rocks at distinct heat treatment temperatures while sustaining a constant stress, we selected an axial stress level of $\sigma_1 = 70$ MPa, coupled with a confining pressure of $\sigma_3 = 5$ MPa. This choice was made based on the intersection of the ranges of axial stress σ_1 values when rock undergoes creep failure at different temperatures, i.e., $\sigma_{1\text{ci}} \leq \sigma_1 < \sigma_{1\text{peak}}$.

Figure 12.11 depicts creep curves for rocks subjected to various heat treatment temperatures and exhibiting different trends in friction coefficients. Notably, these curves illustrate that as the heat treatment temperature rises, the time to failure decreases. Figure 12.12, on the other hand, illustrates the evolution of wing crack length during creep for different friction coefficient trends post heat treatment.

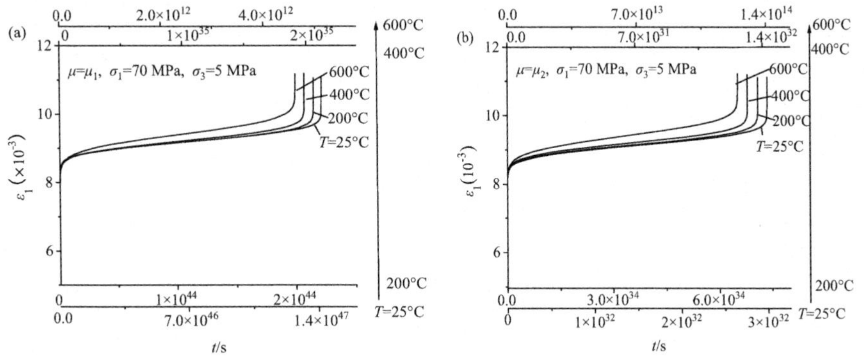

Fig. 12.11 Creep curves at different temperatures for **a** $\mu = \mu_1$, **b** $\mu = \mu_2$, after heat treatment

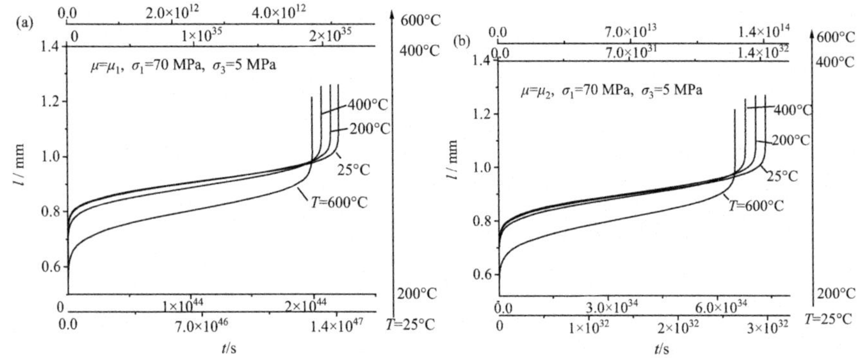

Fig. 12.12 Curves of crack length with time at different temperatures for **a** $\mu = \mu_1$, **b** $\mu = \mu_2$, after heat treatment

Utilizing the crack length–time evolution results derived from Eq. (12.12) and substituting them into Eq. (5.4), we can deduce the time evolution equation for the crack tip stress intensity factor, denoted as $K_{\text{I}}(t,T)$, for the rock. When $K_{\text{I}}(t,T)$ is normalized by the fracture toughness $K_{\text{IC}}(T)$, it yields a time evolution graph illustrating the normalized crack tip stress intensity factor, as portrayed in Fig. 12.13. Notably, during steady-state creep, $K_{\text{I}}(t,T)/K_{\text{IC}}(T)$ consistently remains below 1, signifying that the crack tip stress intensity factor can propagate even when it's less than the fracture toughness K_{IC} throughout the creep process. This normalized strength factor's time evolution can be categorized into three stages: an initial slow decrease, followed by stabilization, and ultimately a rapid increase upon reaching the failure time. Importantly, these three stages correspond to the crack length evolution displayed in Fig. 12.12, which undergoes a deceleration, stabilizes, and then experiences a rapid increase over time.

In the context of long-term strength, we define the long-term strength as the axial stress σ_1 at which the creep curve transitions to accelerated creep as σ_1 gradually

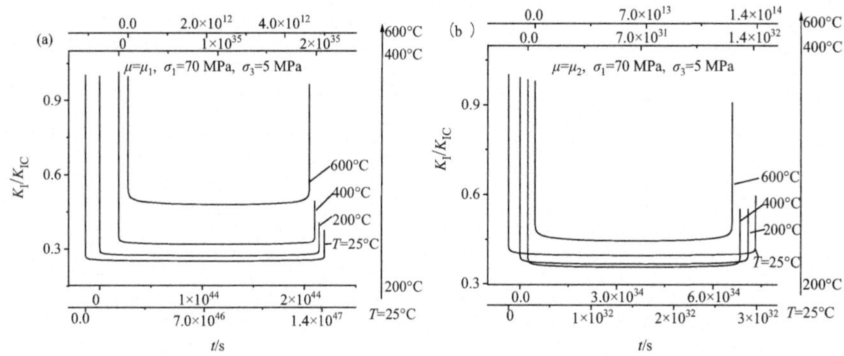

Fig. 12.13 Normalized stress intensity factor evolution curves with time at different temperatures for **a** $\mu = \mu_1$, **b** $\mu = \mu_2$, after heat treatment

increases. When time t approaches infinity (in our study, we used a time value of 1.6×10^{200} s), continuous adjustments to the axial stress enable us to determine the long-term strength. It's worth noting that under specific conditions, even when the axial stress σ_1 reaches its minimum value at $\sigma_1 = \sigma_{1ci}$, an accelerated creep phase may still be observed in the creep curve. However, when $\sigma_1 < \sigma_{1ci}$, the rock does not undergo creep failure. In such scenarios, we can regard σ_{1ci} as the critical value at which the creep curve enters the accelerated creep stage. Consequently, the long-term strength equates to the crack initiation strength, elucidating why both crack initiation strength and long-term strength exhibit similar trends with increasing temperature, as discussed earlier.

Figure 12.14 illustrates the variation in long-term strength with temperature for different friction coefficient trends. When $\mu = \mu_1$, it's evident that higher confining pressure leads to a broader range of long-term strength variation with temperature, characterized by a continuous decrease. Conversely, when $\mu = \mu_2$, increasing confining pressure alters the long-term strength trend with temperature from an initial increase to a subsequent decrease.

For a more in-depth exploration of how different friction coefficient trends impact long-term strength, Fig. 12.15 presents the difference in long-term strength between cases with friction coefficients μ_1 and μ_2, respectively, as a function of heat treatment temperature. Notably, as the temperature increases, the difference in long-term strength initially decreases before starting to rise again. It reaches its minimum value when the temperature is near the threshold temperature. This observation suggests that as the heat treatment temperature rises, the distinctions in long-term strength due to varying friction coefficients gradually diminish. When the heat treatment temperature approaches the threshold temperature, the influence of different friction coefficient trends on long-term strength becomes less pronounced.

Figure 12.16 illustrates the normalized variation of failure time with temperature after heat treatment, considering different friction coefficient trends (normalized with respect to the first data point). Notably, confining pressure has no discernible impact

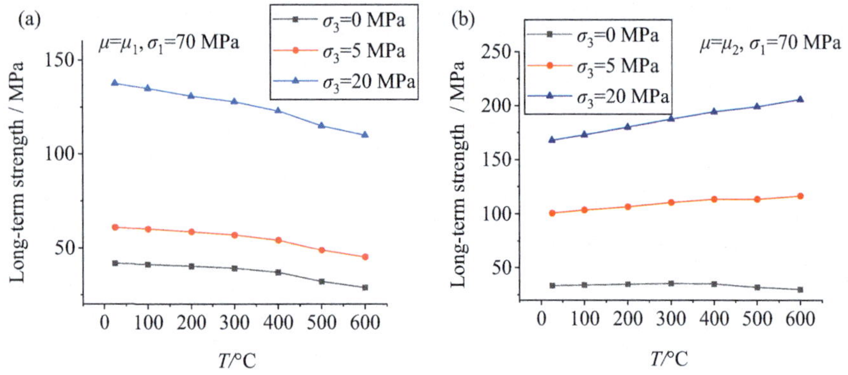

Fig. 12.14 Long-term strength variation curves with temperature for **a** $\mu = \mu_1$, **b** $\mu = \mu_2$, after heat treatment

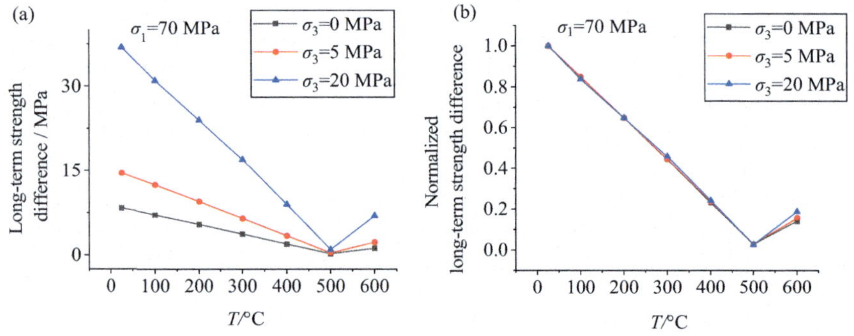

Fig. 12.15 Long-term strength difference variation curves with temperature for **a** $\mu = \mu_1$, **b** $\mu = \mu_2$, after heat treatment

on the trend of failure time with temperature. When $\mu = \mu_1$, there is a consistent and monotonous decrease in failure time as the temperature rises. Conversely, when $\mu = \mu_2$, the failure time initially experiences an increase followed by a decrease with increasing temperature. Furthermore, the range of variation expands with higher confining pressure.

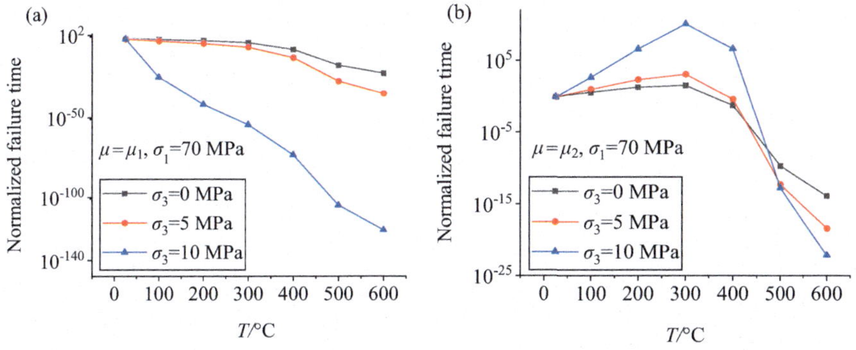

Fig. 12.16 Failure time variation curves with temperature for **a** $\mu = \mu_1$, **b** $\mu = \mu_2$, after treatment

12.3 Conclusion

This section developed a comprehensive macro-microscopic mechanical model to analyze the creep behavior of brittle rocks, taking into consideration the influence of heat treatment temperature. This model has enabled us to explore the interconnected impacts of heat treatment temperature and confining pressure on various mechanical properties of brittle rocks, including initiation stress, peak strength, long-term strength, and the time to long-term creep failure, all while accounting for different trends in the friction coefficient with temperature.

(1) Confining Pressure Amplifies Temperature Effects: Under two distinct friction coefficient trends with temperature, i.e., monotonically decreasing and decreasing, increasing confining pressure significantly widens the spectrum of variation in both short-term and long-term mechanical properties of rocks as temperature rises.
(2) Short-term Mechanical Properties: For short-term mechanical properties, when the friction coefficient (μ) equals μ_1, elevated heat treatment temperature consistently results in a decline in both peak strength ($\sigma_{1\text{peak}}$) and crack initiation stress ($\sigma_{1\text{ci}}$). Confining pressure (σ_3) does not exert any discernible influence on these trends. Conversely, for $\mu = \mu_2$, the trend in $\sigma_{1\text{peak}}$ with temperature remains consistent regardless of σ_3, featuring an initial increase followed by a decrease. However, the behavior of $\sigma_{1\text{ci}}$ evolves with increasing σ_3, transitioning from an initial increase followed by a decrease to a monotonic increase.
(3) Long-term Mechanical Properties: Concerning long-term mechanical properties, when $\mu = \mu_1$, increasing heat treatment temperature leads to a continuous reduction in long-term strength and failure time across various confining pressures. In contrast, for $\mu = \mu_2$, the trend in long-term strength shifts from an initial increase followed by a decrease to a monotonic increase with increasing σ_3, while the behavior of failure time remains unchanged, displaying an initial increase followed by a decrease.

(4) Threshold Temperature: Notably, when the heat treatment temperature reaches a specific threshold value, the disparity in long-term strength between the two friction coefficient trends becomes minimal. This suggests that, at this threshold temperature, the influence of the friction coefficient on rock behavior aligns.

References

Chen S, Yang C, Wang G. Evolution of thermal damage and permeability of Beishan granite. Appl Thermal Eng. 2017;110:1533–42.

Fan LF, Gao JW, Wu ZJ, Yang SQ, Ma GW. An investigation of thermal effects on micro-properties of granite by X-ray CT technique. Appl Thermal Eng. 2018;140:505–19.

Feng G, Wang XC, Kang Y, Luo SG, Hu YQ. Effects of temperature on the relationship between mode-I fracture toughness and tensile strength of rock. Multidisc Digital Publ Inst. 2019;9(7):1326.

Huang YH, Yang SQ, Tian WL, Zhao J, Ma D, Zhang CS. Physical and mechanical behavior of granite containing pre-existing holes after high temperature treatment. Arch Civ Mech Eng. 2017;17(4):912–25.

Li D, Ma J, Wan Q, Zhu Q, Han Z. Effect of thermal treatment on the fracture toughness and subcritical crack growth of granite in double-torsion test. Eng Fract Mech. 2021;253: 107903.

Liu JF, Luo X, Feng G, Zhou L, Yang X, Lu G. Creep characteristics of thermally-stressed Beishan granite under triaxial compression. Int J Rock Mech Mining Sci. 2023;162: 105302.

Liu S, XU J. An experimental study on the physico-mechanical properties of two post-high-temperature rocks. Eng Geol. 2015;185:63–70.

Meng F, Song J, Wong LNY, Wang Z, Zhang C. Characterization of roughness and shear behavior of thermally treated granite fractures. Eng Geol. 2021;293:106287.

Nasseri MHB, Schubnel A, Young RP. Coupled evolutions of fracture toughness and elastic wave velocities at high crack density in thermally treated Westerly granite. Int J Rock Mech Mining Sci. 2007;44(4):601–16.

Sha S, Rong G, Chen ZH, Li B, Zhang Z. Experimental evaluation of physical and mechanical properties of geothermal reservoir rock after different cooling treatments. Rock Mech Rock Eng. 2020;53:4967–91.

Shao S, Wasantha PLP, Ranjith PG, Chen BK. Effect of cooling rate on the mechanical behavior of heated Strathbogie granite with different grain sizes. Int J Rock Mech Mining Sci. 2014;70:381–7.

Srinivasan V, Tripathy A, Gupta T, Singh TN. An investigation on the influence of thermal damage on the physical, mechanical and acoustic behavior of Indian Gondwana shale. Rock Mech Rock Eng. 2020;53:2865–85.

Tang ZC. Experimental Investigation on Temperature-dependent Shear Behaviors of Granite Discontinuity. Rock Mech Rock Eng. 2020;53(9):4043–60.

Tang ZC, Zhang Y. Temperature-dependent peak shear-strength criterion for granite fractures. Eng Geol. 2020;269: 105552.

Tang ZC, Sun M, Peng J. Influence of high temperature duration on physical, thermal and mechanical properties of a fine-grained marble. Appl Therm Eng. 2019;156:34–50.

Wu XH. Study on physical and mechanical properties and damage mechanism of granite at different heating temperatures after water-cooling [Ph.D. diss]. Beinjing: University of Science and Technology Beijing, 2022. (in Chinese))

Xu XL, Zhang ZZ. Acoustic emission and damage characteristics of granite subjected to high temperature. Adv Mater Sci Eng. 2018;8149870.

Zhu Z, Tian H, Mei G, Jiang G, Dou B, Xiao P. Experimental investigation on mechanical behaviors of Nanan granite after thermal treatment under conventional triaxial compression. Environ Earth Sci. 2021;80:46.

Uncited References

Preisig G, Eberhardt E, Smithyman M, Preh A, Bonzanigo L. Hydromechanical rock mass fatigue in deep-seated landslides accompanying seasonal variations in pore pressures. Rock Mech Rock Eng. 2016;49(6):2333–51.

Sagong M, Bobet A. Coalescence of multiple flaws in a rock-model material in uniaxial compression. Int J Rock Mech Mining Sci. 2002;39(2):229–41.

Zivaljevic S, Tomanovic Z. Experimental research of the effects of preconsolidation on the time-dependent deformations—creep of marl. Mech Time-Dependent Mater. 2015;19:43–59.

The manufacturer's authorised representative in the EU is Springer Nature Customer Service Centre GmbH, Europaplatz 3, 69115 Heidelberg, Germany. If you have any concerns regarding our products, please contact ProductSafety@springernature.com

Printed and bound by CPI Group (UK) Ltd, Croydon, CR0 4YY
01/10/2026
02225900-0003